METHODS IN MOLECULAR BIOLOGY

Series Editor
John M. Walker
School of Life and Medical Sciences
University of Hertfordshire
Hatfield, Hertfordshire, AL10 9AB, UK

For further volumes:
http://www.springer.com/series/7651

Acinetobacter baumannii

Methods and Protocols

Edited by

Indranil Biswas

Department of Microbiology, Molecular Genetics and Immunology, University of Kansas Medical Center, Kansas City, KS, USA

Philip N. Rather

Department of Microbiology and Immunology, Emory University, Atlanta, GA, USA

Humana Press

Editors
Indranil Biswas
Department of Microbiology
Molecular Genetics and Immunology
University of Kansas Medical Center
Kansas City, KS, USA

Philip N. Rather
Department of Microbiology
and Immunology
Emory University
Atlanta, GA, USA

ISSN 1064-3745 ISSN 1940-6029 (electronic)
Methods in Molecular Biology
ISBN 978-1-4939-9117-4 ISBN 978-1-4939-9118-1 (eBook)
https://doi.org/10.1007/978-1-4939-9118-1

Library of Congress Control Number: 2018968407

This Humana Press imprint is published by the registered company Springer Science+Business Media, LLC, part of Springer Nature.
The registered company address is: 233 Spring Street, New York, NY 10013, U.S.A.

Preface

Acinetobacter baumannii is a Gram-negative bacterium that has become a serious human pathogen. This bacterium is responsible for infections of the lung, skin/soft tissue, bloodstream, and urinary tract. Although *A. baumannii* is primarily an opportunistic pathogen, infections in the community have been increasing steadily, and this worrisome event suggests that strains with enhanced virulence have arisen. The ability to treat *A. baumannii* infections has become challenging due to the extensive array of antibiotic resistance mechanisms that are present. Resistance to "last resort" antibiotics such as the carbapenems or colistin is now commonly seen, and pan-resistant strains have been reported with increasing frequency. This has prompted the World Health Organization to list *A. baumannii* as a number one priority for the research and development of new antibiotics. The importance of this bacterium has led to an explosion of studies to address antibiotic resistance, virulence, and basic aspects of physiology and metabolism. This book will serve as a useful reference for all *A. baumannii* researchers, as it contains an extensive compilation of state-of-the-art techniques to study this important pathogen.

Kansas City, KS, USA ***Indranil Biswas***
Atlanta, GA, USA ***Philip N. Rather***

Contents

Contributors

LUIS A. ACTIS • *Department of Microbiology, Miami University, Oxford, OH, USA*

YONAS A. ALAMNEH • *Wound Infections Department, Bacterial Diseases Branch, Walter Reed Army Institute of Research, Silver Spring, MD, USA*

AMJAD ALI • *Atta-ur-Rahman School of Applied Biosciences (ASAB), National University of Sciences and Technology (NUST), Islamabad, Pakistan*

SHALINI ANANDAN • *Department of Clinical Microbiology, Christian Medical College, Vellore, Tamil Nadu, India*

SARAH E. ANDERSON • *Department of Microbiology and Immunology, Emory University, Atlanta, GA, USA; Emory Antibiotic Resistance Center, Emory University, Atlanta, GA, USA*

SANKALP ARYA • *Department of Biotechnology, Indian Institute of Technology-Roorkee, Roorkee, Uttarakhand, India; Division of Agricultural and Environmental Sciences, University of Nottingham, Nottingham, UK*

SHIFA TARIQ ASHRAF • *Atta-ur-Rahman School of Applied Biosciences (ASAB), National University of Sciences and Technology (NUST), Islamabad, Pakistan*

RAZIYA FATHIMA ASLAM • *Department of Clinical Microbiology, Christian Medical College, Vellore, Tamil Nadu, India*

YAMUNA DEVI BAKTHAVACHALAM • *Department of Clinical Microbiology, Christian Medical College, Vellore, Tamil Nadu, India*

JAIDEEP BANERJEE • *Wound Infections Department, Bacterial Diseases Branch, Walter Reed Army Institute of Research, Silver Spring, MD, USA*

INDRANIL BISWAS • *Department of Microbiology, Molecular Genetics and Immunology, University of Kansas Medical Center, Kansas City, KS, USA*

DARREN J. CREEK • *Drug Delivery, Disposition and Dynamics, Monash Institute of Pharmaceutical Sciences, Monash University, Parkville, VIC, Australia*

BRYAN W. DAVIES • *Department of Molecular Biosciences, University of Texas at Austin, Austin, TX, USA; Institute for Cellular and Molecular Biology, University of Texas at Austin, Austin, TX, USA; Center for Systems and Synthetic Biology, University of Texas at Austin, Austin, TX, USA; John Ring LaMontagne Center for Infectious Diseases, Institute for Cellular and Molecular Biology, University of Texas at Austin, Austin, TX, USA*

EMMANUELLE DÉ • *Normandie University, Unirouen, Rouen, France; Polymers, Biopolymers, Surfaces Laboratory, CNRS, UMR 6270, UFR des Sciences et Techniques, Mont-Saint-Aignan, France*

P. MALAKA DE SILVA • *Department of Microbiology, University of Manitoba, Winnipeg, MB, Canada*

SAMANDRA T. DEMONS • *Wound Infections Department, Bacterial Diseases Branch, Walter Reed Army Institute of Research, Silver Spring, MD, USA*

SARA DOMINGUES • *Faculty of Pharmacy, University of Coimbra, Coimbra, Portugal*

VINEET DUBEY • *Department of Biotechnology, Indian Institute of Technology-Roorkee, Roorkee, Uttarakhand, India*

KALEIGH DUCAS-MOWCHUN • *Department of Microbiology, University of Manitoba, Winnipeg, MB, Canada*

Shweta Fulsundar • *Institute of Bioinformatics and Biotechnology, Savitribai Phule Pune University, Pune, India*

Larry A. Gallagher • *Department of Genome Sciences, University of Washington, Seattle, WA, USA*

Anindya S. Ghosh • *Department of Biotechnology, Indian Institute of Technology Kharagpur, Kharagpur, West Bengal, India*

Erin R. Green • *Department of Pathology, Microbiology, and Immunology, Vanderbilt University Medical Center, Nashville, TN, USA; Vanderbilt Institute for Infection, Immunology and Inflammation, Vanderbilt University Medical Center, Nashville, TN, USA*

Anja Heins • *I&L Biosystems France SAS, Lyon, France*

Carmen M. Herrera • *Department of Infectious Diseases, University of Georgia, Athens, GA, USA*

Ayush Kumar • *Department of Microbiology, University of Manitoba, Winnipeg, MB, Canada; Manitoba Chemosensory Biology Group, University of Manitoba, Winnipeg, MB, Canada*

Binesh Lal • *Department of Clinical Microbiology, Christian Medical College, Vellore, Tamil Nadu, India*

Jian Li • *Laboratory of Antimicrobial Systems Pharmacology, Department of Microbiology, Biomedicine Discovery Institute, Monash University, Melbourne, VIC, Australia*

Alexandra Machen • *Department of Biochemistry and Molecular Biology, University of Kansas Medical Center, Kansas City, KS, USA*

Mohd Hafidz Mahamad Maifiah • *International Institute for Halal Research and Training, International Islamic University Malaysia, Gombak, Selangor, Malaysia; Drug Delivery, Disposition and Dynamics, Monash Institute of Pharmaceutical Sciences, Monash University, Parkville, VIC, Australia*

D. Scott Merrell • *Department of Microbiology and Immunology, Uniformed Services University of the Health Sciences, Bethesda, MD, USA*

Joshua Mettlach • *Department of Microbiology, Molecular Genetics and Immunology, University of Kansas Medical Center, Kansas City, KS, USA*

Sarah L. J. Michel • *Department of Pharmaceutical Sciences, University of Maryland, Baltimore, MD, USA*

Bertrand Naudin • *Normandie University, Unirouen, Rouen, France; Polymers, Biopolymers, Surfaces Laboratory, CNRS, UMR 6270, UFR des Sciences et Techniques, Mont-Saint-Aignan, France*

Heather M. Neu • *Department of Pharmaceutical Sciences, University of Maryland, Baltimore, MD, USA*

Marion Nicol • *Normandie University, Unirouen, Rouen, France; Polymers, Biopolymers, Surfaces Laboratory, CNRS, UMR 6270, UFR des Sciences et Techniques, Mont-Saint-Aignan, France*

Kaare Magne Nielsen • *Department of Life Sciences and Health, OsloMet—Oslo Metropolitan University, Oslo, Norway*

Shilpa Pal • *Department of Biotechnology, Indian Institute of Technology Kharagpur, Kharagpur, West Bengal, India*

Lauren D. Palmer • *Department of Pathology, Microbiology, and Immunology, Vanderbilt University Medical Center, Nashville, TN, USA; Vanderbilt Institute for Infection, Immunology and Inflammation, Vanderbilt University Medical Center, Nashville, TN, USA*

RANJANA PATHANIA • *Department of Biotechnology, Indian Institute of Technology-Roorkee, Roorkee, Uttarakhand, India*

RAKESH PATIDAR • *Department of Microbiology, University of Manitoba, Winnipeg, MB, Canada*

AIMEE R. PAULK TIERNEY • *Department of Microbiology and Immunology, Emory University School of Medicine, Emory University, Atlanta, GA, USA*

WILLIAM F. PENWELL • *Department of Biology and Marine Science, Jacksonville University, Jacksonville, FL, USA*

STÉPHANE PINHAL • *I&L Biosystems France SAS, Lyon, France*

MATTHEW J. POWERS • *Department of Infectious Diseases, University of Georgia, College of Veterinary Medicine, Athens, GA, USA; Department of Microbiology, University of Georgia, Athens, GA, USA*

AGILA K. PRAGASAM • *Department of Clinical Microbiology, Christian Medical College, Vellore, Tamil Nadu, India*

JOHN A. J. PRAKASH • *Department of Clinical Microbiology, Christian Medical College, Vellore, Tamil Nadu, India*

PHILIP N. RATHER • *Department of Microbiology and Immunology, Emory University, Atlanta, GA, USA*

NITHIN SAM RAVI • *Department of Clinical Microbiology, Christian Medical College, Vellore, Tamil Nadu, India*

HERBERT P. SCHWEIZER • *Department of Molecular Genetics and Microbiology, College of Medicine, Emerging Pathogens Institute, University of Florida, Gainesville, FL, USA*

DEEPAK SEN • *Department of Biotechnology, Indian Institute of Technology-Roorkee, Roorkee, Uttarakhand, India*

FATIMA SHAHID • *Atta-ur-Rahman School of Applied Biosciences (ASAB), National University of Sciences and Technology (NUST), Islamabad, Pakistan*

ATIN SHARMA • *Department of Biotechnology, Indian Institute of Technology-Roorkee, Roorkee, Uttarakhand, India*

JONATHAN P. SHEARER • *Wound Infections Department, Bacterial Diseases Branch, Walter Reed Army Institute of Research, Silver Spring, MD, USA*

JESSICA R. SHELDON • *Department of Pathology, Microbiology, and Immunology, Vanderbilt University Medical Center, Nashville, TN, USA; Vanderbilt Institute for Infection, Immunology and Inflammation, Vanderbilt University Medical Center, Nashville, TN, USA*

EDGAR X. SHERMAN • *Department of Microbiology and Immunology, Emory University, Atlanta, GA, USA; Emory Vaccine Center, Atlanta, GA, USA; Division of Infectious Diseases, Department of Medicine, Emory University School of Medicine, Atlanta, GA, USA; Emory Antibiotic Resistance Center, Atlanta, GA, USA*

ERIC P. SKAAR • *Department of Pathology, Microbiology, and Immunology, Vanderbilt University Medical Center, Nashville, TN, USA; Vanderbilt Institute for Infection, Immunology and Inflammation, Vanderbilt University Medical Center, Nashville, TN, USA*

EVELYN SKIEBE • *Robert Koch Institute, Wernigerode, Germany*

MITCHELL G. THOMPSON • *Department of Plant and Microbial Biology, University of California Berkeley, Berkeley, CA, USA*

KYLE A. TIPTON • *Department of Microbiology and Immunology, Emory University School of Medicine, Emory University, Atlanta, GA, USA*

M. STEPHEN TRENT • *Department of Infectious Diseases, University of Georgia, College of Veterinary Medicine, Athens, GA, USA; Department of Microbiology, University of Georgia, Athens, GA, USA*
ASHLEY T. TUCKER • *Department of Molecular Biosciences, University of Texas at Austin, Austin, TX, USA*
BALAJI VEERARAGHAVAN • *Department of Clinical Microbiology, Christian Medical College, Vellore, Tamil Nadu, India*
TONY VELKOV • *Department of Pharmacology and Therapeutics, The University of Melbourne, Parkville, VIC, Australia*
SARANYA VIJAYAKUMAR • *Department of Clinical Microbiology, Christian Medical College, Vellore, Tamil Nadu, India*
XUN WANG • *Department of Molecular Biosciences, University of Texas at Austin, Austin, TX, USA*
DAVID S. WEISS • *Emory Vaccine Center, Atlanta, GA, USA; Division of Infectious Diseases, Department of Medicine, Emory University School of Medicine, Atlanta, GA, USA; Emory Antibiotic Resistance Center, Atlanta, GA, USA; Research Service, Atlanta VA Medical Center, Decatur, GA, USA*
GOTTFRIED WILHARM • *Robert Koch Institute, Wernigerode, Germany*
CAITLIN L. WILLIAMS • *Department of Microbiology and Immunology, Uniformed Services University of the Health Sciences, Bethesda, MD, USA*
JESSIE E. WOZNIAK • *Department of Microbiology and Immunology, Emory University, Atlanta, GA, USA; Emory Vaccine Center, Atlanta, GA, USA; Division of Infectious Diseases, Department of Medicine, Emory University School of Medicine, Atlanta, GA, USA; Emory Antibiotic Resistance Center, Atlanta, GA, USA*
SÜLEYMAN YILDIRIM • *Department of Medical Microbiology, International School of Medicine, Istanbul Medipol University, Istanbul, Turkey*
DANIEL V. ZURAWSKI • *Wound Infections Department, Bacterial Diseases Branch, Walter Reed Army Institute of Research, Silver Spring, MD, USA*

Chapter 1

Specimen Collection, Processing, Culture, and Biochemical Identification of *Acinetobacter* spp.

Binesh Lal, Saranya Vijayakumar, Shalini Anandan, and Balaji Veeraraghavan

Abstract

Specimen collection and processing is an important aspect of clinical microbiology laboratory. The reports are dependent on the quality of the specimen and the time between the collection and processing. Appropriate methodology needs to be followed for the collection, amount, type, labeling, transportation, and processing of the specimens especially for organism like *Acinetobacter* species. Various biochemical tests are used for identification of various organisms. Such identification depends on the ability of organisms to produce certain enzymes or to utilize certain compound to be identified by biochemical tests.

Key words *Acinetobacter*, Specimen collection, Culture identification, Biochemical identification

1 Introduction

The value and reliability of microbiological reports are directly affected by the quality of the specimen received by the laboratory and the length of time between its collection and processing [1]. The laboratory should issue written instructions like the amount and type of specimen required, the container to use, the need for any preservative or transport medium, the best time to collect a specimen, the aseptic and safe methods of collection to avoid contamination and accidental infection, the labeling of the specimen container, and the conditions in which specimens need to be kept prior to and during their transport to the laboratory [2].

Biochemical tests are very important in the identification of bacterial isolates and different bacterial species. Different types of biochemical reactions each organism undergoes act as a thumbprint for its identification. Bacterial physiology differs from one species to the other. These differences in carbohydrate metabolism, protein metabolism, fat metabolism, production of certain enzymes, and ability to utilize a particular compound help them to be identified

Indranil Biswas and Philip N. Rather (eds.), *Acinetobacter baumannii: Methods and Protocols*, Methods in Molecular Biology, vol. 1946, https://doi.org/10.1007/978-1-4939-9118-1_1, © Springer Science+Business Media, LLC, part of Springer Nature 2019

by the biochemical tests [3, 4]. The following chapter describes in brief about the various specimen collections, processing methods for *Acinetobacter* species, and biochemical tests used in the identification of *Acinetobacter* species.

2 Materials

2.1 Common Materials

1. Glass slide.
2. Permanent marker.
3. Inoculation loop.
4. Bunsen burner.
5. 5 mL syringe.
6. Needle (22 gauge).
7. Petri plates (75 mm diameter).
8. Test tubes (glass 12 × 100 mm).

2.2 Sample Collection

2.2.1 Blood Specimen

1. BacT/ALERT bottle (*see* **Note 1**).
2. Tourniquet.
3. 2% alcoholic chlorhexidine.
4. Culture tubes (18 × 150 mm).
5. Alcohol pad.

2.2.2 Respiratory Specimen

1. Sterile red-capped 40 mL wide-mouthed screw-capped container.
2. Saline.
3. Aspirator.
4. Suction catheter.

2.3 Gram Staining

1. 1% crystal violet.
2. 2% iodine.
3. Acetone.
4. 0.34% safranin.

2.4 Culture Growth and Identification

1. Nutrient agar: 5 g peptic digest of animal tissue, 5 g sodium chloride, 1.5 g beef extract, 1.5 g yeast extract, 15 g agar, 1000 mL water, pH 7.4 (*see* **Note 2**).
2. MacConkey agar: 3 g peptone (meat and casein), 17 g pancreatic digest of gelatin, 10 g lactose monohydrate, 1.5 g bile salts, 5 g sodium chloride, 0.001 g crystal violet, 0.03 g neutral red, 13.5 g agar, 1000 mL water, pH (at 25 °C) 7.1 ± 0.2 (*see* **Note 3**).
3. Blood agar: 10 g beef heart peptone, 10 g tryptose, 5 g sodium chloride, 15 g agar, 1000 mL water, pH 7.3. Add 5% v/v defibrinated blood (*see* **Note 4**).

2.5 Cytochrome Oxidase Test (See Note 5)

1. 1% tetramethyl-*p*-phenylene-diamine dihydrochloride (oxidizing reagent).
2. Whatman No. 1 filter paper.
3. Clean glass rod or loop.
4. *Pseudomonas aeruginosa* (ATCC 27853) as positive control.
5. *Escherichia coli* (ATCC 25922) as negative control.

2.6 Mannitol Motility Medium (See Note 6)

1. Agar.
2. Peptone.
3. Mannitol.
4. Potassium nitrate.
5. 1% phenol red solution.
6. Glass test tubes (12 × 100 mm).
7. Straight needle (stab loop—Nichrome wire).
8. *P. aeruginosa* (ATCC 27853) or *E. coli* (ATCC 25922) as positive control.
9. *Klebsiella pneumoniae* (ATCC 700603) as negative control.

2.7 Triple Sugar Iron Agar Test (See Note 7)

1. Sodium chloride.
2. Yeast extract.
3. Peptone.
4. Agar.
5. Lactose.
6. Sucrose.
7. Dextrose.
8. Sodium thiosulfate.
9. Ferrous sulfate.
10. *E. coli* (ATCC 25922) as positive control.
11. *P. aeruginosa* (ATCC 27853) as negative control.

2.8 Indole Test (See Note 8)

1. Test tube with tryptone broth.
2. Platinum inoculation loop/wooden stick.
3. Burner.
4. Culture plate.
5. *p*-Dimethyl-aminobenzaldehyde.
6. *E. coli* (ATCC 25922) as positive control.
7. *K. pneumoniae* (ATCC 700603) as negative control.

2.9 Citrate Utilization Test (See Note 9)

1. Simmons' citrate medium, pH 6.9: 0.2 g magnesium sulfate, 1 g ammonium dihydrogen phosphate, 1 g dipotassium phosphate, 2 g sodium citrate, 5 g sodium chloride, 15 g agar, 0.08 g bromothymol blue, 1000 mL distilled water.
2. *K. pneumoniae* (ATCC 700603) as positive control.
3. *E. coli* (ATCC 25922) as negative control.

2.10 Oxidative-Fermentative Glucose (OF Glucose) Medium Growth (See Note 10)

1. OF medium of Hugh-Leifson: 2 g peptone, 5 g sodium chloride, 10 g glucose, 0.03 g bromothymol blue, 0.3 g dipotassium phosphate, 3 g agar, 1000 mL distilled water, pH 7.1.
2. Stab loop (Nichrome wire).
3. Test tubes (12 × 100 mm).
4. Burner.
5. Culture plates (75 mm).
6. Paraffin oil.
7. *P. aeruginosa* (ATCC 27853) as positive control (oxidation).
8. *Acinetobacter lwoffii* (clinical strain) as negative control (oxidation).
9. *Staphylococcus aureus* (ATCC 25923) as positive control (fermentation).
10. *P. aeruginosa* (ATCC 27853) as negative control (fermentation).

2.11 Arginine Decarboxylase Test (See Note 11)

1. Moeller decarboxylase broth: 5 g peptone, 5 g beef extract, 0.5 g glucose, 0.005 g pyridoxal, 10 g L-arginine, 0.005 cresol red, 0.01 bromocresol purple, 1000 mL distilled water; adjust pH—6.0.
2. Mineral oil.
3. *Aeromonas* sp. as positive control.
4. *Vibrio cholerae* as negative control.

2.12 Nitrate Reduction Assay (See Note 12)

1. Nitrate broth or nitrate agar: 3 g beef extract, 5 g peptone, 1 g potassium nitrate, 12 g agar (nitrite-free), 1000 mL distilled water.
2. Reagent A: 5 g alpha-naphthylamine, 1000 mL 30% acetic acid (5 N).
3. Reagent B: 8 g sulfanilic acid, 1000 mL 30% acetic acid (5 N).
4. *E. coli* (ATCC 25922) as positive control.
5. *Acinetobacter* spp. as negative control.
6. Durham's tubes (6 × 50 mm).

2.13 Urease Test (See Note 13)

1. Stuart's urea broth, pH 6.8: 0.1 g yeast extract, 9.1 g monopotassium phosphate, 9.5 g disodium phosphate, 20 g urea, 0.01 g phenol red, 1000 mL distilled water.
2. Christensen's urea agar, pH 6.8: 1 g peptone, 1 g glucose, 5 g sodium chloride, 2 g monopotassium phosphate, 20 g urea, 0.012 g phenol red, 15 g agar, 1000 mL distilled water.
3. *Proteus mirabilis* as positive control.
4. *E. coli* (ATCC 25922) as negative control.

3 Methods

3.1 Specimen Collection (Blood) (See Fig. 1)

1. Collect blood by strict aseptic technique.
2. Select the site of venipuncture. Wash the intended site with soap and water prior to venipuncture.
3. Apply tourniquet 3–4 in. above the site of venipuncture.
4. Wear an examination glove.
5. Scrub the venipuncture site gently but firmly with the 2% alcoholic chlorhexidine cotton beginning in the center and continuing in an outward direction circularly for an area of 4–5 in. in diameter.
6. Repeat this procedure twice and leave it for air-dry.
7. Drape the site with sterile towels.
8. Allow the area to air-dry completely. Do not use gauze to dry.
9. Even if the area appears to dry sooner, wait at least 3 full minutes (contact time for maximum effect to kill the organism) before performing venipuncture.
10. Don't touch the site after cleaning.
11. Instruct the patient to clench and unclench the fist.
12. Perform phlebotomy using the needle and syringe (5–8 mL and 2–4 mL per bottle, recommended for adult and pediatric patients, respectively).
13. Release the tourniquet and withdraw the needle.
14. Apply pressure to the site of venipuncture, and place a bandage over the puncture site.
15. Clean the rubber stopper of the blood culture bottle with alcohol pad.
16. Inoculate blood into blood culture bottles.
17. Label the blood culture bottles with the date, name, and hospital number of the patient and send the specimens to the laboratory (*see* **Notes 14** and **15**).
18. Keep inoculated bottles at 37 °C or at room temperature.

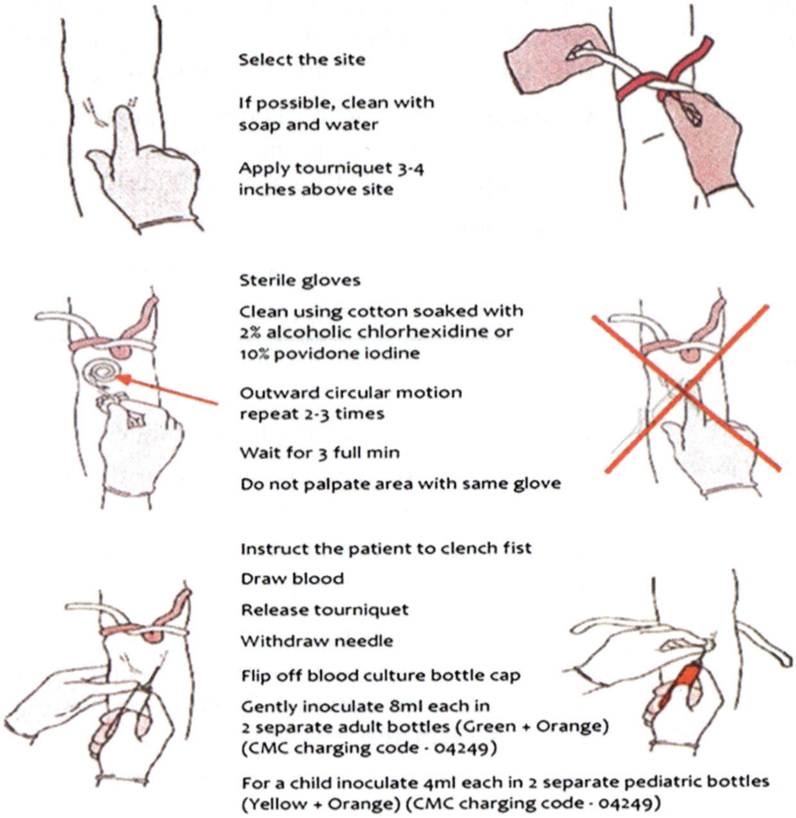

Fig. 1 Schematic diagram of specimen collection from blood

3.2 Specimen Collection (Sputum)

1. Give patient a sterile wide-mouthed screw-capped container.
2. Instruct the patient to inhale deeply 2–3 times and cough out deep from the chest.
3. Open the container and spit the sputum into the bottle.
4. Avoid saliva or nasal secretions.
5. Close the container for further processing.

3.3 Bronchoalveolar Lavage (BAL) and Bronchial Washings

1. Collect bronchial samples by injecting a variable volume of saline through a bronchoscope channel and aspirate in 3–4 aliquots.
2. Collect the aspirate in sterile disposable red-capped 40 mL wide-mouthed container or sterile bottles.
3. Collect bronchial washings by aspirating small amounts of instilled saline from the large airways of the respiratory tract.
4. Store the samples in sterile disposable 40 mL wide-mouthed containers or if from ICU in sterile 100 mL bottles.

3.4 Collection of Endotracheal (ET) Aspirate

1. Follow aseptic precaution—apron, glove, and mask.
2. If the patient is conscious, explain the procedure.
3. Verify patient assessment by measuring vital signs, auscultate, position the patient, and know the abnormal side.
4. Before sample collection, ensure adequate saturation and other vitals are normal.
5. Hyperoxygenate with 100% oxygen and hyperventilate (few breaths).
6. Connect the suction catheter and mucus extractor appropriately.
7. With the help of another staff, collect the sample properly.
8. Flush the catheter with 2–3 mL of sterile water for irrigation.
9. Continue 100% oxygen for few more minutes and then revert back to the original settings.
10. Label the container properly and arrange to send the sample for further processing.

3.5 Gram Staining (See Note 16)

1. Take a clean, grease-free glass slide.
2. Mark the slide and make a smear with the collected samples.
3. Fix the smear (sample) by heating over a Bunsen burner (*see* **Note 17**).
4. Place the slide on a staining tray and cover the smear with crystal violet. Allow to stain for 1 min.
5. Tilt the slide and gently rinse with distilled water until the stain is removed.
6. Cover the smear with Gram's iodine and allow to sit for 1 min.
7. Tilt the slide and gently rinse with distilled water. Tilt the slide and let 2–3 drops of acetone run over the slide. If the last drop is still purple, continue decolorizing, 2–3 drops at a time, until the decolorizer runs clear. Rinse with distilled water (*see* **Note 18**).
8. Cover the smear with safranin, and stain for 1 min.
9. Interpret the results. Gram-positive bacteria will appear as purple/violet color, while Gram-negative bacteria will appear as pink color.

3.6 Preliminary Identification of Acinetobacter spp. by Culture Method and Propagation of Specimen Cultures

1. Take samples from the collected tubes which showed Gram-negative coccobacilli in the Gram stain.
2. Streak samples with a sterile disposable inoculation loop on nutrient agar, MacConkey agar, or blood agar plates.
3. Incubate the inoculated plates at 37 °C under aerobic condition.

4. Colony morphology on nutrient agar, blood agar, and MacConkey agar plates has to be observed and documented (*see* **Note 19**).
5. Inoculate the biochemical reactions for preliminary identification (*see* **Note 20**).
6. Inoculate the biochemicals for species identification (*see* **Note 21**).
7. Interpret the results based on positive and negative controls.

3.7 Cytochrome Oxidase Test

3.7.1 Direct Plate Method

1. Prepare fresh solution of 1% solution of tetramethyl-*p*-phenylene-diamine dihydrochloride (*see* **Note 22**).
2. Directly pour the solution onto the plate with streaked colonies on the nutrient agar so as to cover the surface and then decant.
3. Observe the plate for color change within 10 s (*see* **Note 23**).

3.7.2 Indirect Filter Paper Method

1. Soak a portion of Whatman No. 1 filter paper in a petri dish with few drops of 1% solution of tetramethyl-*p*-phenylene-diamine dihydrochloride.
2. Pick isolated colonies grown on nutrient agar with platinum loop or clean glass rod and smear over the moist area (*see* **Note 24**).
3. Observe the inoculated area of the filter paper for color change (*see* **Note 23**).

3.8 Mannitol Motility Test

1. Touch a straight needle to a colony of 18–24 cultures growing on agar medium.
2. Stab once to a depth of 1/3–1/2 in. in the middle of the tube.
3. Make sure to stab the medium with an inoculating needle in the center of the tube (*see* **Note 25**).
4. Incubate at 35–37 °C and examine after 24 h.
5. Interpret the results based on positive and negative controls (*see* **Note 26**).

3.9 Triple Sugar Iron Agar Test

1. Using an inoculation needle, touch a well-isolated colony from the nutrient agar plate.
2. Inoculate the triple sugar iron agar, first by stabbing to the bottom of the tube through the center of the medium and then by streaking the surface of the slant.
3. Leave the cap loosely and incubate the tube at 35–37 °C for 18–24 h.
4. Interpret the results based on control strains (*see* **Note 27**).

3.10 Indole Test

1. Inoculate tryptone broth with suspected colonies from the nutrient agar plate.

2. Incubate at 35 °C for 18–24 h under aerobic condition.
3. Add 15 drops of *p*-dimethyl-aminobenzaldehyde reagent to the inner wall of the tube.
4. Interpret the results based on control strains (*see* **Note 28**).

3.11 Citrate Utilization Test

1. Inoculate Simmons' citrate agar on the slant by touching 18–24-h-old culture from the nutrient agar (*see* **Note 29**).
2. Incubate at 35–37 °C for up to 7 days under aerobic condition.
3. Observe the growth and the development of blue color denoting alkalinization (*see* **Note 30**).

3.12 Growth on OF Medium

1. Inoculate two tubes of OF medium with fresh culture of the test organism from the nutrient plate to a depth of 1 cm.
2. After inoculation, one tube is overlaid with sterile liquid paraffin oil to form a 5 mm layer.
3. Incubate both the tubes at 37 °C in ambient air for up to 7 days.
4. Interpret the results based on controls (*see* **Note 31**).

3.13 Arginine Decarboxylase Test

1. Using a light inoculum from an 18–24 h pure culture, inoculate the tube.
2. Overlay the tube with sterile mineral oil to cover about 1 cm of the surface.
3. Incubate at 35 °C for 18–24 h.
4. Interpret the results based on controls (*see* **Note 32**).

3.14 Nitrate Reduction Assay

1. Inoculate nitrate broth with culture of the test organism from the nutrient agar plate.
2. Incubate for 48 h at 35–37 °C. Test these cultures 24 h after growth is detected or after a maximum of 7 days.
3. After the incubation period, test the nitrate broth culture for the presence of gas, reduction of nitrate, and reduction of nitrite according to the following.
4. Observe the Durham tube (inverted small test tube) for the presence of gas which is indicated by bubbles inside the tube. Add five drops each of nitrate reagents A and B.
5. Observe for 3 min for a red color to develop.
6. If no color develops, further test with zinc. Dip a wooden applicator stick into the zinc, and transfer the amount to the nitrate broth culture to which reagents A and B have been added.
7. Observe for 3 min for a red color to develop.
8. Interpret the results based on controls (*see* **Note 33**).

3.15 Urease Test

1. Streak the surface of a urea agar with well-isolated colony from the nutrient agar plate.
2. Leave the cap loosely and incubate the tube at 35–37 °C for 48 h to 7 days.
3. Interpret the results based on the controls (*see* **Note 34**).

3.16 Final Interpretation

1. Using the flowchart shown in Fig. 2, determine the identity of the bacterial isolates from the clinical samples.

4 Notes

1. BacT/ALERT blood culture bottle color codes: yellow, pediatric culture bottle; green, adult culture bottle; and orange, anaerobic organism culture bottle.
2. Nutrient agar is used for the cultivation of less fastidious microorganisms and can be enriched with blood or other biological fluids. Peptic digest of animal tissue, beef extract, and yeast extract provide the necessary nitrogen compounds, carbon, vitamins, and also some trace ingredients necessary for the growth of bacteria. Sodium chloride maintains the osmotic equilibrium of the medium.
3. MacConkey agar is a selective and differential medium used to isolate and differentiate members of the *Enterobacteriaceae* based on the ability to ferment lactose. Bile salts and crystal violet inhibit the growth of Gram-positive bacteria. Neutral red dye is a pH indicator that is colorless above a pH of 6.8 and red at a pH below 6.8. The acid accumulating from lactose fermentation turns the dye red.
4. Blood agar is a nutritious medium. Beef extract and tryptose provide carbon, nitrogen, amino acids, and vitamins. Sodium chloride helps in maintaining the osmotic equilibrium of the medium. Addition of blood makes the medium more nutritious by providing additional growth factors required by fastidious organisms. It also helps in visualizing the hemolytic reactions.
5. The cytochromes are hemoproteins which contain iron and act as the link in the chain of aerobic respiration which results in the formation of water by transferring electrons to oxygen. Oxidase test depends on the presence of certain oxidases in bacteria which will catalyze the transport of electrons between electron donors in the bacteria and a redox dye – tetramethyl-*p*-phenylene-diamine.
6. Mannitol motility agar medium is made by dissolving 0.3 g agar and 2 g peptone in 100 mL distilled water and adjusting

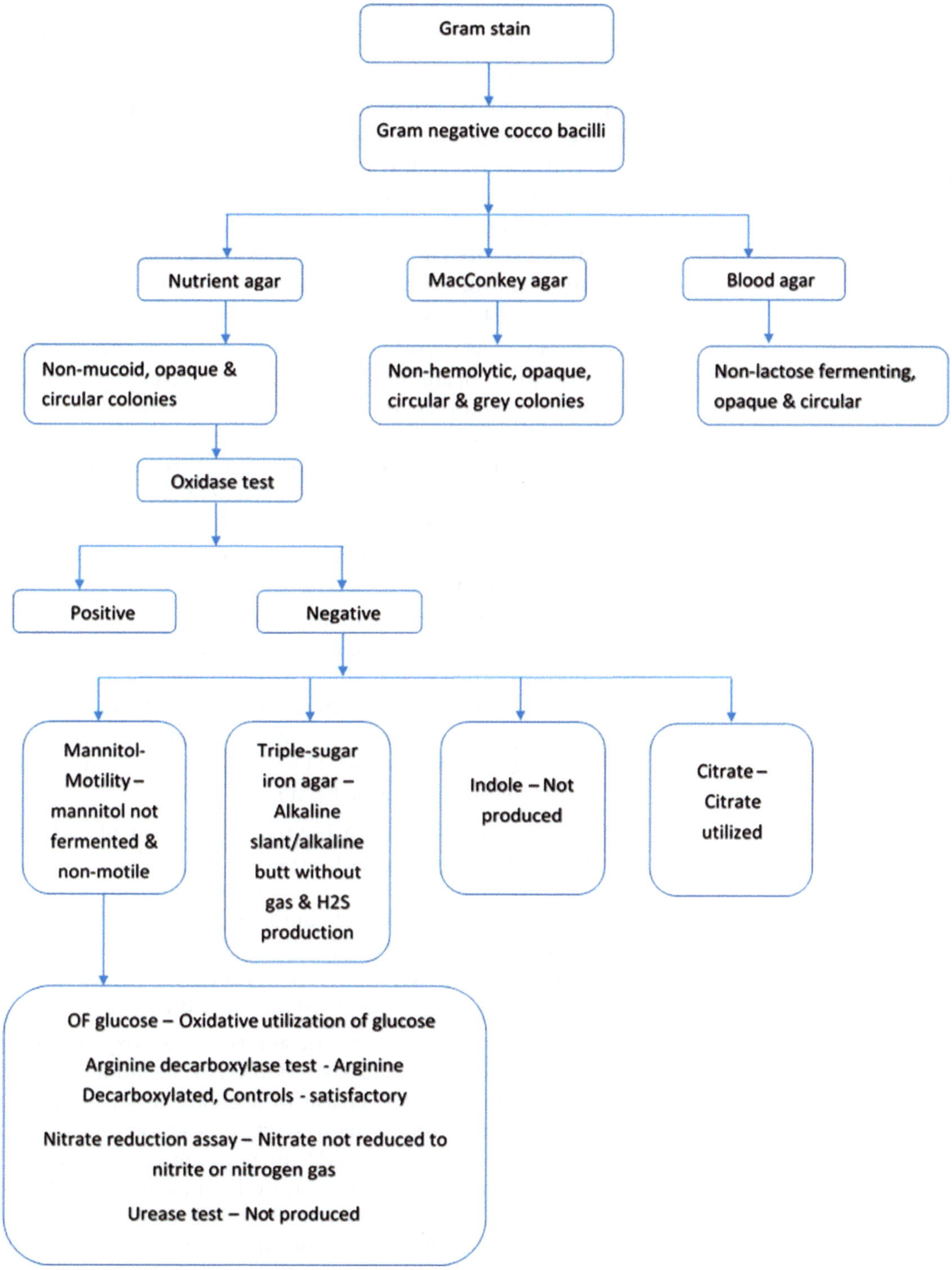

Fig. 2 Algorithm for the identification of *Acinetobacter baumannii-calcoaceticus* complex

the pH with diluted HCl. Then add 0.2 g mannitol, 0.1 g potassium nitrate, 0.4 mL 1% phenol red solution. Distribute 4 mL in test tubes and autoclave.

7. Add 0.5 g sodium chloride, 2 g peptone, and 1.5 g agar in 100 mL distilled water. Dissolve by keeping in a boiling water bath, and add the following: 1 g lactose, 1 g sucrose, 0.1 g dextrose, 0.03 g sodium thiosulfate, and 0.02 g ferrous sulfate. Adjust the pH to 7.6. Then add 0.024 g phenol red. Distribute into small test tubes in 4 mL quantities and autoclave. Keep the tubes in a slanting position immediately after that, so as to get a deep butt and a short slant.
8. Indole is a benzyl pyrrole which is one of the metabolic degradation products of amino acid tryptophan. Bacteria that possess the enzyme tryptophanase are capable of hydrolyzing and deaminating tryptophan with the production of indole, pyruvic acid, and ammonia. Indole is then tested for a colorimetric reaction using *p*-dimethyl-aminobenzaldehyde. The development of bright fuchsia red color at the interface of the reagent and the broth within seconds indicates a positive reaction and the presence of indole.
9. Bacteria which have the ability to grow in this medium produce an enzyme, citrate permease, which converts citrate to pyruvate. Pyruvate then enters the organism's metabolic cycle for the production of energy. Bacteria use the citrate and convert ammonium phosphate into ammonia and ammonium hydroxide, creating an alkaline pH. The change in pH turns the bromothymol blue indicator from green to blue.
10. The OF medium contains a high concentration of carbohydrates (1%) with low concentration of peptone (0.2%) to overcome the possibility of an aerobic organism utilizing peptone, thereby producing an alkaline condition which neutralizes any slight acidity produced by an oxidative organism. Dipotassium phosphate promotes fermentation and acts as buffer to control the pH. Bromothymol blue is used as an indicator.
11. This test measures the enzymatic ability (decarboxylase) of an organism to decarboxylate (or hydrolyze) an amino acid to form an amine. The conversion of arginine to citrulline is a dihydrolyze, rather than a decarboxylase, reaction, in which an amine group (NH_2) is removed from arginine as a first step and citrulline to ornithine which then undergoes decarboxylation to form putrescine. Moeller decarboxylase medium contains both glucose and amino acid. During the initial stage of incubation, the glucose is fermented and produces yellow color. If the arginine is decarboxylated, alkaline amines are formed, and the medium reverts to its original purple color.

12. This test is used to differentiate the ability of the organism to reduce nitrate to nitrite using anaerobic respiration. The organisms produce nitrate reductase, which converts nitrate to nitrite, and the reduction is determined by adding sulfanilic acid and alpha-naphthylamine. The sulfanilic acid and nitrite react to form a diazonium salt which then couples with alpha-naphthylamine to produce a red, water-soluble azo dye. If no color change occurs, the organism did not reduce nitrate or further reduced to NH_3, NO, or N_2O_2. At this point, zinc is added, and if nitrate remains, the zinc will reduce the compound to nitrite, and the reaction will turn positive indicating a negative test result. If there is no color change after the addition of zinc, this indicates that the organism reduced nitrate to one of the nitrogen compounds.
13. This test is used to determine an organism's ability to produce the enzyme urease, which hydrolyzes urea. Urea is the product of decarboxylation of amino acids. Hydrolysis of urea produces ammonia and carbon dioxide. Ammonia formation alkalinizes the medium, and the change in pH is detected by the color change of phenol red from light orange (pH 6.8) to magenta pink (pH 8.1).
14. Do not refrigerate samples. If there is a delay in transporting the specimen, keep it at room temperature or in an incubator at 37 °C.
15. Good venipuncture technique will avoid false-positive result that can be due to skin flora. In addition, certain percentage of the flora may overgrow the pathogen. The pathogen will be missed, i.e., once any organism multiplies either pathogen or skin flora, the machine flags positive for that blood culture.
16. In Gram stain, the primary stain is crystal violet, and all cells take up the purple crystal violet stain. Gram's iodine is applied to the bacterial smears. The iodine acts as a mordant, binds with crystal violet, and locks it into the peptidoglycan of bacteria. It also intensifies the purple color. The decolorizing agent, acetone, is a lipid solvent that melts the Gram-negative outer membrane and leads to decolourization of Gram-negative cells. It also dehydrates proteins, helping the primary stain to remain in Gram-positive cell walls. Finally, the counterstain safranin is applied, which stains the decolourized gram negative cells pink. Thus, at the end of the staining procedure, Gram-positive cells are purple, and Gram-negative cells are pink.
17. Do not despair if the stained smear is not visible to the naked eye; this may happen with a Gram-negative reaction.
18. Take care to make an even smear; otherwise, alcohol will continue to wash the violet/purple color from thick parts of the smear, while thin parts are being over-decolorized.

19. On nutrient agar, non-mucoid, opaque, and circular colonies will be observed. On blood agar, nonhemolytic, opaque, circular, and gray color colonies will be observed. On MacConkey agar, non-lactose, fermenting, opaque, and circular colonies will be observed.
20. Cytochrome oxidase test, mannitol motility medium, triple sugar iron agar, indole test, citrate utilization test, and urease test.
21. Oxidative-fermentative medium, arginine decarboxylase test, and nitrate reduction test.
22. The oxidase reagent is unstable and has to be freshly prepared.
23. When present, the cytochrome oxidase oxidizes the reagent tetramethyl-*p*-phenylene-diamine to indophenol, a deep blue or purple color product. When the enzyme is not present, the reagent remains in the reduced state and colorless.
24. The bacterial colonies must be transferred to the filter paper using clean platinum loop or glass rod, since rusted loop will give a false-positive reaction. Colonies from MacConkey agar will give a pink-violet color due to carry-over from the medium and show a false-positive reaction.
25. Rough stabbing or using a loop may give a false appearance of turbidity which is due to mechanical splitting of the medium.
26. Motile organisms will spread out into the medium from the site of inoculation where non-motile organisms remain. Mannitol fermenting organisms reduce the pH of the medium and turn the color from red to yellow, whereas mannitol non-fermenting organisms show no color change.
27. Alkaline slant/alkaline butt (K/K)—glucose, lactose, and sucrose non-utilizer.

    Alkaline slant/acid butt (K/A)—glucose fermentation only.

    Acid slant/acid butt (A/A)—glucose, sucrose, and/or lactose fermenter.

    Black precipitate in the butt indicates ferrous sulfide and hydrogen sulfide gas production. Bubbles or cracks in the tube indicate carbon dioxide or hydrogen production.
28. A positive indole test is indicated by the formation of pink to red color in the reagent layer on top of the medium within seconds of adding the reagent. If a culture is indole negative, the reagent will remain yellow.
29. Heavy inoculum may give false-positive result. Diluting the inoculum in saline prior to inoculate citrate is recommended to avoid carry-over of other carbon sources.

30. A positive result indicates growth on the medium, with a color change of the indicator. Growth results in the bromothymol blue indicator turning from green to blue. Negative result indicates absence of growth.
31. Acid production is required in the medium by the appearance of a yellow color. In the case of oxidative organisms, color production may be first noted near the surface of the medium. The following are the reaction patterns. Oxidative metabolism will produce acid (yellow) in open tube and alkaline (green) in covered tubes. Fermentative metabolism will produce acid in both the tubes, while non-saccharolytic metabolism will produce alkaline.
32. Decarboxylase test may show two layers of different colors, yellow and purple. Shake the tube gently before attempting to make an interpretation. The control tube containing no amino acid should remain yellow after 18–24 h of incubation, denoting that only glucose was fermented. A positive control (purple) invalidates all the decarboxylase tests, and no interpretation can be made. Change of purple color indicates a positive reaction; no color change indicates a negative reaction.
33. Development of red color within 30 s after the addition of reagents indicates presence of nitrite and positive reaction for nitrate reduction. If no color develops, this indicates either the nitrates have not been reduced or they have been reduced to products other than nitrite and a negative reaction for nitrate reduction.
34. Change in the color of slant from light orange to magenta indicates a positive reaction, while no color change indicates a negative reaction.
35. Manually, the genus- and species-level identification will take 48 h and 72 h, respectively. With automated systems like VITEK MS, the identification will be done within 3 h [5].

References

1. Winn WC (2006) Koneman's color atlas and textbook of diagnostic microbiology. Lippincott Williams & Wilkins, Philadelphia, PA
2. Tille P (2015) Bailey & Scott's diagnostic microbiology. Elsevier Health Sciences. C. V. Mosby Co, St. Louis
3. Collee JG, Fraser AG, Marmion BP, Simmons A (1996) Mackie & McCartney practical medical microbiology. Churchill Livingstone, New York, NY
4. MacFaddin JF (2000) Biochemical tests for identification of medical bacteria. Lippincott Williams & Wilkins, Philadelphia, PA
5. Chatzigeorgiou K-S, Sergentanis TN, Tsiodras S, Hamodrakas SJ, Bagos PG (2011) Phoenix 100 versus Vitek 2 in the Identification of Gram-Positive and Gram-Negative Bacteria: a Comprehensive Meta-Analysis. J Clin Microbiol 49(9):3284–3291

Chapter 2

Genotyping of *Acinetobacter baumannii* in Nosocomial Outbreak and Surveillance

Saranya Vijayakumar, Balaji Veeraraghavan, Agila K. Pragasam, and Yamuna Devi Bakthavachalam

Abstract

Acinetobacter baumannii is considered to be an important nosocomial pathogen responsible for various outbreaks that have resulted in a need for effective epidemiological typing methods. Different typing methods are available for *A. baumannii* epidemiological studies. Currently, the phenotypic typing methods are not being used and replaced by various molecular methods. In this chapter, two important epidemiological typing methods, pulsed-field gel electrophoresis (PFGE) and multi-locus sequence typing (MLST), are discussed.

Key words *Acinetobacter baumannii*, Epidemiological typing, PFGE, MLST

1 Introduction

Acinetobacter baumannii is an opportunistic gram-negative bacterium which has emerged as a clinically relevant pathogen causing a wide range of nosocomial infection outbreaks, community-acquired infections, or war- and natural disaster-related infections. The ability of *A. baumannii* to acquire resistance to all or almost all available antimicrobial agents and its ability to persist for a long time in a clinical environment are considered to be the main features behind its epidemic behavior and thus its successful emergence [1]. Initially, phenotypic typing methods were used to decipher the clonal nature of *A. baumannii* isolates. However, currently numerous genotypic typing methods are in place. There are a wide range of genotypic methods available for epidemiological typing of *A. baumannii* clinical isolates [1]. The most gold standard epidemiological method is pulsed-field gel electrophoresis followed by multi-locus sequence typing [2–4]. In this chapter, the abovementioned two methods will be discussed with respect to *A. baumannii*.

Indranil Biswas and Philip N. Rather (eds.), *Acinetobacter baumannii: Methods and Protocols*, Methods in Molecular Biology, vol. 1946, https://doi.org/10.1007/978-1-4939-9118-1_2,

2 Materials

2.1 Pulsed-Field Gel Electrophoresis (PFGE) (See Note 1)

1. NDSK buffer: 0.5 M EDTA, 1% *N*-lauroylsarcosine, proteinase K (pH 9.5). Solution may be heated to 50 °C to dissolve components.
2. TNE buffer: 10 mM Tris–HCl, 200 mM NaCl, 100 mM EDTA, pH 7.2.
3. EC lysis buffer: 6 mM Tris–HCl, pH 7.6, 1 M NaCl, 100 mM EDTA, 0.5% Brij-58, 0.2% deoxycholate, 0.5% *N*-lauroylsarcosine, lysozyme, RNase.
4. TE buffer: 10 mM Tris, 1 mM EDTA, pH 8.0.
5. 0.5 M EDTA.
6. Phosphate-buffered saline (PBS).
7. 1× TAE buffer: 40 mM Tris-acetate, 1 mM EDTA.
8. 0.5× TBE buffer: 44.5 mM Tris-borate, 1 mM EDTA.
9. 0.1 M phenylmethanesulfonyl fluoride (PMSF) solution (*see* **Note 2**).
10. Triton X-100, membrane research grade.
11. Restriction enzymes and corresponding 10× reaction buffer.
12. SeaKem standard (LE) or high gelling temperature (HGT) agarose for standard gels.
13. SeaPlaque LMP agarose for preparative gels.
14. Electrophoresis setup.
15. Commercially purchased PFGE system or custom-constructed PFGE system.
16. Gel casting tray and comb (custom made or commercial).
17. DNA insert mold (custom made or purchased) (*see* **Note 3**).

2.2 Multi-Locus Sequence Typing (MLST) (See Note 4)

1. Bacterial genomic DNA.
2. PCR buffer with enzyme.
3. Molecular grade water.
4. Forward and reverse primer (*see* **Note 5**).
5. Pipettes.
6. Thermal cycler.

3 Methods

3.1 Pulsed-Field Gel Electrophoresis

1. Determine the shape and volume of the desired inserts. The inserts should correspond to the dimensions of the wells in the final agarose gel. Alternatively, agarose inserts can be prepared

by filling 1 mL syringe or trough with the agarose-cell mixture and cutting the resulting brick into slices.

2. Obtain sample materials from the desired source (such as clinical isolates).
3. Separate the cells from storage or growth media, and wash to remove the debris and any lysed cells.
4. Pellet cells by centrifuging at 2000 × *g* for 15 min at 4 °C.
5. Discard the supernatant and wash the cells in TNE buffer twice.
6. Quantitate the cells using spectrophotometry (*see* **Note 6**).
7. Heat a 1.2% solution of LMP agarose in distilled water, and equilibrate to 42–50 °C in a water bath. This solution can be held until ready for use.
8. Mix the cells and agarose together. Mix the cells with an equal amount of agarose (*see* **Note 7**).
9. Pipette the agarose-cell mixture into a casting mold, taking care of bubbles. Allow the molds to set at 4 °C until the agarose has gelled.
10. Unmold the solidified inserts. If the agarose-cell suspension was poured as a brick, slice the brick into appropriate pieces.
11. Incubate the insert in EC lysis buffer overnight at 37 °C to create spheroplasts, disrupt membranes, and digest other cellular debris.
12. Drain the buffer from the tube and replace with NDSK.
13. Incubate an additional two nights in NDSK at 50 °C.
14. Place the inserts in TE containing 0.01 mM. Incubate overnight at room temperature (20–25) with gentle shaking.
15. Dialyze the inserts into a larger volume of TE. Gently shake or rock the inserts at 4 °C, and make several buffer changes over the course of at least 24 h.
16. Prepare agarose "chops" by mincing an insert into smaller (0.5–1 mm) pieces using a sterile glass cover slip.
17. Prepare digestion mixture, total volume 200 μL by combining 50 U of restriction enzyme, 20 μL of appropriate 10× restriction buffer, BSA, 0.02% Triton X-100, agarose chops, and sterile water to make up the volume. Gently shake the insert at the recommended temperature overnight (*see* **Note 8**).
18. To stop the reactions, remove the digestion mixture, and replace with 1 mL NDSK. Shake on ice for 2 h.
19. Determine the size and volume of the gel. Select the type of agarose to use. Standard agarose and HGT agarose are both used for PFGE. "Pulsed-field grade" agarose can also be purchased but is costly.

20. Decide an appropriate percentage of agarose. Commonly, 1% agarose gels are used for resolving DNA molecules up to 2 Mb. Lower concentrations of agarose can also be used if short running times are desired; however, higher gel concentrations reduce smearing and yield sharper bands.
21. Decide on a suitable buffer. Gels can be run in 1× TAE or 0.5× TBE. If the sample will be excised for subsequent enzymatic reactions, we recommend TAE buffer.
22. Add the appropriate amount of agarose to buffer, and dissolve using a hot plate stirrer. Ensure that agarose particles are completely dissolved before pouring, and cool solution to at least 55–60 °C before pouring.
23. Load the samples onto the gel. Fill the electrophoresis tank with buffer, and remove any air bubbles if present. Add loading dye very carefully. Load appropriate markers or ladders. Load appropriate control samples from every batch of DNA to ensure that the genomic molecules are intact.
24. Fill the electrophoresis tank with running buffer, and the running temperature should be set at 8.5 °C. Gently sink the gel into the tank.
25. Perform electrophoresis. Stain or document the results obtained. Pulsed-field gels can be stained and visualized using dyes and protocols for standard DNA agarose gels (ethidium bromide).

3.2 Multi-Locus Sequence Typing

1. Prepare master mix using PCR buffer with enzyme, molecular grade water, forward primer, and reverse primer for seven housekeeping genes separately.
2. Aliquot the master mix into seven respective tubes.
3. Then add genomic DNA into seven tubes containing master mix for seven housekeeping genes, respectively.
4. Set the respective cycling condition for MLST PCR in the machine. Keep the seven tubes into the machine carefully.
5. After PCR, remove the PCR tubes carefully, and store at 4 °C until the gel run.
6. Once the gel run is completed and the base pair for all the seven housekeeping genes found satisfactory, the remaining PCR products are given for targeted sequencing.
7. After targeted sequencing, the allelic sequence obtained for each of the seven genes was BLAST matched against the respective MLST database (PubMLST or Pasteur MLST) to find the allelic profile of seven housekeeping genes.
8. The allelic profile was submitted to the database to find the respective sequence type.

4 Notes

1. Pulsed-field gel electrophoresis (PFGE) is still considered as the gold standard for epidemiological investigations of *A. baumannii*. The genome is cut by a restriction enzyme, producing a few large fragments that are separated by electrophoresis technique, periodically changing the direction of an electric field, and allowing the resolution of large fragments up to 1000 kb. The choice of restriction enzymes is a key factor in PFGE process. The most widely used restriction enzymes is *Apa*I followed by *Sma*I. PFGE has got high discriminatory power which makes it the technique of choice for typing *A. baumannii* outbreaks. PFGE has also proven its role in retrospective local studies, in national and international surveillance of multidrug-resistant (MDR) *A. baumannii*.
2. Prepare a 0.1 mM solution of PMSF in isopropyl alcohol. Store at −20 °C. PMSF is a toxic chemical. Wear gloves and face protection.
3. Alternatively, 1 mL plastic syringe with tip cut off or similarly shaped item such as Teflon tubing.
4. MLST is based on the multi-locus enzyme electrophoresis but directly relies on DNA sequence comparison of internal fragments of protein-encoding housekeeping genes. For each gene fragment, the different sequences are assigned as distinct alleles, and each isolate is defined by the alleles at each of the housekeeping loci (the allelic profile or sequence type [ST]). Two MLST schemes are available for *A. baumannii*: PubMLST or Oxford scheme and Pasteur MLST.
5. For Oxford or PubMLST scheme—seven housekeeping genes (*gltA*, *gyrB*, *gdhB*, *recA*, *cpn60*, *gpi*, *rpoD*). For Pasteur scheme—seven housekeeping genes (*cpn60*, *fusA*, *gltA*, *pyrG*, *recA*, *rplB*, *rpoB*).
6. Resuspension of pellets after final wash and/or count, it should be noted that this solution will be diluted to 1:1 into agarose. Therefore, the concentration and the resuspension volume should be calculated accordingly.
7. The cell suspension should be briefly warmed in a water bath at 37 °C before mixing with agarose. Mixing should be done carefully to avoid damaging the cells and shearing the DNA.
8. All the components of the digestion mixture should be checked for endogenous activity. This can be performed by setting up mock digestions with chopped inserts and restriction buffer. Additional reaction component can be added to separate

sample and incubated for few hours at proper temperature. When PFGE is run on these samples, the nuclease activity will be revealed as smeared bands or smear of high-molecular-weight DNA entering the gel matrix.

References

1. Rafei R, Kempf M, Eveillard M, Dabboussi F, Hamze M, Joly-Guillou ML (2014) Current molecular methods in epidemiological typing of *Acinetobacter baumannii*. Future Microbiol 9(10):1179–1194
2. Herschleb J, Ananiev G, Schwartz DC (2007) Pulsed-field gel electrophoresis. Nat Protoc 2 (3):677
3. Bartual SG, Seifert H, Hippler C, Luzon MAD, Wisplinghoff H, Rodríguez-Valera F (2005) Development of a multilocus sequence typing scheme for characterization of clinical isolates of *Acinetobacter baumannii*. J Clin Microbiol 43(9):4382–4390
4. Diancourt L, Passet V, Nemec A, Dijkshoorn L, Brisse S (2010) The population structure of *Acinetobacter baumannii*: expanding multi-resistant clones from an ancestral susceptible genetic pool. PLoS One 5(4):e10034

Chapter 3

Antimicrobial Susceptibility Testing Methods for *Acinetobacter* spp.

Balaji Veeraraghavan, Saranya Vijayakumar, Agila K. Pragasam, Yamuna Devi Bakthavachalam, and John A. J. Prakash

Abstract

Serial twofold dilution methods are most commonly used to identify the antimicrobial activities of antibiotics. This can be achieved by different methods like broth dilution or agar dilution. Though these methods are simple, they can be influenced by various experimental factors and result in discrepancy. The following protocol has been validated for *Acinetobacter* species, including *A. baumannii*. It is important to include appropriate control strains to determine the minimum inhibitory concentration values and to compare the experiment results.

Key words *Acinetobacter* spp., Antimicrobials, Susceptibility, Disk diffusion, MIC

1 Introduction

Acinetobacter species are ubiquitous in nature and the most common gram-negative organisms found. They have the ability to develop resistance to a variety of antimicrobial agents and to cause infection in debilitated hosts. Isolates that are clinically significant must be tested for antimicrobial susceptibility in order to provide appropriate therapy [1]. In *Acinetobacter* species, especially the important nosocomial pathogen, *Acinetobacter baumannii*, the increase in drug resistance can be due to intrinsic or chromosomal and acquired or plasmid-mediated mechanisms. Bacterial strains causing nosocomial infections can be defined with terms ranging from multidrug (MDR) to extensively drug resistant (XDR) to pan-drug resistant (PDR), depending upon their extent of drug resistance. Clinically, a strain can be categorized as "susceptible," "intermediate," or "resistant" based on their minimum inhibitory concentration (MIC) of a particular antimicrobial agent. MIC is defined as the lowest drug concentration at which no bacterial growth is visible to the naked eye. International agencies like

Indranil Biswas and Philip N. Rather (eds.), *Acinetobacter baumannii: Methods and Protocols*, Methods in Molecular Biology, vol. 1946, https://doi.org/10.1007/978-1-4939-9118-1_3,

CLSI and EUCAST regularly evaluate the breakpoint concentrations and also provide drug susceptibility testing protocols as well as results interpretation. Breakpoints are being revised annually, if required.

Various factors may influence MIC values; therefore, it is important to validate the experiments by testing appropriate reference strains as controls [2–6]. In addition, the composition of media used for MIC determination needs special attention, as it varies from one manufacturer to the other or from one batch to another. Also the resistance level of a strain to an antibiotic results from the interplay of various mechanisms; the physiological condition of the bacteria to be tested may have a huge impact on the MIC of a drug. Therefore, to avoid uninterpretable results, a fresh overnight grown culture must be used for antimicrobial susceptibility testing.

We describe different methods for testing antimicrobial susceptibility. This includes macro broth dilution, micro broth dilution, agar dilution methods, checkerboard assay, and time kill assay.

2 Materials

2.1 Common Materials

1. Cation-adjusted Mueller-Hinton broth—CAMHB (*see* **Note 1**).
2. Sterile Pasteur pipettes.
3. Overnight culture of test and control organisms.
4. Required antibiotic in powder form (either from the manufacturer or standard laboratory accompanied by a statement of its activity in mg/unit or per mL. Clinical preparations should not be used for reference technique).
5. Required solvent for the antibiotic.
6. Sterile distilled water—500 mL.
7. Sterile normal saline.
8. 0.5 McFarland turbidity standard.
9. Multichannel pipettes—50 μL and 10 μL with sterile disposable tips.

2.2 Macro Broth Dilution Method

1. Sterile graduated pipettes—10 mL, 5 mL, 2 mL, and 1 mL.
2. Sterile capped 75 × 12 mm tubes or small screw-capped bottles.
3. Test tube rack to hold 22 tubes in two rows, i.e., 11 tubes in each row.

2.3 Micro Broth Dilution Method

1. U-bottom or V-bottom 96-well microtiter plates.
2. Micropipettes 20–200 μL, 10 μL with sterile tips.

2.4 Agar Dilution Method

1. Cation-adjusted Mueller-Hinton Agar (CAMHB)..
2. 90 mm diameter petri dishes.

2.5 Time Kill Assay

1. Sterile saline dispensed in 0.9 mL and 9.9 mL aliquots.
2. Sterile 1 mL, 5 mL, and 10 mL pipettes.
3. Nutrient agar plates.

3 Methods

3.1 Preparation of Antimicrobial Stock Solution

1. Purchase antibiotic powders from pharmaceutical companies or suitable suppliers. Store the antibiotic powders according to the manufacturer's recommendations (*see* **Notes 2–5**).
2. Prepare antimicrobial agent stock solutions according to the potency mentioned in the technical information sheet.
3. The range of concentration to be tested for each antibiotic is to be chosen based on the quality control ranges of the QC strain against the test organism.
4. Use the following formula to prepare stock solutions of antibiotic:

$$\text{Volume(mL)} = \frac{\text{Weight(mg)} \times \text{Potency } (\mu g/\text{mg})}{\text{Concentration } (\mu g/\text{mL})}$$

5. Dissolve the antibiotic powders using an appropriate diluent. Aliquot the stock solution and store. Aliquots must be thawed on the same day of testing and discarded.
6. For time kill assay:

$$\text{Volume(mL)} = \frac{\text{Weight(mg)} \times \text{Potency } (\mu g/\text{mg})}{\text{Concentration } (\mu g/\text{mL})}$$

Prepare stock solution in a suitable solvent, so that the concentration is 100 times the X concentration that will be tested in time kill test, as 0.1 mL of the stock concentration is added into the test medium of 9.9 mL CAMHB (1/100 dilution).

3.2 Inoculum Preparation

1. Inoculate 2–3 morphologically similar colonies of overnight grown culture from a nutrient agar plate/suitable media into 1.5 mL nutrient broth (*see* **Note 6**).
2. Incubate at 37 °C for 2 h and then adjust the culture to 0.5 McFarland standards.

3.3 Macro Broth Dilution (BMD)

1. Prepare serial twofold dilution of the antimicrobial range to be tested in broth.
2. A minimum final dilution of 1 mL of each dilution is needed for the test.
3. Prepare inoculum as per standard method mentioned before. Dilute the 0.5 McFarland suspension 1:150 in a tube (1×10^6 CFU/mL).
4. Within 15 min of preparation of the standardized inoculum, add 1 mL into tubes containing 1 mL of the antimicrobial dilution (final inoculum of 5×10^5 CFU/mL).
5. Include a growth control tube with the standard inoculum in broth not containing any antimicrobial agent.
6. Include a sterility control tube with the broth alone with no antimicrobials or inoculum.
7. Subculture the inoculum into nonselective media for purity check.
8. Verify the inoculum as follows:
   (a) Take 0.01 mL of the growth control and dilute in 10 mL sterile saline.
   (b) Plate 0.1 mL onto a nonselective plate. Incubate overnight at 37 °C.
   (c) Count colonies after overnight incubation.
   (d) Approximate count of 50 indicates an inoculum density of 5×10^5 CFU/mL.
9. Incubate the tubes at 37 °C for 16–20 h in ambient air.
10. Refer to **Note 11** for interpretation of results.

3.4 Micro Broth Dilution

1. Prepare antimicrobial stock and serial twofold dilution of the antimicrobial range to be tested in broth as in macro broth dilution.
2. Add 0.1 mL of the antibiotic into the wells of 96-well microtiter plate.
3. Include one sterility well (only broth, no antimicrobial/inoculum) and one growth control well (broth with inoculum, no antimicrobial) in each tray.
4. Add 0.01 mL of the standardized inoculum preparation (0.5 McFarland) in all the wells except the sterility well using prongs or with micropipettes.
5. Seal the plate with a plastic tape or in a plastic bag.
6. Check purity of inoculum by subculturing onto a nonselective agar plate.
7. Verify the inoculum as follows:

(a) Take 0.01 mL of the growth control and dilute in 10 mL sterile saline.

(b) Plate 0.1 mL onto a nonselective plate. Incubate overnight at 37 °C.

(c) Count colonies after overnight incubation.

(d) Approximate count of 50 indicates an inoculum density of 5×10^5 CFU/mL.

8. Incubate the trays stacked no higher than four plates at 37 °C for 16–20 h in ambient air.
9. Refer to **Note 12** for interpretation of results.

3.5 Agar Dilution Method

1. Prepare antimicrobial stock and serial twofold dilution of the antimicrobial range to be tested in broth as in macro broth dilution. The antimicrobial solution should have 10 times the needed concentration on plate (e.g., required concentration on plate is 4 μg/mL; prepare 40 μg/mL solution: 2 mL of 40 μg/mL solution is added with 18 mL of Mueller-Hinton agar (MHA) for plate preparation).
2. Label petri dishes with the antimicrobial concentration to be poured.
3. Add 2 mL of the antimicrobial solution with 18 mL of melted MHA at 55 °C and mix thoroughly.
4. Pour the mixture onto the petri dish with appropriate label and allow to set.
5. Include one plate without antimicrobial agents to serve as growth control.
6. Dry the plates at 37 °C with their lids tipped for 20–30 min in an incubator.
7. Prepare inoculum as per standard method mentioned before. Dilute 1:10 in either sterile broth (nutrient/MHB) or saline.
8. Inoculate either with a multiple inoculator as spots or with a wire loop or a platinum loop calibrated to deliver 0.001 mL and spread over a 5 mm to 8 mm spot on the MHA containing antimicrobial agents.
9. Allow the plates to stand at room temperature until moisture in inoculum spots has been absorbed in the agar.
10. Incubate at 37 °C for 16–20 h in ambient air.
11. Refer to **Note 13** for interpretation of results.

3.6 Micro Broth Checkerboard Assay (See Appendix 1)

1. Determine the minimum inhibitory concentration (MIC) of single agent for the test isolate.
2. Prepare serial doubling dilution of antimicrobial solution two times the needed concentration in the well (*see* Fig. 1).

Agent A →

Agent B ↓

GC	0.015X MIC	0.03X MIC	0.06X MIC	0.125X MIC	0.25X MIC	0.5 X MIC	1X MIC	2X MIC	4X MIC	8X MIC	16X MIC
0.06X MIC											
0.125X MIC											
0.25X MIC											
0.5X MIC					E6						
1X MIC											
2X MIC											
4X MIC											**SC**

Fig. 1 Template for preparation of checkerboard panel. First row and first column contain single antimicrobial agents A and B, respectively. Other wells contain combination of antimicrobials except A1 and H12. Example: Well E6 contains antimicrobial agent A at 0.25 times and antimicrobial agent B at 0.5 times of MIC. *GC* growth control; *SC* sterility control

3. Include concentration ranging from four to eight times the expected MIC to at least 1/8–1/16 times the expected MIC. As an example, to test an isolate for antimicrobials A and B with MICs 32 and 4 μg/mL, respectively, test concentrations 2–256 μg/mL for agent A and concentrations 0.25–32 μg/mL for agent B.
4. Dispense 50 μL of the agents and 50 μL of CAMHB along the first row and column. Leave well A1 for growth control as shown in the template below. To the remaining wells, add 50 μL each of agents A and B as in the template. Leave well H12 for sterility control.
5. Add 100 μL of CAMHB on wells A1 and H12.
6. Prepare inoculum as per standard method mentioned before:
   (a) Dilute the suspension 1:31 by adding 0.8 mL into 25 mL of water diluent.
   (b) Inoculate 10 μL into all wells except sterility check well.
7. Seal the plate with plastic cover and incubate for 16–20 h at 37 °C in ambient air.
8. Refer to **Note 14** for interpretation of results.

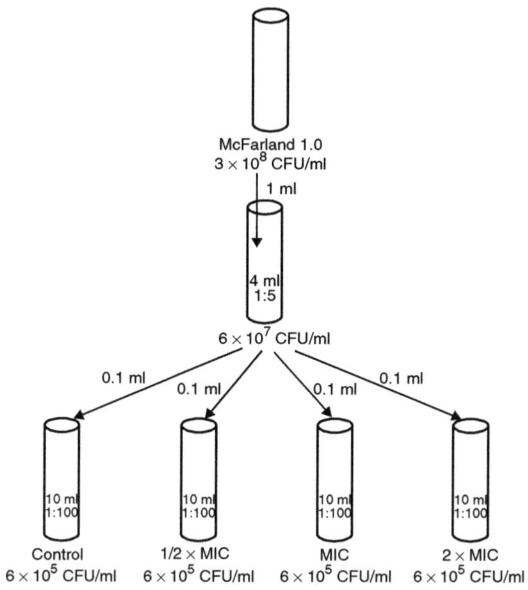

Fig. 2 Schematic diagram of preparation of inoculum tubes for time kill assay

3.7 Time Kill Assay (See Appendix 1)

1. Prepare inoculum as per standard method mentioned before:
   (a) Transfer 0.1 mL of this suspension to 5 mL of CAMHB and incubate at 37 °C till mid-logarithmic growth (2–3 h).
   (b) Adjust turbidity to 1.0 McFarland standard.
   (c) Dilute the suspension 1:5 by adding 1 mL of the suspension to 4 mL of CAMHB (6×10^7 CFU/mL) (*see* Fig. 2).
2. Follow these steps for inoculation:
   (a) Label tubes for single agents, combination, sterility control, and growth control (*see* Fig. 3).
   (b) For individual agents, add 0.1 mL of the antimicrobial solution to 9.9 mL of CAMHB.
   (c) For combination agents, add 0.1 mL of each agent to 9.8 mL of CAMHB.
   (d) Add 10 mL of CAMHB to the growth control and the sterility check tubes.
   (e) Inoculate 0.1 mL of the standardized inoculum into the tubes (6×10^5 CFU/mL), except the sterility check control tube.
   (f) Check purity of inoculum by plating 10 μL onto nonselective media.
   (g) Immediately after inoculation, remove 0.1 mL from each tube to perform serial dilution bacterial count for time zero hours (*see* Fig. 4).
3. Incubate the tubes and the purity plates at 37 °C.

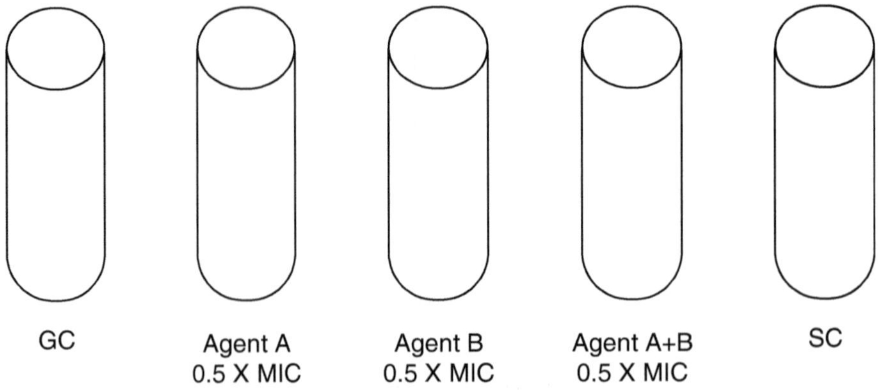

Fig. 3 Tube markings and labelling for testing. *GC* growth control; *SC* sterility control

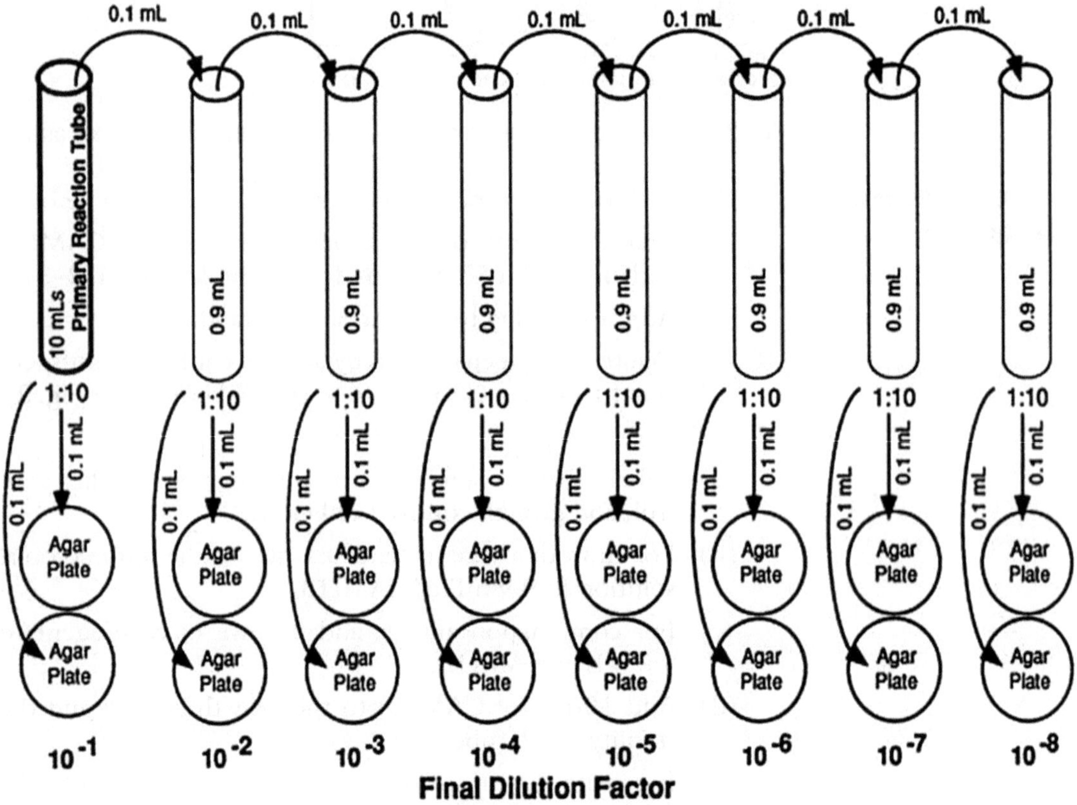

Fig. 4 Dilution scheme for evaluating bacterial colony count during time kill assay

4. Subsequent sampling times are determined by the mode of action of the agents. For concentration-dependent drugs, sample at short time intervals every 1–2 h to 24 h. For time-dependent drugs, sample at longer time intervals every 4–24 h.
5. Incubate at 37 °C immediately after sampling.
6. Refer to **Note 15** for interpretation of results.

4 Notes

1. Generally, MIC determinations are done using Mueller-Hinton agar (MHA). However, MIC results for aminoglycosides and polymyxins vary according to Mg^{2+} and Ca^{2+} concentrations. It is recommended to use cation-adjusted Mueller-Hinton agar (CAMHA).
2. Microbial contamination of antibiotic powders is unusual. However, it is necessary to sterilize the stock solutions by ultrafiltration.
3. Appropriate multichannel pipettes should be used to deliver accurate volumes of bacterial suspension.
4. Freezing stock solution more than once is not recommended. Antibiotics like β-lactams can undergo degradation when thawed and refrozen. Individual aliquots of stock solutions must be prepared and used.
5. The pipette tip should be changed whenever preparing a different dilution.
6. The volume of MHA should be prepared according to the number of strains and antibiotics tested.
7. The plates with antibiotics should be prepared on the same day. If they contain stable antibiotic, they can be stored at 4–8 °C for up to 3 days [7].
8. For subculture of test strains, drug-free medium should be used [7].
9. Use of commercial antimicrobial susceptibility testing (cAST) methods. In the post-antibiotic era, very few agents like tetracycline (minocycline and tigecycline) and colistin are available for the management of drug-resistant infections. Majority of the clinical microbiology laboratories perform automation-based commercial antimicrobial susceptibility testing (cAST) methods. However, caution must be taken while reporting susceptibility to *Acinetobacter* spp. when cAST is performed. A brief note on these cAST pitfalls is summarized below [8].

   Colistin: Micro broth dilution is the reference method for determining susceptibility to colistin. CLSI and/or EUCAST do not recommend the diffusion-based assays and cAST for colistin testing.

   Tigecycline: No breakpoints are available for clinical interpretation of susceptible or non-susceptible by either CLSI or EUCAST guidelines. Testing is critical, as there are no FDA-approved cAST available as of today. BMD is to be performed, as it is the reference method. Tests other than BMD yield very major error rates of 7% for Microscan, 36% for Vietk2, and 60% for E-test. Hence, caution must be taken while reporting the test results for these agents based on the testing method.

10. Antimicrobials not to be tested and reported for *Acinetobacter* spp. *A. baumannii* is inherently resistant to few antimicrobials, for which susceptibility should not be tested and reported. It includes ampicillin, amoxicillin, amoxicillin/clavulanate, aztreonam, ertapenem, trimethoprim, chloramphenicol, and fosfomycin [7].
11. Results and interpretation for macro broth dilution: Check purity of inoculum (should be homogeneous). Also check sterility control tube; it should be clear/non-turbid. Verify growth control tube for adequate turbidity (for macro broth dilution). Always run a standard strain of known MIC value with the test as the control to check the reagents and conditions. MIC is expressed as the lowest dilution, which inhibited growth judged by lack of turbidity in the tube.
12. Results and interpretation for micro broth dilution: Check purity of inoculum (should be homogeneous). Also check sterility control tube; it should be clear/non-turbid. Check growth control for adequate growth of at least 2 mm button formation. MIC is the lowest concentration of antimicrobial agent that completely inhibits growth of the organism in the microdilution wells as detected by the unaided eye. When single skipped well is seen, read the highest MIC. Do not report results if more than one skipped well is present.
13. Results and interpretation for agar dilution: Place the plates on a dark, nonreflecting surface and record MIC as the lowest concentration of antimicrobial agent that completely inhibits growth. Disregard a single colony or faint haze caused by inoculum. If two or more colonies persist in concentrations beyond an obvious end point, or there is no growth at lower concentration and growth at higher concentration, check purity and repeat if required.
14. Results and interpretation for checkerboard assay: Check purity of inoculum by subculturing onto a nonselective agar plate. Verify count by transferring 10 μL from the growth control well to 10 mL sterile saline. Mix thoroughly and plate 100 μL onto a nonselective agar plate and incubate at 37 °C overnight. A colony count of 30–70 indicates adequate inoculum of 4–5 $\times$ 10^5 CFU/mL. Check the growth control well for heavy turbidity. Record MIC of agents A and B from the first row and column, respectively. Examine each combination well and record growth or no growth for each well in the template mentioned below (*see* Fig. 5). For all clear wells in the turbid/non-turbid border , calculate the fractional inhibitory concentration (FIC) of each agent as follows:

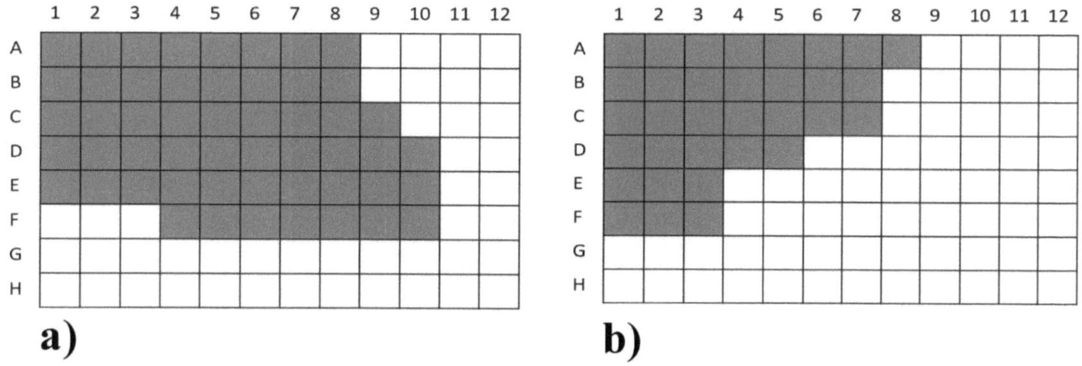

Fig. 5 Representation of checkerboard assay. Test results showing the pattern of turbid versus non-turbid wells for the interpretation of synergistic or antagonistic effect of the antimicrobial combinations tested. (**a**) Antagonistic effect and (**b**) synergistic effect. Gray box, turbid (indicates growth); white box, non-turbid (indicates no growth)

$$\text{FIC of agent A} = \frac{\text{MIC of agent A in combination}}{\text{MIC of agent A alone}}$$

$$\text{FIC of agent B} = \frac{\text{MIC of agent B in combination}}{\text{MIC of agent B alone}}$$

Calculate the Σ FIC index by adding FIC of agent A and agent B and record the minimum and the maximum Σ FIC index. Interpret the results as follows:

Synergism is indicated by Σ FIC index ≤ 0.5.

Indifference by Σ FIC index 0.5–4.

Antagonism by Σ FIC index >4.

15. Results and interpretation for time kill assay: When synergism is expected for the combination, the concentration most likely to produce synergism as detected by previous checkerboard assay is used. A standard of $0.5\times$ MIC is generally tested to demonstrate synergy of any combinations. Alternatively, antimicrobial concentration achievable in blood after usual dosage and administration route can be used. Antimicrobial concentration 100 times the required volume is prepared. Each antibiotic solution should be checked by testing with known QC strains. After 24 h, inspect the tubes for turbidity and final sampling for count. Determine the colony count at time zero and at other sampling times. Re-incubate plates if the colonies appear small to count. Average the counts from plates with 30–300 colonies. Do not use plates with <30 or >300 colonies unless these are the only counts available for that time period. Convert to actual CFU/mL by multiplying average raw counts by diluting

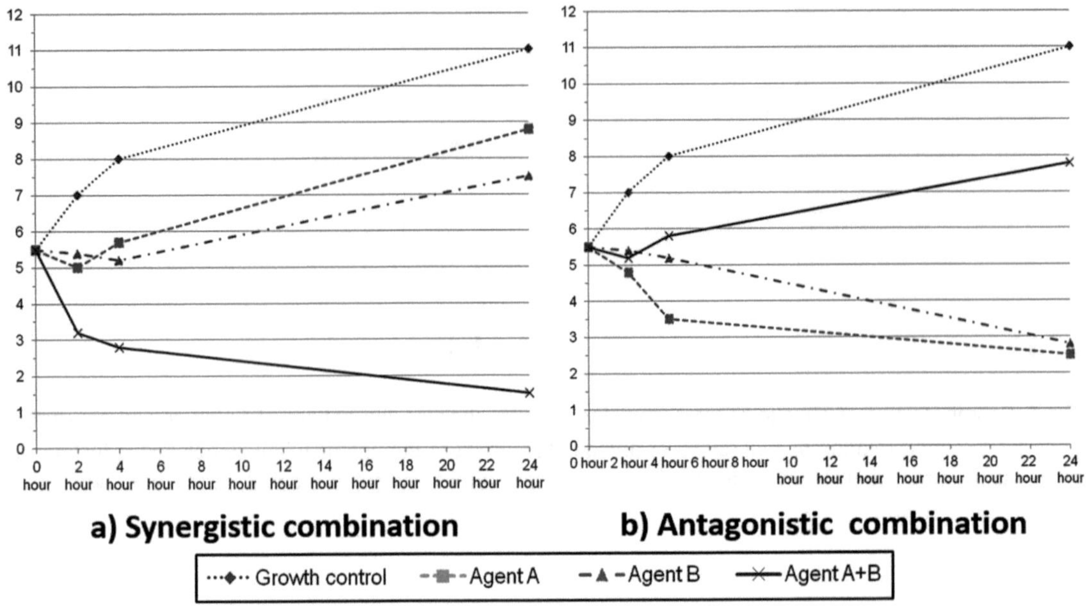

Fig. 6 Representation of growth kill curve of synergistic and antagonistic combination by time kill assay

factors. Convert CFU/mL to $\log_{10}$ data. The bacterial counts at each time period can also be expressed as line graph with the $\log_{10}$ count in the x-axis and time in the y-axis (as shown in Fig. 6). Determine the $\log_{10}$ CFU/mL between the most active single agent and the combination of agents. Synergy is defined as ≥ 2 $\log_{10}$ CFU/mL decrease by the combination compared to the most active single agent. Antagonism is defined as ≥ 2 $\log_{10}$ CFU/mL increase by the combination compared to the most active single agent. Bactericidal effect is indicated by change of ≥ 3 $\log_{10}$ decrease at 24 h from the initial inoculum.

Appendix 1: Templates to Record the Test Results

A. Micro-broth Checkerboard Assay Worksheet

Date: **Combination drug:** **Test isolate ID:**

	1	2	3	4	5	6	7	8	9	10	11	12
A	GC											
B												
C												
D												
E												
F												
G												
H												SC

Test isolate ID		Sterility Control		Purity Check		
Clear Wells		FIC (A)		FIC (B)	ΣFIC	Interpretation

Remarks:

B. Time-kill Assay Worksheet

Date:
Agent (A):
MIC (A):

Sample ID/year:
Organism:
Agent (B):
MIC (B):

Antibiotic conc. tested (μg/mL)	Time (h)	Actual CFU count at respective dilution (0.1 mL plated in duplicates)								Actual CFU		Log_{10} CFU	
		10^{-1}	10^{-2}	10^{-3}	10^{-4}	10^{-5}	10^{-6}	10^{-7}	10^{-8}	Avg. CFU/mL	Δ CFU/mL*	Avg. CFU/mL	Δ CFU/mL*

*ΔCFU/ml = Control – Drug

Sterility Control
Purity check
Smear report:

Growth Control	**At 24 h**
Agent A	
Agent B	
Agent A + Agent B	

Re incubation Reading after 48 h

Remarks:

Signature

	0 h, 3 h, 6 h (Day 1)	24 h (Day 2)	
1.			
2.			

References

1. Swenson JM, Killgore GE, Tenover FC (2004) Antimicrobial susceptibility testing of *Acinetobacter spp.* by NCCLS broth microdilution and disk diffusion methods. J Clin Microbiol 42 (11):5102–5108
2. Laishram S, Pragasam AK, Bakthavatchalam YD, Veeraraghavan B (2017) An update on technical, interpretative and clinical relevance of antimicrobial synergy testing methodologies. Indian J Med Microbiol 35(4):445
3. Li X, Wang L, Zhang XJ, Yang Y, Gong WT, Xu B, Zhu YQ, Liu W (2014) Evaluation of meropenem regimens suppressing emergence of resistance in Acinetobacter baumannii with human simulated exposure in an in vitro intravenous-infusion hollow-fiber infection model. Antimicrob Agents Chemother 58 (11):6773–6781
4. NCCLS (2003) Performance standards for antimicrobial disk susceptibility tests, 8th edn. Approved standard M2-A8. NCCLS, Wayne, PA
5. NCCLS (2003) Methods for dilution antimicrobial susceptibility tests for bacteria that grow aerobically, 6th edn. Approved standard M7-A6. NCCLS, Wayne, PA
6. Truant AL (ed) (2002) Manual of commercial methods in clinical microbiology. ASM Press, Washington, DC
7. CLSI (2018) Performance standards for antimicrobial susceptibility testing 28th edn. CLSI, Wayne, PA
8. Eric Wenzler, Debra A. Goff, Romney Humphries, Ellie J. C. Goldstein, (2017) Anticipating the Unpredictable: A Review of Antimicrobial Stewardship and Acinetobacter Infections. Infectious Diseases and Therapy 6 (2):149–172

Chapter 4

Methods to Evaluate Colistin Heteroresistance in *Acinetobacter baumannii*

Edgar X. Sherman, Jessie E. Wozniak, and David S. Weiss

Abstract

The nosocomial pathogen *Acinetobacter baumannii* is a growing threat to public health due to its increasing resistance to antibiotics including the last-line polymyxin, colistin. Heteroresistance to colistin has been described in *A. baumannii*, wherein a resistant subpopulation of cells coexisting with a majority susceptible subpopulation actively grows in the presence of antibiotic and can cause treatment failure. The shortcomings of diagnostic tests in detecting colistin heteroresistance are especially worrisome as they may lead to clinicians unknowingly prescribing an ineffective antibiotic, leading to increased patient morbidity and mortality.

Several techniques can be used to detect heteroresistance, and the purpose of this chapter is to outline effective methods for identifying, quantifying, and analyzing heteroresistance to colistin in *A. baumannii*. We will highlight the advantages and disadvantages of techniques including population analysis profile (PAP), Etest, and disc diffusion, as well as additional methods to distinguish heteroresistance from other forms of resistance. While the scope of this chapter will focus on colistin heteroresistance in *A. baumannii*, these techniques can be adapted for the study of heteroresistance to other antibiotics and in other bacteria with slight modifications.

Key words Heteroresistance, Resistant subpopulations, Population analysis profile, Antibiotic susceptibility testing, Phenotypic antibiotic resistance, Polymyxin, Colistin

1 Introduction

Acinetobacter baumannii is a Gram-negative opportunistic nosocomial pathogen causing multiple types of infections including ventilator-associated pneumonia (VAP) and bacteremia, primarily in immunocompromised patients [1, 2]. This pathogen is a growing global health concern due to its increasing resistance to antibiotics. The Centers for Disease Control and Prevention have categorized multidrug-resistant *A. baumannii* as a serious threat to public health, and the World Health Organization listed

Edgar X. Sherman and Jessie E. Wozniak contributed equally to this work.

Indranil Biswas and Philip N. Rather (eds.), *Acinetobacter baumannii: Methods and Protocols*, Methods in Molecular Biology, vol. 1946, https://doi.org/10.1007/978-1-4939-9118-1_4, © Springer Science+Business Media, LLC, part of Springer Nature 2019

A. baumannii as the top priority antibiotic-resistant pathogen for new drug development [3, 4]. Unfortunately, isolates resistant to all available antibiotics have emerged, making this organism a significant obstacle to treat in the clinic.

To combat the growing threat of antibiotic resistance, we must address remaining gaps in our understanding of resistance mechanisms. Heteroresistance is an underappreciated and understudied phenomenon wherein a small subpopulation of bacterial cells resist an antibiotic, while the majority of the population is susceptible [5]. This is in contrast to "conventional" resistance in which the entire population is resistant to the antibiotic. During drug treatment of a heteroresistant isolate, the susceptible population is killed, while the resistant cells survive and can rapidly expand. This rapid expansion makes this phenomenon distinct from persistence, in which a metabolically quiescent population of cells remains dormant and thereby resists antibiotics [6]. Heteroresistance was first described in 1947 and has since been identified in both Gram-negative and Gram-positive organisms against multiple classes of antibiotics [7–14].

The frequency of the resistant subpopulation in heteroresistant strains can vary, and when it is exceedingly low (less than ~1 in 10,000 cells), this form of resistance is often undetected by current diagnostic tests. Recently, Band et al. described strains of *Enterobacter cloacae* exhibiting heteroresistance (either detected or undetected) to the last-line polymyxin antibiotic colistin that caused colistin treatment failure and subsequent lethal infections in mice [15]. These results clearly demonstrated that heteroresistance can mediate failure of antibiotics in vivo. The shortcomings of diagnostic tests in detecting colistin heteroresistance are especially worrisome as this may lead to clinicians unknowingly prescribing ineffective antibiotics, causing increased patient morbidity and mortality.

Heteroresistance to colistin has been reported in *A. baumannii* [16–21]. The purpose of this chapter is to detail effective methods of detecting, quantifying, and analyzing colistin heteroresistance in *A. baumannii*.

2 Materials

Lysogeny broth (LB)—*Approximate formula per liter: tryptone (10.0 g), yeast extract (5.0 g), sodium chloride (10.0 g).*

Lysogeny broth (LB) agar—*Approximate formula per liter: tryptone (10.0 g), yeast extract (5.0 g), sodium chloride (10.0 g), agar (15.0 g).*

Sterile 100 mm × 15 mm polystyrene petri dishes.

LB agar plates.

Sterile phosphate-buffered saline (PBS).
Sterile 5 mL polystyrene tubes.
Sterile 96-well polystyrene plates.
Blank 6 mm filter paper discs.
Colistin sulfate (*Sigma-Aldrich*).
Sterile 1 μL inoculating loops.
Colistin Etest strips (*bioMerieux*).
Bacterial strains.

3 Methods

Unless otherwise mentioned, all steps are to be performed at room temperature and in a biosafety cabinet. Working stocks of colistin should be prepared fresh for each experiment and diluted in sterile water.

3.1 Population Analysis Profile (PAP)

Population analysis profile (PAP) is the gold standard method for detecting heteroresistance. Briefly, this technique involves the quantification of the proportion of resistant cells existing within a culture at a variety of antibiotic concentrations. PAP is the only current method for identifying heteroresistance that also quantitatively determines the frequency of the resistant subpopulation of cells. While it is the most reproducible and reliable method for detecting heteroresistance, it is also more time-consuming and requires more materials than other methods [22].

Day 1

1. Streak out isolates to be tested from freezer stocks on lysogeny broth (LB) agar plates, and incubate overnight at 37 °C.

   Prepare lysogeny broth agar plates as described:

   (a) Prepare LB agar in sterile H_2O on a magnetic stir plate.

   (b) Autoclave LB agar.

   (c) Cool LB agar in a water bath (60 °C) for 1 h.

   (d) Remove agar from the water bath and place on a spinning magnetic stir plate.

   (e) When agar has cooled (hot, but not uncomfortable to the touch), add colistin.

   (f) Final concentrations of colistin in LB agar should be 0, 1, 2, 4, 8, 16, 32, or 128 μg/mL.

   (g) Pour agar into petri dishes and let it stand overnight or until plates harden.

   (h) Store plates at 4 °C—Do not use plates that are more than 10 days old.

- Agar plates should have a depth of 4 mm (approximately 20 mL in a 100 mm circular plate).
- Plates should be dried before use with no visible drops of water on the agar surface or lid [5].
- Plates may be stored at 4 °C but must be properly dried before use to remove excess moisture.

Day 2

1. Pick an isolated colony with a sterile loop from the LB plates, and inoculate into 2 mL sterile lysogeny broth, and grow at 37 °C shaking at 225 rpm overnight (~16 h).

Day 3

1. Dilute 10 μL of the overnight culture into 90 μL sterile PBS in a sterile 96-well polystyrene plate. Make 10× serial dilutions ranging in concentration from 10^{-1} to 10^{-7}.
2. Plate dilutions ranging from 10^{-2} to 10^{-8} in 10 μL drops on agar plates (Fig. 1) of each colistin concentration, 0–128 μg/mL (*see* **Note 1**).
3. Allow plates to dry with covers removed for 15 min or until no liquid is visible on plates (*see* **Note 2**).
4. Incubate plates at 37 °C for 16 h.

Day 4

Analysis of data should be as follows:

Count the number of colonies at the highest dilution factor at which there is growth (*see* **Note 3**). For statistical interpretation, only count dilutions wherein the countable colonies are between 10 and 100. If it appears there are fewer than 10 colonies at your lowest dilution factor, we recommend plating a sample from this dilution in a higher volume on a single plate. With these data, the proportion of colistin resistant cells at each concentration can be quantified using the following formula:

$$\frac{(\text{number of colonies on colistin plate} \times \text{dilution factor})}{(\text{number of colonies on antibiotic free plate} \times \text{dilution factor})}$$

Interpret results as follows:

Based on our laboratory's experience performing this assay, as well as commentary in the literature, we recommend the following interpretation of population analysis profile results (*see* **Note 4**) [5, 15]:

Susceptible: Proportion of surviving bacteria at 2–128 μg/mL is below 10^{-7} (the limit of detection).

Resistant: Proportion of surviving bacteria at a concentration above 2 μg/mL is above 50%.

Heteroresistant: Proportion of surviving bacteria at a concentration at least four fold above the antibiotic's breakpoint is between 10^{-7} and 50% (Fig. 2).

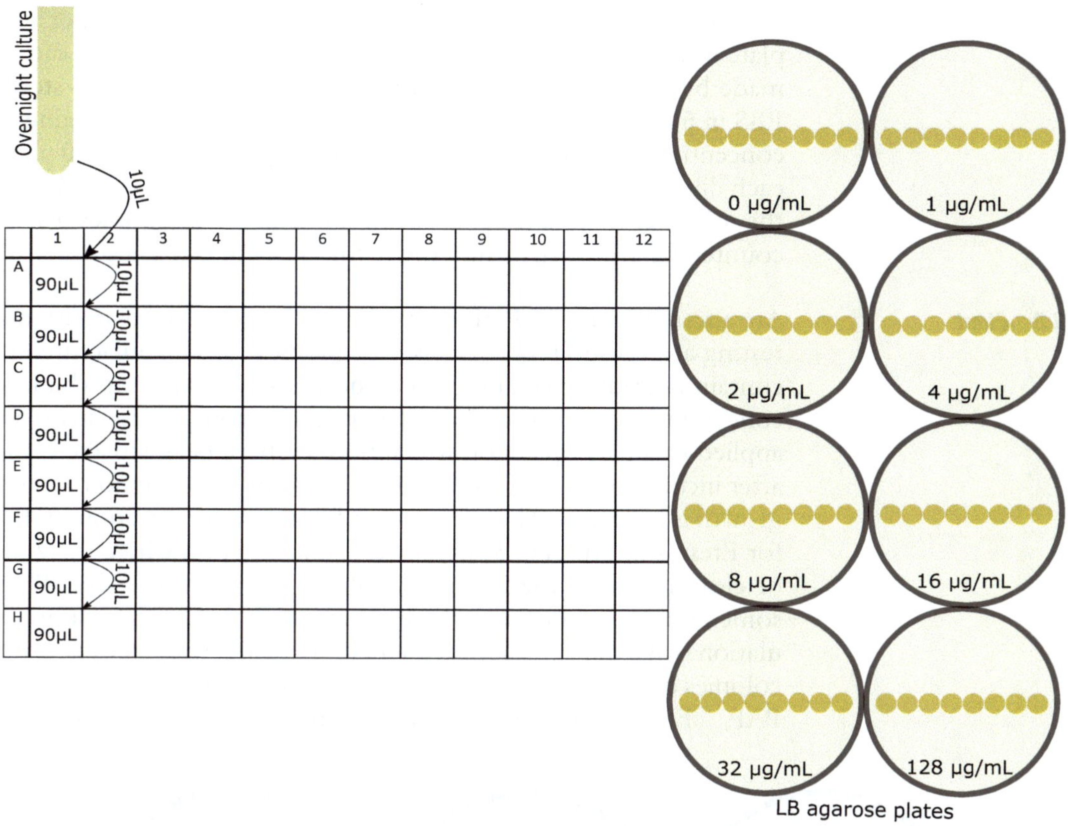

Fig. 1 Sample population analysis profile (PAP) procedure using 96-well plate. Note—while the authors prefer to plate eight serial dilutions in 10 μL drops per plate, there are multiple ways to achieve these results (*see* **Note 1**). Up to six rows can be spot plated on a 100 mm circular agar petri dish plate

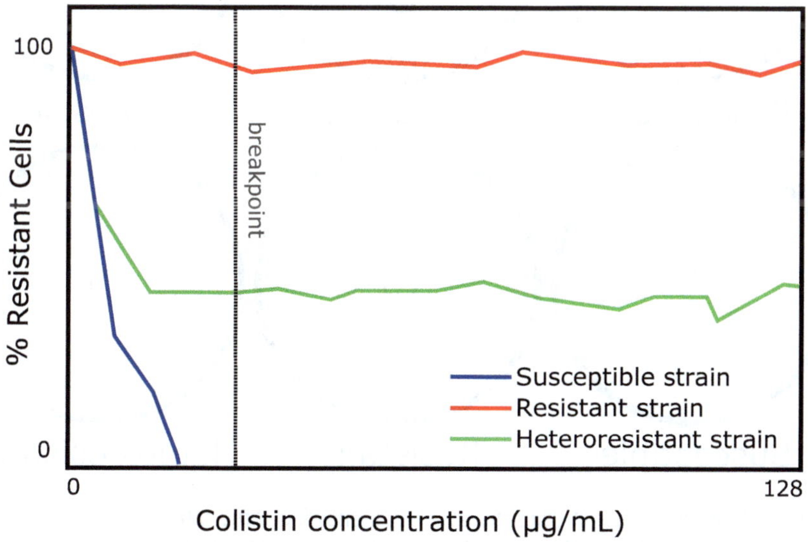

Fig. 2 Example of graph of a typical population analysis profile with three strains of varying susceptibility

Alternatively, dilutions can be plated with one sample per agar plate instead of the spot method. To do this, serial dilutions can be made by diluting 100 μL of overnight culture into 900 μL sterile PBS in 5 mL culture tubes, making 10× serial dilutions ranging in concentration from 10^{-1} to 10^{-7}. Once prepared, spread 100 μL of each dilution onto an individual agar plate. To enumerate CFU, use the dilution that allows between 30 and 300 colonies to be counted, and quantify using the methods mentioned above.

3.2 Etest

An alternative to PAP that can be utilized for heteroresistance testing are Etests (Fig. 3). Etests are plastic strips manufactured to contain a gradient of antibiotic concentrations from top (highest concentration) to bottom (lowest concentration). These strips are applied to an agar plate on which bacteria have been streaked, and after incubation, a "zone of clearing" appears in which no bacterial growth is observed. The minimum inhibitory concentration (MIC) for Etest is interpreted as the lowest concentration where the zone of clearing meets the Etest strip along the bottom edge. When some heteroresistant isolates (those in which the resistant subpopulation is at a high enough frequency) are plated with an Etest strip, colonies can be observed within the zone of clearing. Compared to PAP, Etests are a nonquantitative method to observe

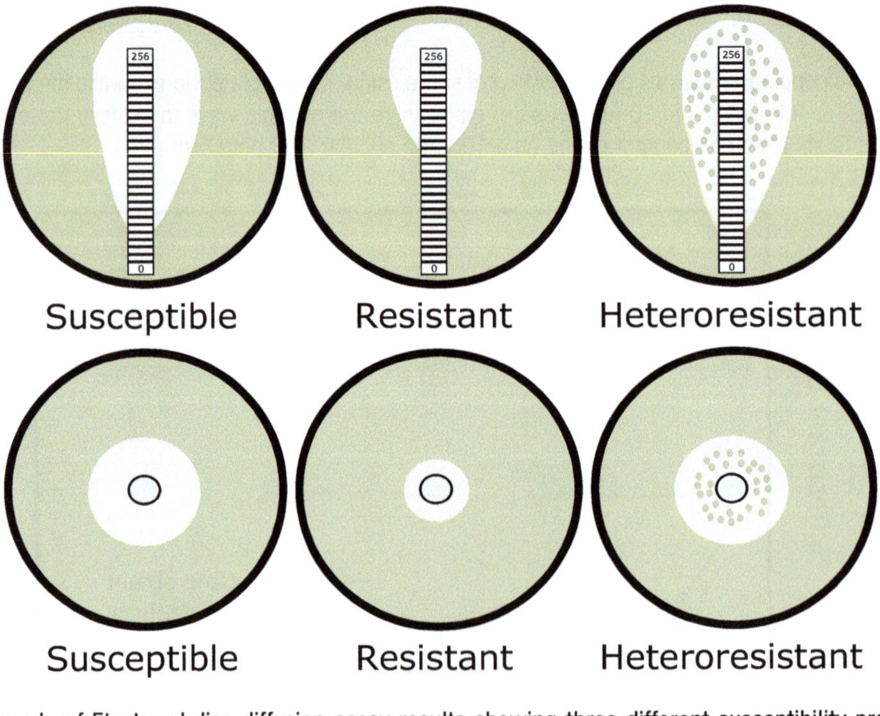

Fig. 3 Example of Etest and disc diffusion assay results showing three different susceptibility profiles. The minimum inhibitory concentration (MIC) for Etest (top row) is interpreted as the lowest concentration where the zone of clearing meets the Etest strip along the bottom edge. The MIC for disc diffusion (bottom row) is interpreted by comparing the diameter of the zone of clearing, is drug and dose dependent, and can be accessed through EUCAST guidelines [24]

heteroresistance as the appearance of colonies within the zone of clearing will depend on the frequency of the resistant subpopulation of cells (for heteroresistant isolates with a low frequency resistant subpopulation, colonies will not appear in the zone of clearing, and thus these strains will not be classified as heteroresistant using Etest). Another shortcoming is that Etest strips can be expensive and may not be ideal to screen numerous strains for heteroresistance. This method has been adapted and modified from the agar disc diffusion method standard described by CLSI [23] and is outlined below:

Day 1

1. Streak out bacterial strains from freezer stocks on lysogeny broth agar plates, and grow overnight in a 37 °C incubator.

Day 2

1. Pick an isolated colony with a sterile inoculating loop from the LB plates, and transfer into 2 mL sterile lysogeny broth. Subsequently grow at 37 °C shaking at 225 rpm overnight (~16 h).

Day 3

1. Prepare LB agar plates for inoculation (*see* instructions and notes in Subheading 3.1. LB agar plates used for Etest should not contain antibiotics).
2. Remove Etest strips from 4 °C, and allow them to reach room temperature before application (this takes about 30 min).
3. Dilute culture to be tested to between 1×10^8 and 2×10^8 CFU/mL (~0.5 McFarland Standard) (*see* **Note 5**), and use the suspension within 15 min (*see* **Note 6**).
4. Dip a sterile swab in the diluted culture tubes, and rub off excess culture along the side of the tube. Swab along the surface of a dry petri dish (no visible water spots), rotate the plate 60°, and then repeat these steps two more times. Allow excess moisture to dry before Etest strip application, but do not exceed 15 min after inoculation (*see* **Note 7**).
5. Apply Etest strip to the surface of the agar plate with the label side up. Make sure the strip does not shift during placement.
6. Within 15 min of Etest strip application, invert plates, and incubate at 37 °C for 16–20 h.

Day 4

Examine plates for zones of clearing, and then examine zones for any resistant colonies that may have appeared. Colonies can appear uniformly throughout the zone or close to the edge of the zone. Colonies may be spontaneous mutants rather than the resistant subpopulation of a heteroresistant strain and should thus be tested through the methods described in Subheading 3.4 below.

3.3 Disc Diffusion

Disc diffusion is used to determine an organism's MIC to an antibiotic using discs concentrated with the drug (Fig. 3). Similar to Etest, this method is nonquantitative but can detect heteroresistant isolates that harbor a resistant subpopulation present at a high enough frequency. Unlike Etest which contains a gradient of antibiotic concentrations, in disc diffusion, each disc is loaded with a specific concentration. The discs are placed on agar plates after bacteria have been spread out in a lawn. The antimicrobial diffuses radially from the disc creating gradients of antibiotic concentrations. After overnight incubation, the MIC for disc diffusion is interpreted by comparing the diameter of the zone of clearing and is drug and dose dependent. This method can be adapted to detect resistant subpopulations by observing colonies that may grow within the zone of clearing. While this method is not as sensitive as PAP, it can be useful to screen multiple strains for heteroresistance. Discs preloaded with antibiotic can be purchased commercially but can also be prepared in a laboratory. Preparing discs can be more cost-effective when testing multiple isolates and can be used to work within a specific range of antibiotic concentrations for which commercial discs are not available. Results can vary if disc preparation or application is not carefully controlled. This method has been adapted and modified from the agar disc diffusion method standard described by EUCAST [24] and is outlined below:

Day 1

1. Streak out bacterial strains from freezer stocks on lysogeny broth agar plates (*see* instructions and notes in Subheading 3.1. LB agar plates used for disc diffusion should not contain antibiotics), and grow overnight in a 37 °C shaking incubator.

Day 2

1. Pick an isolated colony with a sterile inoculating loop from the LB plates, and transfer into 2 mL sterile lysogeny broth, and then grow at 37 °C shaking at 225 rpm overnight (~16 h).

Day 3

1. Prepare antibiotic discs (or antibiotic discs can be purchased from a manufacturer).
   (a) Prepare colistin sulfate stock solution (concentrations are based on total μg/disc so calculate accordingly, e.g., 128 μg disc would require 12.8 μL of 10 mg/mL stock solution).
   (b) Add 5–20 μL of antibiotic/solution to blank filter paper discs (*see* **Note 8**).
   (c) Allow discs to air-dry for at least 15–20 min before application to plates.
2. Dilute culture to be tested to between 1 × 10^8 and 2 × 10^8 CFU/mL (~0.5 McFarland Standard) (*see* **Note 5**), and use the suspension within 15 min (*see* **Note 6**).

3. Dip a sterile swab into the diluted culture tubes, rub off excess culture along the side of the tube, and swab along the surface of a dry petri dish (no visible water spots). Rotate the plate 60° and repeat, and then rotate the plate an additional 60° and repeat. Allow excess moisture to dry before application of the disc, but do not exceed 15 min after inoculation (*see* **Note 7**).
4. Apply disc to the surface of the agar plate with the label side up. Make sure the strip does not shift during placement.
5. Within 15 min. of disc application, invert plates and incubate at 37 °C for 16–20 h.

Day 4

1. Examine plates for zones of clearing, and then examine zones for any resistant colonies. Colonies can appear uniformly throughout the zone or close to the edge of the zone. These colonies can be examined for potential heteroresistance with the methods described below.

To verify that heteroresistance observed by PAP, Etest, or disc diffusion is not due to a stable mutation in a small proportion of the total population or due to persister cells, further examination is necessary. Two methods can be utilized to distinguish between heteroresistance and these other forms of resistance [15].

3.4 Time-Kill Assay

The purpose of this assay is to distinguish the resistant subpopulation in heteroresistance from persisters. Canonically, persisters are a subset of the population of a strain of bacteria that resist antibiotics by entering a state of quiescence with no or very limited growth. Heteroresistance differs from persistence in that the resistant subpopulation of a heteroresistant strain is able to grow and rapidly expand in the presence of antibiotics.

Day 1

1. Streak out isolates from freezer stocks on lysogeny broth agar plates, and grow overnight in a 37 °C incubator.

Day 2

1. Inoculate 2 mL LB with a single colony from overnight plates, and grow at 37 °C shaking at 225 rpm.

Day 3

2. Dilute 10 μL overnight culture into 10 mL LB without colistin and 10 mL LB with colistin (10 μg/mL).
3. Serially dilute and plate these cultures on LB agar plates with (10 μg/mL) and without colistin. This is time zero.
4. Every hour for 8 h, serially dilute and plate both cultures on LB agar plates with (10 μg/mL) and without colistin overnight (~16 h).

Day 4

Analysis of data should be as follows:

Colony-forming units counted from LB agar plates without colistin (total CFU) and colony-forming units counted from LB agar plates with colistin (resistant CFU) should be plotted on a graph over the 8-h duration of this experiment. An increase of resistant CFU is expected to be observed over time in colistin heteroresistant strains of *Acinetobacter baumannii*.

3.5 Resistant Colony Restreak

A single colony picked from a population analysis profile plate or from within the zone of clearing on an Etest or disc diffusion assay plate (*see* **Note 9**) can be grown overnight in broth in the absence of colistin (several days of subculture if necessary), and the methods used to detect heteroresistance can be repeated to determine whether the cells maintain or lose their resistance phenotype (*see* **Note 10**). If the strains are heteroresistant, there should be a decrease in the frequency of the resistant subpopulation after culture in the absence of colistin. If the colonies now exhibit stable resistance, the frequency of the resistant subpopulation should not decrease after a few days of subculture in media without colistin.

4 Notes

1. Lysogeny broth agar plates containing colistin may result in difficulty spot plating. If this becomes a problem during population analysis profile assays, we recommend pouring thinner plates (15 mL or less) and allowing plates to dry in the biosafety cabinet for 10–15 min prior to plating. While it is our recommendation to plate 10 μL per dilution so as to conserve both time and plates, it is possible to use one LB plate per dilution.
2. Plates can be dried in a biosafety cabinet for 30 min with the lid removed, 20–25 °C overnight, or at 37 °C with the lid removed for 15 min.
3. For all assays, we recommend counting colonies at 24 h after plating and leaving plates at 37 °C for an additional 24 h to observe slow-growing colonies within the bacterial population.
4. To control for the variance that appears by creating population analysis profile plates, it is advised to use a control strain where percent survival can be verified for each set of PAP plates that are prepared to ensure consistency with results.
5. Alternatively, a direct colony suspension method can be used in which morphologically similar colonies are picked directly from a plate and resuspended in saline and adjusted to a turbidity resulting in a $1–2 \times 10^8$ CFU/mL suspension.

6. Do not allow the suspension to sit for more than 60 min to prevent overgrowth and incorrect interpretation if MIC is also desired.
7. If inoculated plates are left sitting for prolonged periods of time before the disc is applied, the bacteria may begin growing in the absence of antibiotic, leading to erroneous results.
8. An empty sterile petri dish may be used to prepare discs.
9. If colonies are present.
10. It may require several passages before the subpopulation frequency begins to lower, if after several passages the percent resistance does not decrease then the increase may be due to conventionally resistant bacteria.

References

1. Antunes LC, Visca P, Towner KJ (2014) Acinetobacter baumannii: evolution of a global pathogen. Pathog Dis 71(3):292–301. https://doi.org/10.1111/2049-632x.12125. Epub 2014/01/01
2. Dijkshoorn L, Nemec A, Seifert H (2007) An increasing threat in hospitals: multidrug-resistant Acinetobacter baumannii. Nat Rev Microbiol 5(12):939–951. https://doi.org/10.1038/nrmicro1789. Epub 2007/11/17
3. Badave GK, Kulkarni D (2015) Biofilm producing multidrug resistant Acinetobacter baumannii: an emerging challenge. J Clin Diagn Res 9(1):Dc08–Dc10. https://doi.org/10.7860/jcdr/2015/11014.5398. Epub 2015/03/05
4. Bergogne-Berezin E, Towner KJ (1996) Acinetobacter spp. as nosocomial pathogens: microbiological, clinical, and epidemiological features. Clin Microbiol Rev 9(2):148–165. Epub 1996/04/01
5. El-Halfawy OM, Valvano MA (2015) Antimicrobial heteroresistance: an emerging field in need of clarity. Clin Microbiol Rev 28 (1):191–207. https://doi.org/10.1128/cmr.00058-14. Epub 2015/01/09
6. Maisonneuve E, Gerdes K (2014) Molecular mechanisms underlying bacterial persisters. Cell 157(3):539–548. https://doi.org/10.1016/j.cell.2014.02.050. Epub 2014/04/29
7. Bianco C, Arena F, Rossetti B, Tordini G, Migliorini L, Galluzzi P, Cerase A, De Luca A, Rossolini GM, Montagnani F (2014) First report of spondylodiscitis due to vancomycin heteroresistant Staphylococcus capitis in immunocompetent host. J Infect Chemother 20(10):639–642. https://doi.org/10.1016/j.jiac.2014.04.004. Epub 2014/08/19
8. Chen Y, Hu D, Zhang Q, Liao XP, Liu YH, Sun J (2017) Efflux pump overexpression contributes to tigecycline heteroresistance in Salmonella enterica serovar Typhimurium. Front Cell Infect Microbiol 7:37. https://doi.org/10.3389/fcimb.2017.00037. Epub 2017/03/07
9. Cherkaoui A, Diene SM, Renzoni A, Emonet S, Renzi G, Francois P, Schrenzel J (2017) Imipenem heteroresistance in nontypeable Haemophilus influenzae is linked to a combination of altered PBP3, slow drug influx and direct efflux regulation. Clin Microbiol Infect 23(2):118 e9–118e19. https://doi.org/10.1016/j.cmi.2016.10.009. Epub 2016/10/21
10. Gosciniak G, Biernat MM, Binkowska A, Kus A, Iwanczak B (2017) Frequency of infection with Helicobacter pylori isolates of different antimicrobial profiles in children and adolescents: a preliminary study. Adv Clin Exp Med 26(2):263–268. https://doi.org/10.17219/acem/67716. Epub 2017/08/10
11. Napier BA, Band V, Burd EM, Weiss DS (2014) Colistin heteroresistance in Enterobacter cloacae is associated with cross-resistance to the host antimicrobial lysozyme. Antimicrob Agents Chemother 58(9):5594–5597. https://doi.org/10.1128/AAC.02432-14
12. Sun JD, Huang SF, Yang SS, Pu SL, Zhang CM, Zhang LP (2015) Impact of carbapenem heteroresistance among clinical isolates of invasive Escherichia coli in Chongqing, southwestern China. Clin Microbiol Infect 21(5):469 e1–469 10. https://doi.org/10.1016/j.cmi.2014.12.013. Epub 2015/02/05
13. Zheng C, Li S, Luo Z, Pi R, Sun H, He Q, Tang K, Luo M, Li Y, Couvin D, Rastogi N,

Sun Q (2015) Mixed infections and rifampin heteroresistance among Mycobacterium tuberculosis clinical isolates. J Clin Microbiol 53 (7):2138–2147. https://doi.org/10.1128/jcm.03507-14. Epub 2015/04/24

14. Alexander HE, Leidy G (1947) mode of action of streptomycin on type b hemophilus influenzae: II. Nature of resistant variants. J Exp Med 85(6):607–621. Epub 1947/05/31
15. Band VI, Crispell EK, Napier BA, Herrera CM, Tharp GK, Vavikolanu K, Pohl J, Read TD, Bosinger SE, Trent MS, Burd EM, Weiss DS (2016) Antibiotic failure mediated by a resistant subpopulation in Enterobacter cloacae. Nat Microbiol 1(6):16053. https://doi.org/10.1038/nmicrobiol.2016.53. Epub 2016/08/31
16. Hawley JS, Murray CK, Jorgensen JH (2008) Colistin heteroresistance in acinetobacter and its association with previous colistin therapy. Antimicrob Agents Chemother 52 (1):351–352. https://doi.org/10.1128/aac.00766-07. Epub 2007/10/24
17. Moosavian M, Shoja S, Nashibi R, Ebrahimi N, Tabatabaiefar MA, Rostami S, Peymani A (2014) Post neurosurgical meningitis due to colistin heteroresistant Acinetobacter baumannii. Jundishapur J Microbiol 7(10):e12287. https://doi.org/10.5812/jjm.12287. Epub 2015/01/30
18. Rodriguez CH, Bombicino K, Granados G, Nastro M, Vay C, Famiglietti A (2009) Selection of colistin-resistant Acinetobacter baumannii isolates in postneurosurgical meningitis in an intensive care unit with high presence of heteroresistance to colistin. Diagn Microbiol Infect Dis 65(2):188–191. https://doi.org/10.1016/j.diagmicrobio.2009.05.019. Epub 2009/09/15
19. Rodriguez CH, Nastro M, Fiorilli G, Dabos L, Lopez Calvo J, Farina ME, Vay C, Famiglietti A (2016) Trends in the resistance profiles of Acinetobacter baumannii endemic clones in a university hospital of Argentina. J Chemother 28(1):25–27. https://doi.org/10.1179/1973947814y.0000000213. Epub 2014/10/01
20. Yau W, Owen RJ, Poudyal A, Bell JM, Turnidge JD, Yu HH, Nation RL, Li J (2009) Colistin hetero-resistance in multidrug-resistant Acinetobacter baumannii clinical isolates from the Western Pacific region in the SENTRY antimicrobial surveillance programme. J Infect 58(2):138–144. https://doi.org/10.1016/j.jinf.2008.11.002. Epub 2008/12/09
21. Li J, Rayner CR, Nation RL, Owen RJ, Spelman D, Tan KE, Liolios L (2006) Heteroresistance to colistin in multidrug-resistant Acinetobacter baumannii. Antimicrob Agents Chemother 50(9):2946–2950. https://doi.org/10.1128/aac.00103-06. Epub 2006/08/31
22. Satola SW, Farley MM, Anderson KF, Patel JB (2011) Comparison of detection methods for heteroresistant vancomycin-intermediate Staphylococcus aureus, with the population analysis profile method as the reference method. J Clin Microbiol 49(1):177–183. https://doi.org/10.1128/jcm.01128-10. Epub 2010/11/05
23. CLSI (2015) Methods for dilution antimicrobial susceptibility tests for bacteria that grow aerobically; approved standard, 10th edn. CLSI document M07-A10. Clinical and Laboratory Standards Institute, Wayne PA
24. Matuschek E, Brown DF, Kahlmeter G (2014) Development of the EUCAST disk diffusion antimicrobial susceptibility testing method and its implementation in routine microbiology laboratories. Clin Microbiol Infect 20(4): O255–O266. https://doi.org/10.1111/1469-0691.12373. Epub 2013/10/18

Chapter 5

Testing Metal Sensitivity of *A. baumannii* Strains: Survival in Copper-Supplemented Liquid Media and on Copper-Containing Surfaces

Caitlin L. Williams and D. Scott Merrell

Abstract

This chapter describes protocols for determining *A. baumannii* isolates' overall levels of sensitivity to heavy metals; copper is used as a model heavy metal. Measurements of the ability of strains to grow in the presence of various concentrations of copper in liquid media and on copper-containing surfaces are described.

Key words *Acinetobacter*, *A. baumannii*, Copper, Heavy metal, Surface, Coupon

1 Introduction

Acinetobacter baumannii is notoriously resistant to most antibiotics; therefore, many research efforts focus on identification of novel antimicrobial targets and on development of new treatment options for patients. Heavy metals, including copper, are essential for bacterial life but are also inherently toxic to the cell when present in excess quantities [1]. Thus, there is potential for novel metal-based drug targets and therapeutics. Indeed, bacterial strains with mutations in crucial copper homeostasis mechanisms have been shown to be attenuated in in vivo infection models, highlighting the importance of these systems for survival and virulence [2]. Copper also has significant potential for use as a decontaminant. *A. baumannii* are able to persist and spread among patients in healthcare environments because they survive on dry surfaces; nosocomial outbreaks can be caused by contamination of surfaces in hospital rooms [3, 4]. Covering high-touch surfaces with copper dramatically reduces the level of contamination and, consequently, reduces the number of infections acquired by the patients in the room [5–9]. Protocols presented describe methods to assess strains' overall sensitivity level to a given metal, e.g., copper, when present in liquid media or on metal surfaces.

Indranil Biswas and Philip N. Rather (eds.), *Acinetobacter baumannii: Methods and Protocols*, Methods in Molecular Biology, vol. 1946, https://doi.org/10.1007/978-1-4939-9118-1_5, © Springer Science+Business Media, LLC, part of Springer Nature 2019

2 Materials

1. 5× M9 salts: Dissolve 64 g of $Na_2HPO_4{\cdot}7H_2O$, 15 g of KH_2PO_4, 2.5 g of NaCl, and 5 g of NH_4Cl in 1 L of deionized distilled H_2O (di-dH_2O) and autoclave [10]. Store at room temperature.
2. 20% glucose: Dissolve 10 g of D-glucose in 50 mL of di-dH_2O and filter sterilize with a 0.2 μm filter. Store at room temperature.
3. 1 M $MgSO_4$: Dissolve 12.32 g of $MgSO_4{\cdot}7H_2O$ in 50 mL of di-dH_2O and filter sterilize. Store at room temperature.
4. 1 M $CaCl_2$: Dissolve 1.47 g of $CaCl_2{\cdot}2H_2O$ in 10 mL of di-dH_2O and filter sterilize. Store at room temperature.
5. Bacto™ Casamino Acids (BD Biosciences).
6. 1 M $CuSO_4$: Dissolve 1.60 g of anhydrous copper sulfate in 10 mL of di-dH_2O. Store at 4 °C (*see* **Note 1**).
7. LB agar: Add 40 g/L to di-dH_2O. Autoclave, and cool to approximately 55 °C. Pour approximately 16–20 mL of agar into each petri dish. Let cool and solidify at room temperature. Store at 4 °C in closed bags. Before use, dry the plates by incubating at 37 °C for approximately 10 min with the lids open.
8. Petri dishes.
9. 1× PBS: Add 10 mL of 10× PBS liquid concentrate (Millipore Sigma) to 90 mL of di-dH_2O and autoclave or filter sterilize. Store at room temperature.
10. Stainless steel 304 sheet, 28 gauge (Online Metals, Seattle) (*see* **Note 2**).
11. Copper 110 H02 sheet, 25 gauge (Online Metals).
12. Cartridge Brass 260 sheet, 26 gauge (Online Metals).
13. Bullet Storm 5 blender (Next Advance, Averill Park, NY).
14. 5 mL Screw Cap Transport Tubes.
15. Acetone.
16. Pure Ethanol 200 Proof.

3 Methods

3.1 Copper Exposure in Liquid

1. Prepare 250 mL of growth medium: In a large beaker on a stir plate, combine 194 mL of di-dH_2O, 50 mL of 5× M9 Salts, 0.5 mL of 1 M $MgSO_4$, 5 mL of 20% glucose, 25 μL of 1 M $CaCl_2$, and 0.25 g of Casamino Acids [10]. Sterilize with a 0.2 μm filter. Use immediately or store at 4 °C for up to 7 days (*see* **Notes 3** and **4**).

2. Inoculate 4 mL of growth medium in a large glass test tube with a single bacterial colony (*see* **Note 5**).
3. Incubate the cultures at 37 °C, shaking at 190 rpm overnight (*see* **Note 6**).
4. Pipette 10 mL of growth medium into 25 mL glass flasks. Add the required volume of copper sulfate to achieve desired concentration (*see* **Notes 7–9**). Swirl immediately to dissolve.
5. Measure the OD_{600} of the overnight cultures, and determine the volume required to start 10 mL subcultures at an OD_{600} of 0.05. Pipette the required volume into 1.5 mL tubes, and pellet the bacteria by spinning at max speed (21,130 × *g*) for 1 min in a tabletop centrifuge (*see* **Notes 10** and **11**).
6. Discard the supernatant, resuspend each bacterial pellet in growth medium taken directly from one particular 10 mL subculture, and then add the resuspended bacteria back to that same subculture. Swirl to evenly mix the bacteria. Start a timer to record the length of exposure (*see* **Note 12**).
7. Immediately remove T0 samples for measurement by OD_{600} and plating for CFU: Remove 0.5 mL of culture to a cuvette to measure the OD_{600}. Remove 10 μL of culture to a microtiter plate well containing 90 μL of PBS to be used for plating.
8. Immediately incubate flasks at 37 °C, 190 rpm.
9. Use the cells placed in the microtiter plate from **step 7** to make serial tenfold dilutions to obtain 10^{-1} through 10^{-7} dilutions (*see* **Note 13**).
10. Spot 10 μL of the required dilutions onto an LB agar plate, and tilt the plate to create four or five streaks down the plate. Allow the plates to dry before inverting them (*see* **Notes 14** and **15**) (*see* Fig. 1).
11. For the next 6 h, take samples for growth measurement every hour (0–6), and plate for CFU every other hour (0, 2, 4, 6) (*see* **Note 16**).
12. Incubate the agar plates at 37 °C overnight and count CFU the next day.
13. Calculate the CFU/mL of each culture (*see* **Note 17**).
14. Plot the data as OD_{600} vs. time and CFU/mL vs. time.

3.2 Copper Exposure on Metal Surfaces

1. Prepare the growth medium: In a large beaker on a stir plate, combine 194 mL of di-dH_2O, 50 mL of 5× M9 salts, 0.5 mL of 1 M $MgSO_4$, 5 mL of 20% glucose, 25 μL of 1 M $CaCl_2$, and 0.25 g of Casamino Acids [10]. Sterilize with a 0.2 μm filter. Use immediately or store at 4 °C for up to 7 days (*see* **Notes 3** and **4**).

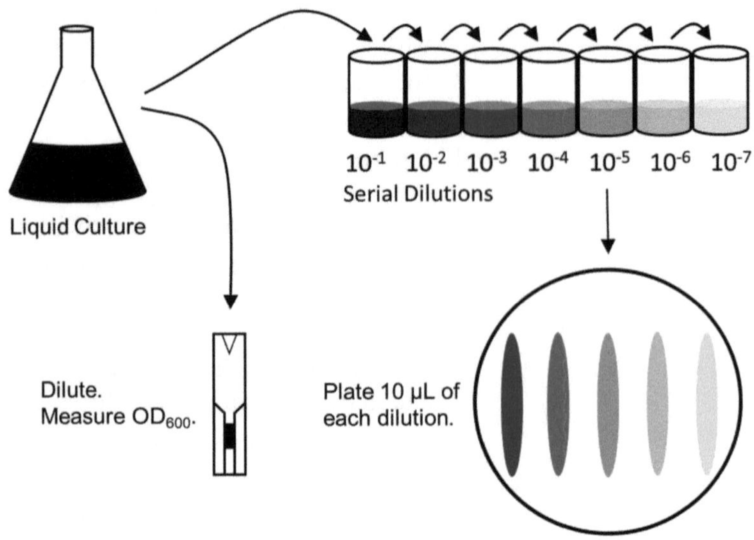

Fig. 1 Techniques to measure bacterial growth. Over the time course, growth is assessed both by culture turbidity and by quantifying bacterial numbers. Culture turbidity is measured by the OD_{600} of a sample from the culture; the sample may need to be diluted, so that the measured value is within the linear range of the spectrophotometer (*see* **Note 11**). To enumerate bacterial numbers, seven tenfold dilutions are performed in PBS to obtain 10^{-1} through 10^{-7} dilutions; 10 μL aliquots of each dilution are streaked down an LB agar plate (*see* **Notes 13** and **14**). After overnight incubation at 37 °C, single colonies are counted, and the CFU/mL of the original culture is calculated (*see* **Note 17**)

2. Prepare the metal coupons: Cut metal sheets into coupons (1 × 1 × 0.04–0.05 cm). Degrease the coupons by vortexing in ~20 mL of acetone for 30 s. Store immersed in absolute ethanol. Prior to use, sterilize each coupon by flaming and then place them in sterile petri dishes (*see* **Note 18**) [11–13].
3. Inoculate 4 mL of growth medium in a large glass test tube with a single bacterial colony (*see* **Note 5**).
4. Incubate the cultures at 37 °C, shaking at 190 rpm overnight (*see* **Note 6**).
5. Pipette 10 mL of growth medium into 25 mL glass flasks.
6. Measure the OD_{600} of the overnight cultures, and determine the volume required to start 10 mL subcultures at an OD_{600} of 0.05. Pipette the required volume into 1.5 mL tubes, and pellet the bacteria by spinning at max speed (21,130 × *g*) for 1 min in a tabletop centrifuge (*see* **Notes 10** and **11**).
7. Discard the supernatant, resuspend each bacterial pellet in growth medium taken directly from one particular 10 mL subculture, and then add the resuspended bacteria back to

that same subculture. Swirl to evenly mix the bacteria. Start a timer to record the length of exposure (*see* **Note 12**).

8. Incubate flasks immediately at 37 °C, 190 rpm.
9. Grow until the desired growth phase is reached (*see* **Note 19**).
10. Take a 10 μL sample of the culture and perform serial tenfold dilutions in PBS. Plate 10 μL streaks of the required dilutions to determine the number of bacteria in the 5 μL inoculum (*see* **Notes 13–15**).
11. Label the petri dishes containing the coupons with the times at which each will be inoculated and the end time of the incubation (*see* **Notes 20** and **21**).
12. Spot 5 μL of the culture onto each of the coupons at 30 s intervals (*see* **Notes 22** and **23**).
13. Incubate the closed petri dishes on the bench for the desired time, e.g., 15–75 min.
14. Pick up the coupon with sterile tweezers approximately 5–10 s before the end of incubation period. When the timepoint is reached, drop it into a bead beater tube containing 3 mL of PBS. Close the cap and place it into the bead beater. Continue collecting the rest of the coupons for that timepoint, e.g., all of the coupons with 15 min incubations.
15. Bead beat the group of tubes for 1 min on power 1 (*see* **Note 24**).
16. Remove 10 μL of each resuspension to a microtiter plate well containing 90 μL of PBS.
17. Make serial tenfold dilutions to obtain 10^{-1} through 10^{-4} dilutions (*see* **Notes 13** and **25**).
18. Spot 10 μL of the required dilutions onto an LB agar plate, and tilt the plate to create four or five streaks down the plate. Allow the plates to dry before inverting them (*see* **Note 14**) (*see* plating portion of Fig. 1).
19. Incubate the agar plates at 37 °C overnight and count CFU the next day.
20. Calculate the CFU/coupon of each culture (*see* **Note 26**).
21. Plot data as CFU/coupon vs. time.

4 Notes

1. We have noticed differences between fresh stock solutions of $CuSO_4$ and solutions stored at room temperature for many months. We recommend making small batches of fresh solution every 6–12 months and storing at 4 °C.

2. Metal sheets can be ordered from a variety of companies; we chose Online Metals Inc. because we were able to purchase small quantities, e.g., 12 × 12 in. sheets.
3. We have noticed reduced bacterial growth in M9 growth medium that is greater than 7 days old; therefore, we make fresh batches as often as possible and store the media at 4 °C.
4. We chose M9 medium with Casamino Acids for these experiments because it is a defined, minimal growth medium in which *A. baumannii* can still grow. It is important to use a minimal medium to limit sequestration of the added metal ions by components that are typically found in rich media. Metal binding to these components reduces the effective concentration of metal.
5. Some strains of *A. baumannii* undergo phase variation [14]. Our lab carefully checks each plate of colonies struck from a freezer stock and chooses a single, opaque colony for each experiment.
6. Our lab has defined an overnight culture as 16–19 h of growth. This length of time may be changed, but it is important to keep consistent timing between replicate experiments in order to maximize reproducibility.
7. The range and interval of concentrations chosen to achieve a dose response will depend on the growth conditions, metal, overall sensitivity of the strain, etc. To show a dose response, we typically include one low concentration that does not affect growth, the highest soluble concentration to show maximum effect, and at least two intermediate concentrations to look for intermediate effects. For example, strain AB5075 grows normally in 0 and 0.5 mM $CuSO_4$ but is partially inhibited by 1 and 1.5 mM [13]. For the more sensitive strains, we include 0.1 and 0.25 mM as well; at these low concentrations, we see minimal effects, while 0.5, 1, and 1.5 mM completely inhibit growth and may kill the cells.
8. The volume of 1 M copper sulfate added to a 10 mL culture is typically 1–15 μL, resulting in a final concentration of 0.1–1.5 mM. We have found that concentrations above 1.5 mM do not reliably stay in solution in this minimal medium and therefore are not recommended.
9. It is important that the flasks used for the growth curves have been previously acid washed to remove all heavy metals. We submerge flasks in a 10% nitric acid bath overnight, rinse 10× with dH_2O, cover with lids or foil, and autoclave.
10. Our overnight cultures typically have an OD_{600} of 1.0–1.7. Therefore, the volume of culture pelleted for a 10 mL

subculture is typically 300–500 μL. Example: $[(0.05 \times 10 \text{ mL})/1.5 = 333\ \mu\text{L}]$.

11. To reliably measure high OD_{600}, we dilute 1:2–1:5, so that the measured optical density is no higher than 0.7 and is within the linear range of the spectrophotometer. It is also important to account for the OD_{600} of the added $CuSO_4$ and to make the proper adjustments. For example, when measuring the OD_{600} of a sample containing 1 mM $CuSO_4$, you must either blank the spectrophotometer with medium containing 1 mM $CuSO_4$ *or* subtract the OD_{600} of medium with 1 mM $CuSO_4$ after the data is collected. When measuring the OD_{600} of a sample containing 1 mM $CuSO_4$ that was diluted 1:2, you must make adjustments for 0.5 mM $CuSO_4$.
12. We typically process 20 flasks at a time; it takes ~10 min to start each group of ten subcultures. Note the time each group went in to the incubator as T0, and process the flasks in the same order at each timepoint.
13. We perform our serial dilutions in a 96-well plate; the dilutions could be performed in 1.5 mL tubes; however, it would be more laborious and less cost-effective. Fill each well of the plate with 90 μL of PBS before beginning the experiment. To make the 10^{-1} dilution, add 10 μL of the culture to the first well of the row. Mix by pipetting up and down at least 20 times. Take 10 μL, add it to the next well, and discard the tip(s). Continue across the plate until a dilution series from 10^{-1} to 10^{-7} has been created. Using a 12-channel pipette, an entire plate of dilutions can be completed in only a few minutes.
14. Four or five streaks can fit across a standard 100-mm diameter petri plate. We do not recommend spot plating *A. baumannii* because the colonies grow quickly and will be difficult to count accurately if overlapping.
15. For cultures started at an OD_{600} of 0.05, we plate 10 μL of each dilution between 10^{-3} and 10^{-7} when the bacteria are growing. Lower dilutions may be required if the bacteria are static or dying, e.g., 10^{-1} through 10^{-5}. Plating fewer dilutions saves plates; however, depending on the experiment, metal, etc., it may be more informative to plate the complete dilution series to ensure countable colonies are captured at each timepoint.
16. The length of the time course and sampling interval can be adjusted as necessary depending on the chosen metal, desired information, bacterial growth rate, etc. We use a 6-h time course to reach early stationary phase in control cultures without added metal. Over that time, we see approximately 1.5 logs of growth in CFU/mL; therefore, plating every other hour is sufficient to obtain a nice curve and uses less supplies.

17. For example, if 25 colonies grew in the 10^{-6} dilution plated with 10 μL, the CFU/mL would equal 2.5×10^9. $[25 \times (1/10^{-6}) \times (1000/10)] = 2.5 \times 10^9$ CFU/mL.

18. Make sure you purchase heavy duty metal cutters capable of cutting the gauge/thickness of sheets you purchase. Stainless steel is typically the hardest of the metals used, so assess the cutters by the gauge of stainless steel they are able to cut.

19. We found that survival on copper-containing coupons is growth phase dependent for strain AB5075; survival is greatest at stationary phase and lowest at log phase [13]. For our experiments, we grew cultures 2.5 h for log phase, 6 h for early stationary phase, and 24 h for stationary phase.

20. We are able to place spots of bacteria onto metal coupons at 30 s intervals, and the recovery process is continuous but not excessively rushed. Labeling the petri dishes of coupons with all of the exact incubation start and end times before beginning makes it easier to stay on track with the precise incubation times.

21. The number of coupons that can be processed during a single experiment will depend on the shortest incubation that will be assessed. If the first timepoint is 15 min, 30 coupons can be spotted (30 s × 30 coupons = 15 min to place all the spots), and then the first group immediately recovered. We typically assess six conditions at five timepoints using this strategy: steel, brass, and copper; strain A and strain B; 15, 30, 45, 60, and 75 min incubations. If very short incubations are used, e.g., 5 min, it is possible to fully process these samples before spotting bacteria onto the rest of the coupons. However, you must be careful with this strategy, as the inoculum culture will be used at different times and may change over time.

22. This protocol uses wet droplets of media containing bacteria, and we begin timing from the moment the droplet is placed onto the metal. Other methods intentionally dry the spots and may not begin timing until the spot has dried [11, 15, 16].

23. We have found that a 5 μL droplet is relatively consistent in size and shape, which is important for determining the surface area of exposure to the metal surface and will affect the concentration of metal ions that accumulate in the droplet. The droplets do not typically evaporate during the test period.

24. One minute at power 1 is the shortest and gentlest setting on the Bullet Storm 5 blender. We previously determined that this was sufficient to resuspend the bacteria in the PBS for plating. A shorter time could be beneficial if available. Also, vortexing has been used to resuspend bacteria from coupons [11, 15–17]. We found that the bead beater enabled us to process our samples more quickly because it holds up to

12 tubes and can resuspend the bacteria in all samples simultaneously.

25. In our experiments, we have a starting inoculum of 1×10^6–5×10^7 cells. These cells are diluted into 3 mL of PBS. Therefore, we plate dilutions from 10^{-1} to 10^{-4} in order to enumerate surviving bacteria. If survival is low, e.g., at later timepoints, we also plate 1 mL of the total resuspension material on a plate. This lowers the limit of detection for the experiment to 3 CFU/coupon.
26. For example, if 25 colonies grew in the 10^{-3} dilution plated with 10 μL, the CFU/coupon would equal 7.5×10^6. $[25 \times (1/10^{-3}) \times (1000/10) \times 3] = 7.5 \times 10^6$ CFU/coupon.

Acknowledgments

Disclaimer and Funding: The opinions or assertions contained herein are the private ones of the authors and are not to be construed as official or reflecting the views of the Department of Defense, the Uniformed Services University of the Health Sciences, or any other agency of the US Government.

Work in the Merrell lab is supported by funds from the NIH and DoD.

References

1. Ladomersky E, Petris MJ (2015) Copper tolerance and virulence in bacteria. Metallomics 7 (6):957–964. https://doi.org/10.1039/c4mt00327f
2. Djoko KY, Ong CL, Walker MJ, McEwan AG (2015) The role of copper and zinc toxicity in innate immune defense against bacterial pathogens. J Biol Chem 290(31):18954–18961. https://doi.org/10.1074/jbc.R115.647099
3. Peleg AY, Seifert H, Paterson DL (2008) *Acinetobacter baumannii*: emergence of a successful pathogen. Clin Microbiol Rev 21 (3):538–582. https://doi.org/10.1128/CMR.00058-07
4. Halachev MR, Chan JZ, Constantinidou CI, Cumley N, Bradley C, Smith-Banks M, Oppenheim B, Pallen MJ (2014) Genomic epidemiology of a protracted hospital outbreak caused by multidrug-resistant *Acinetobacter baumannii* in Birmingham, England. Genome Med 6(11):70. https://doi.org/10.1186/s13073-014-0070-x
5. Schmidt MG, Attaway HH, Sharpe PA, John J Jr, Sepkowitz KA, Morgan A, Fairey SE, Singh S, Steed LL, Cantey JR, Freeman KD, Michels HT, Salgado CD (2012) Sustained reduction of microbial burden on common hospital surfaces through introduction of copper. J Clin Microbiol 50(7):2217–2223. https://doi.org/10.1128/JCM.01032-12
6. Salgado CD, Sepkowitz KA, John JF, Cantey JR, Attaway HH, Freeman KD, Sharpe PA, Michels HT, Schmidt MG (2013) Copper surfaces reduce the rate of healthcare-acquired infections in the intensive care unit. Infect Control Hosp Epidemiol 34(5):479–486. https://doi.org/10.1086/670207
7. Schmidt MG, Attaway HH, Fairey SE, Steed LL, Michels HT, Salgado CD (2013) Copper continuously limits the concentration of bacteria resident on bed rails within the intensive care unit. Infect Control Hosp Epidemiol 34 (5):530–533. https://doi.org/10.1086/670224
8. Schmidt MG, von Dessauer B, Benavente C, Benadof D, Cifuentes P, Elgueta A, Duran C, Navarrete MS (2016) Copper surfaces are associated with significantly lower concentrations

of bacteria on selected surfaces within a pediatric intensive care unit. Am J Infect Control 44 (2):203–209. https://doi.org/10.1016/j.ajic.2015.09.008

9. von Dessauer B, Navarrete MS, Benadof D, Benavente C, Schmidt MG (2016) Potential effectiveness of copper surfaces in reducing health care-associated infection rates in a pediatric intensive and intermediate care unit: a nonrandomized controlled trial. Am J Infect Control. https://doi.org/10.1016/j.ajic.2016.03.053
10. Sambrook J, Russell DW (2001) Molecular cloning: a laboratory manual, 3rd edn. Cold Spring Harbor Laboratory Press, Cold Spring Harbor, NY
11. Warnes SL, Caves V, Keevil CW (2012) Mechanism of copper surface toxicity in *Escherichia coli* O157:H7 and *Salmonella* involves immediate membrane depolarization followed by slower rate of DNA destruction which differs from that observed for Gram-positive bacteria. Environ Microbiol 14(7):1730–1743. https://doi.org/10.1111/j.1462-2920.2011.02677.x
12. Warnes SL, Green SM, Michels HT, Keevil CW (2010) Biocidal efficacy of copper alloys against pathogenic enterococci involves degradation of genomic and plasmid DNAs. Appl Environ Microbiol 76(16):5390–5401. https://doi.org/10.1128/AEM.03050-09
13. Williams CL, Neu HM, Gilbreath JJ, Michel SL, Zurawski DV, Merrell DS (2016) Characterization of copper resistance in *Acinetobacter baumannii*. Appl Environ Microbiol 82 (20):6174–6188. https://doi.org/10.1128/AEM.01813-16
14. Tipton KA, Dimitrova D, Rather PN (2015) Phase-variable control of multiple phenotypes in *Acinetobacter baumannii* strain AB5075. J Bacteriol 197(15):2593–2599. https://doi.org/10.1128/JB.00188-15
15. Eser OK, Ergin A, Hascelik G (2015) Antimicrobial activity of copper alloys against invasive multidrug-resistant nosocomial pathogens. Curr Microbiol 71(2):291–295. https://doi.org/10.1007/s00284-015-0840-8
16. Espirito Santo C, Taudte N, Nies DH, Grass G (2008) Contribution of copper ion resistance to survival of *Escherichia coli* on metallic copper surfaces. Appl Environ Microbiol 74 (4):977–986. https://doi.org/10.1128/AEM.01938-07
17. Souli M, Galani I, Plachouras D, Panagea T, Armaganidis A, Petrikkos G, Giamarellou H (2013) Antimicrobial activity of copper surfaces against carbapenemase-producing contemporary Gram-negative clinical isolates. J Antimicrob Chemother 68(4):852–857. https://doi.org/10.1093/jac/dks473

Chapter 6

A New Method for Determination of Minimum Biofilm Eradication Concentration for Accurate Antimicrobial Therapy

Nithin Sam Ravi, Raziya Fathima Aslam, and Balaji Veeraraghavan

Abstract

Antimicrobial susceptibility testing (AST) is an important technique to find the susceptibility pattern of clinical isolates in order to administer the appropriate drug. One such technique is minimum inhibitory concentration (MIC), which not only identifies the right drug but also suggests the appropriate concentration necessary to neutralize the organisms in planktonic form. MIC can vary in case of adherent organisms since they form biofilms and activate survival mechanisms like quorum sensing. Here we have strategized a new method which used an inoculator plate, a resazurin dye, and a standard plate to identify minimum biofilm eradication concentration (MBEC) of adherent organisms.

Key words Minimum biofilm eradication concentration (MBEC), Minimum inhibitory concentration (MIC), Colony-forming unit (CFU), Calgary Biofilm Device, Antimicrobial susceptibility testing (AST)

1 Introduction

Minimum inhibitory concentration (MIC) is used to determine the antibiotic susceptibility of microorganisms. Routinely MIC is done using planktonic organisms which serve as the reference only to treat non-adherent organisms. MIC is often ineffective to adherent organisms since they activate survival mechanisms such as quorum sensing [1]. Biofilms are microbial communities that adhere to biotic and abiotic surfaces and are frequently embedded within a matrix of extracellular polymeric substances (slime) or other scaffolds [2]. Many studies have suggested that there is reduced susceptibility to antibiotics in microorganisms in biofilm form when compared to planktonically grown cultures. The dosage and antibiotic diffusion play a vital role in eradicating the biofilm formation [3]. Several techniques were used to study biofilm formation, the oldest being modified Robbins device (MRD) which uses a catheter to grow the biofilm. While the method is effective, it takes a longer

Indranil Biswas and Philip N. Rather (eds.), *Acinetobacter baumannii: Methods and Protocols*, Methods in Molecular Biology, vol. 1946, https://doi.org/10.1007/978-1-4939-9118-1_6,

duration in detecting MBEC [4]. Overcoming the disadvantages of MRD, a new technique was devised by Ceri and colleagues called Calgary Biofilm Device (CBD) [5]. This method provides a good overview about the physiology and antimicrobial susceptibility of biofilm formers. Although it is an effective model for finding MBEC, it is also an expensive method to perform in laboratories which lack resources and funding. Here we have evaluated a new reliable method by which MBEC can be obtained rapidly at an affordable price.

2 Materials

All solutions and media were prepared using autoclaved (at least 45 min at 121 °C, 15 psi per 1 L solution), distilled deionized water. Waste disposal protocols were strictly followed while disposing of contaminated solutions and materials.

1. 96-well U-bottom microtiter plates.
2. Disposable Inoculator Plate.
3. Multichannel pipette: P200, 12 channel.
4. Pipettes: P20, P200, and P1000.
5. Resazurin dye (*see* **Note 1**).
6. Cation-adjusted Mueller-Hinton broth (MHB).
7. Conical flask (250 mL).
8. *Acinetobacter baumannii* strains grown in MHB at 37 °C.
9. Test tubes (20 mL).
10. Antibiotics.
11. Autoclaved distilled water.
12. Squeeze bottle (500 mL).
13. 70% ethanol (for sterilization).
14. Spectrophotometry (to read the culture OD).
15. 1.5 mL centrifuge tubes.

3 Methods

All procedures are performed at room temperature (RT) unless specified otherwise.

1. Grow all isolates in MHB at 37 °C overnight.
2. From which take 100 μL and add to 10 mL of fresh MHB. Incubate the broth at 37 °C for 2–3 h.
3. Adjust the OD to 0.5 McFarland standards using MHB.

4. Add 100 μL of the adjusted inoculum to all the wells in a 96-well microtiter plate, and place the inoculator plate over it and incubate at 37 °C for 16 h (*see* **Note 2**).
5. After the production of biofilm in the inoculator plate, two pegs (F12 and G12) from it were taken using a sterile nose plier and were inoculated in the standard plate (*see* **Note 3**). The rest of the inoculator plate with biofilm was inoculated into a 96-well plate containing the antibiotic panel (*see* **Note 4**). The plate setup was placed at 37 °C for overnight incubation.
6. In the standard plate, the CFU in the peg was estimated by comparing it with the standards. The moment of color change in the wells containing the pegs and the appropriate standard well was noted as shown in Fig. 1. It can be estimated by directly viewing or by spectrophotometry reading at 600 nm for more precision.

a

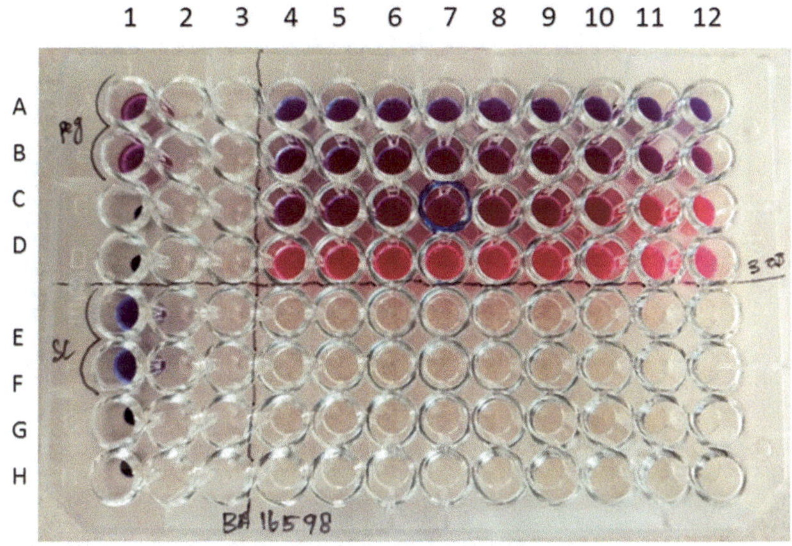

b

	1	2	3	4	5	6	7	8	9	10	11	12
A	Unknown 2.337			STD-1 3.027	STD-2 3.030	STD-3 3.022	STD-4 3.024	STD-5 3.015	STD-6 3.034	STD-7 3.005	STD-8 3.022	STD-9 3.003
B	Unknown 2.284			STD-10 2.991	STD-11 3.005	STD-12 3.000	STD-13 2.980	STD-14 2.834	STD-15 2.942	STD-16 2.942	STD-17 2.771	STD-18 2.109
C				STD-19 2.870	STD-20 2.931	STD-21 2.816	STD-22 2.234	STD-23 2.016	STD-24 2.069	STD-25 1.609	STD-26 0.773	STD-27 0.702
D				STD-28 0.735	STD-29 0.568	STD-30 0.595	STD-31 0.621	STD-32 0.618	STD-33 0.658	STD-34 0.686	STD-35 0.676	STD-36 0.729
E	QC 3.021											
F	QC 3.014											
G												
H												

Fig. 1 (**a**) A standard plate evaluated for *Acinetobacter baumannii* growth using the naked eye. (**b**) The same standard plate read through spectrophotometry at 600 nm for quantitation

7. MIC was done using the planktonic inoculum of the same isolate used for MBEC. The inoculum concentration was prepared based on the concentration obtained from the peg used in the standard plate (*see* **Note 5**). The antibiotic panel used for MIC was exactly similar to the panel used for biofilm susceptibility testing (Fig. 4). The microtiter antibiotic plate was stored at 37 °C overnight.
8. Since the initial inoculum load for both MIC and MBEC was equivalent, the data obtained were directly compared with each other as shown in Fig. 2 (*see* **Note 6**).

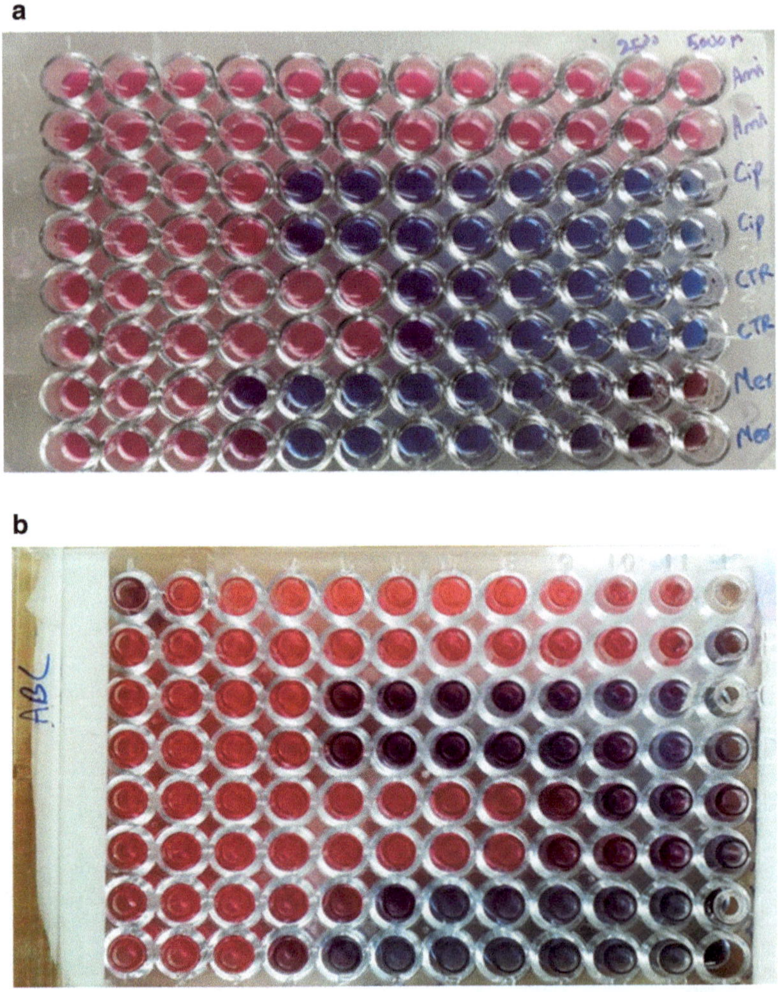

Fig. 2 Comparison of MIC plate (**a**) with MBEC plate (**b**). The pink color represents the growth of the organism, and the purple color represents no growth

4 Notes

1. Resazurin is a dye that changes color from purple to pink based on bacterial growth. The stock concentration is 400 μg in 1 mL of deinonized water (stored at 4 °C). The working concentration is 40 μg/mL.
2. Since all the pegs are dipped in equal concentration of inoculum, there is no significant difference in the biofilm formation between the pegs. It is proved by incubating the inoculum in a 96-well plate with inoculator plate placed over it. After incubating it for 16 h, the inoculator plate was removed and washed using millipore water twice and air-dried. Now the pegs were immersed in 0.1% crystal violet for 5 min and then again washed until the excess crystal violet is removed and air-dried. A fresh 96-well plate was taken and filled with 120 μL of 30% acetic acid in each well, and then the inoculator plate was placed over it for 1 min and then removed. The crystal violet adsorbed by the cell is released in the acetic acid solution, for which the OD at 595 nm is checked using spectrophotometer. Biofilm formed in each peg is compared using one-way analysis of variance (Microsoft Excel data analysis) by applying the Bartlett's test for homogeneity of variance and found no significant difference.
3. Standard plate is used to determine the exact bacterial load formed in the pegs in the inoculator plate. For standard plate preparation, overnight inoculum of the culture for which MBEC has to be determined was collected, and the OD was adjusted to 3 McFarland using MHB. Using this inoculum, the standards are set in four rows, and 9 wells are used in each row of the microtiter plate as shown in Fig. 3. There is onefold difference in antibiotic concentration between the rows and tenfold difference between the columns as shown in Fig. 3. Each well in the standard plate has resazurin, to determine the bacterial growth.
4. Four antibiotics were used in duplicate (one antibiotic in two rows) to increase the accuracy. The concentration of antibiotic ranges from 2.4 μg/mL to 5000 μg/mL as shown in Fig. 4.
5. In Fig. 1b, the OD of the peg containing well corresponds to the well C7 in the standard plate. Since, the innoculum concentrations of all the wells in the standard plate are already known (it is tenfold dilution of 50 μL culture (OD 3) + 50 μL MHB as shown in Fig. 3), that particular dilution can be used for the detection of MIC.
6. The data will be interpreted as shown in Table 1.

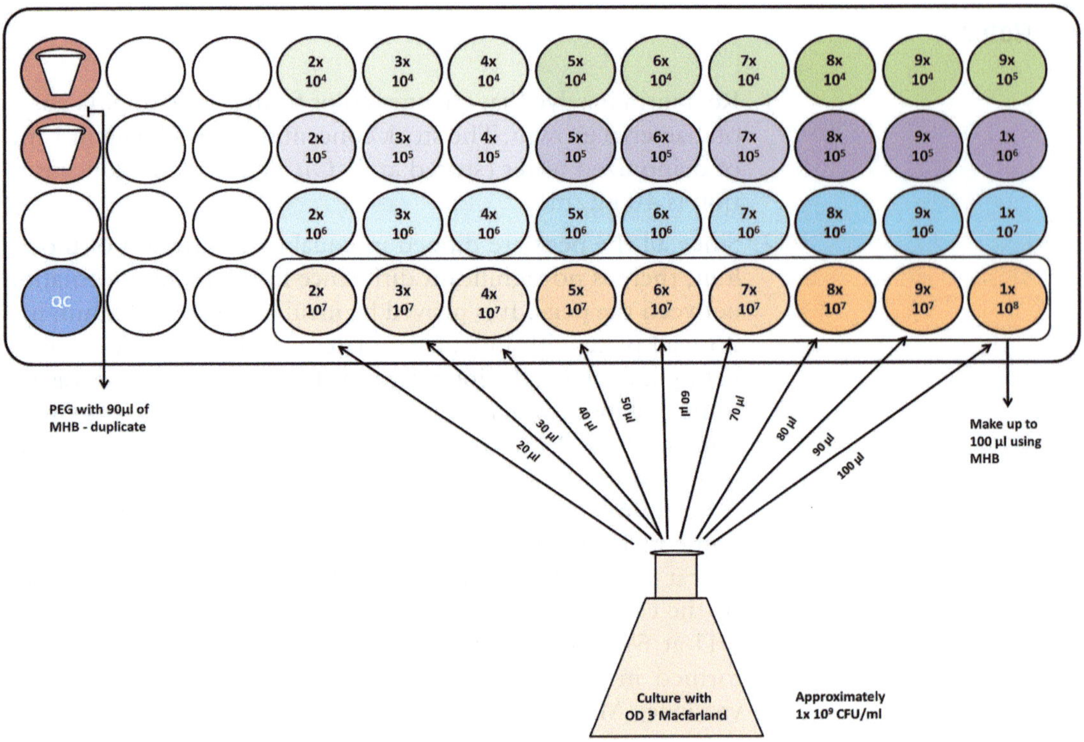

Fig. 3 Layout of a standard plate

	1	2	3	4	5	6	7	8	9	10	11	12	
A	2.4ug/ml	4.8ug/ml	9.7ug/ml	19.5ug/ml	39ug/ml	78ug/ml	156.2ug/ml	312.5ug/ml	625ug/ml	1250ug/ml	2500ug/ml	5000ug/ml	Antibiotic A duplicate
B	2.4ug/ml	4.8ug/ml	9.7ug/ml	19.5ug/ml	39ug/ml	78ug/ml	156.2ug/ml	312.5ug/ml	625ug/ml	1250ug/ml	2500ug/ml	Negative Control	
C	2.4ug/ml	4.8ug/ml	9.7ug/ml	19.5ug/ml	39ug/ml	78ug/ml	156.2ug/ml	312.5ug/ml	625ug/ml	1250ug/ml	2500ug/ml	5000ug/ml	Antibiotic B duplicate
D	2.4ug/ml	4.8ug/ml	9.7ug/ml	19.5ug/ml	39ug/ml	78ug/ml	156.2ug/ml	312.5ug/ml	625ug/ml	1250ug/ml	2500ug/ml	Negative Control	
E	2.4ug/ml	4.8ug/ml	9.7ug/ml	19.5ug/ml	39ug/ml	78ug/ml	156.2ug/ml	312.5ug/ml	625ug/ml	1250ug/ml	2500ug/ml	5000ug/ml	Antibiotic C duplicate
F	2.4ug/ml	4.8ug/ml	9.7ug/ml	19.5ug/ml	39ug/ml	78ug/ml	156.2ug/ml	312.5ug/ml	625ug/ml	1250ug/ml	2500ug/ml	Negative Control	
G	2.4ug/ml	4.8ug/ml	9.7ug/ml	19.5ug/ml	39ug/ml	78ug/ml	156.2ug/ml	312.5ug/ml	625ug/ml	1250ug/ml	2500ug/ml	5000ug/ml	Antibiotic D duplicate
H	2.4ug/ml	4.8ug/ml	9.7ug/ml	19.5ug/ml	39ug/ml	78ug/ml	156.2ug/ml	312.5ug/ml	625ug/ml	1250ug/ml	2500ug/ml	Negative Control	

Fig. 4 Layout of a typical antibiotic panel used for MBEC determination

Table 1
MBEC vs MIC for *Acinetobacter baumannii*

	Acinetobacter baumannii		
Antibiotic	**MBEC (μg/mL)**	**MIC (μg/mL)**	**Fold difference**
Amikacin	>4096	>4096	Not determined
Ceftriaxone	512	128	Four fold increase
Ciprofloxacin	32	32	No change
Meropenem	32	32	No change

Acknowledgments

This work was supported in part by the Department of Clinical Microbiology, CMC, Vellore.

References

1. Li YH, Tian X (2012) Quorum sensing and bacterial social interactions in biofilms. Sensors 12:2519–2538. https://doi.org/10.3390/s120302519
2. Garrett TR, Bhakoo M, Zhang Z (2008) Bacterial adhesion and biofilms on surfaces. Prog Nat Sci 18:1049–1056. https://doi.org/10.1016/j.pnsc.2008.04.001
3. Sepandj F, Ceri H, Gibb A et al (2007) Minimum inhibitory concentration versus minimum biofilm eliminating concentration in evaluation of antibiotic sensitivity of enterococci causing peritonitis [1]. Perit Dial Int 27:464–465
4. Raad I, Darouiche R, Hachem R et al (1995) Antibiotics and prevention of microbial colonization of catheters. These include: antibiotics and prevention of microbial colonization of catheters. Antimicrob Agents Chemother 39:2397–2400. https://doi.org/10.1128/AAC.39.11.2397.Updated
5. Ceri H, Olson ME, Stremick C et al (1999) The Calgary biofilm device: new technology for rapid determination of antibiotic susceptibilities of bacterial biofilms the calgary biofilm device: new technology for rapid determination of antibiotic susceptibilities of bacterial biofilms. J Clin Microbiol 37:1771

Chapter 7

Transformation of *Acinetobacter baumannii*: Electroporation

Mitchell G. Thompson and Süleyman Yildirim

Abstract

Although the pan and the core genome of *Acinetobacter baumannii* and its essential genes are relatively well characterized, functional characterization of these genes has not paralleled the genome-level studies. However, recently developed genetic tools and optimized protocols are poised to accelerate genetic manipulation of *A. baumannii*. Transferring exogenous DNA into the cytosol of bacteria cells is a critical step in genetic characterizations. Conjugation is restricted to the transfer of DNA from one bacterial cell to another, and only a portion of *A. baumannii* clinical isolates are naturally competent. Electroporation, which is thought to transiently create aqueous pores in the membrane, is a preferred method in transferring exogenous DNA as it does not have such limitations. Several factors contribute to efficiency of electroporation and often need to be empirically optimized to maximize efficiency of this procedure. Here we provide an optimized electroporation protocol and guidance for electroporation of clinical MDR isolates of *A. baumannii*.

Key words *Acinetobacter baumannii*, Transformation, Electroporation, Multidrug resistant (MDR)

1 Introduction

Despite the plethora of publically available sequenced genomes of *A. baumannii* strains, genetic tools enabling functional characterization of these genes have only come to the forefront recently [1, 2]. The pan-genome size of *A. baumannii* reaches approximately 12,000 genes; and its core genome includes 2000 genes, many of which are of unknown function [3]. Efficient transformation of the strains of interest is the first and critical step in functional characterization of these genes.

Transformation of bacterial cells in vitro encompasses mainly three fundamental steps: preparation of cells, cell shock, and recovery step. In the preparation step, recipient bacterial cells are conditioned (made competent) so that the cells become receptive to the foreign DNA uptake usually by modifying the cell membrane (s) and weakening the peptidoglycan leading to enhanced

Indranil Biswas and Philip N. Rather (eds.), *Acinetobacter baumannii: Methods and Protocols*, Methods in Molecular Biology, vol. 1946, https://doi.org/10.1007/978-1-4939-9118-1_7,

permeability of DNA through these structures. In the shock step, bacterial cells are treated with heat/electric/sonic shock, which transiently opens pores in the membrane(s) and/or cell wall allowing DNA to permeate into the cytoplasm. Finally, in the recovery step, the transformed cells are incubated in rich medium to repair the shock-induced damages and seal the membrane(s) and/or cell wall [4]. The transformation of bacterial cells can be performed by chemical treatment if the strain is naturally competent or by applying electric field (electroporation). Electroporation is usually more efficient technique than chemical transformation and can be applied to many bacterial species including gram-positive, gram-negative, and acid-fast bacteria [5].

The precise mechanism underlying electro-pore formation is poorly understood. Theoretical and experimental evidences suggest that the application of high electric field to cells results in water wires and water channels across the cell membrane [6]. The voltage across the lipid bilayer under normal physiological conditions is insufficient to cause rearrangement of the phospholipids to create pores. Increasing voltage, however, provides the energy to induce lipid rearrangements ultimately resulting in hydrophilic pores and increased conductivity [7, 8]. But the threshold of this voltage in forming pores is the function of the dipole potential of the membrane and structure of hydrophobic tails of the phospholipids found in the membrane [8, 9]. Besides membrane properties, other factors such as the amount and the purity of DNA, presence of restriction-modification systems in the genome of the strain, temperature, cell density, buffer composition, and the growth phase of the bacterial cells influence the transformation efficiency [4]. Therefore, efficiencies of electrotransformation vary widely due to biochemical differences in cell wall and cell membranes among bacterial species. Hence, electroporation needs to be optimized for bacterial species with distinct cell walls and membranes [10–15]. Recently, electroporation parameters for *A. baumannii* stains have been optimized [11]. Here we describe an optimized electroporation protocol that can be used to electroporate exogenous DNA into the clinical isolates of *A. baumannii.*

2 Materials

Use purified distilled water (18 MΩ/cm at 25 °C) to prepare all solutions and molecular biology grade reagents. Carry out all the experiments at room temperature unless otherwise is indicated.

2.1 Bacterial Growth Media

1. Luria-Bertani (LB) rich medium: 10 g/L tryptone, 5 g/L yeast extract, 10 g/L NaCl, pH 7.4. Sterilize using an autoclave.
2. LB Broth (Lennox): 10 g/L tryptone, 10 g/L yeast extract, 5 g/L NaCl, pH 7.4. Autoclave to sterilize (*see* **Note 1**).

3. Super optimal broth (SOC): 20 g/L peptone, 5 g/L yeast extract, 10 mM NaCl, 2.5 mM KCl, 10 mM $MgCl_2$. Adjust the volume to 1 L with deionized water, and autoclave to sterilize. After autoclaving add 2% (w/v) glucose.
4. LB agar: 10 g/L tryptone, 10 g/L yeast extract, 5 g/L NaCl, 15 g/L agar, pH 7.4. Autoclave to sterilize.

2.2 Electroporation Buffer

Ten percent glycerol diluted in deionized water (volume/volume). Autoclave to sterilize.

2.3 Electroporation Apparatus

Bio-Rad GenePulser Xcell. The following settings were found to be optimal for strain AB5075: 100 Ω, 25 μF, 1.7 kV, for 1 mM gap width cuvettes (*see* **Note 2**).

2.4 Antibiotics

Prepare 1000× stocks of selective antibiotics by dissolving powder antibiotics in deionized water, with the exception of tetracycline which is dissolved in 70% ethanol. Filter sterilize all stock solutions using 0.45 μM filter, and add to the media after autoclaving. Concentrations of the antibiotic stock solutions are as follows: ampicillin (100 mg/mL), tetracycline (10 mg/mL), hygromycin (100 mg/mL), kanamycin (50 mg/mL), and gentamicin (20 mg/mL) (*see* **Note 3**).

3 Methods

1. Three days before electroporation (−3 d), streak out an *A. baumannii* isolate on a freshly prepared LB agar plate, and incubate overnight at 37 °C.
2. From the agar plate (in −2 d), inoculate a single colony into 3 mL of pre-warmed LB media in 10-mL sterile conical tubes (add extra tubes in case bacteria do not grow in the first tube). Grow by shaking the tube at 250 rpm overnight at 37 °C.
3. The next day (−1 d), inoculate 0.5 mL of the overnight culture into 50 mL of fresh LB, pre-warmed to 37 °C. Grow this inoculum under shaking conditions at 250 rpm at 37 °C but for 24 h (*see* **Note 4**).
4. Transfer cells to a 50 mL conical tube, and centrifuge for 10 min at 10,000 × *g* at room temperature. Discard the supernatant, and gently resuspend bacterial pellet in 25 mL of 10% sterile glycerol at room temperature.
5. Spin down the suspended bacteria for 10 min at 10,000 × *g* at room temperature.
6. Decant the supernatant, and again gently resuspend bacterial pellet in 25 mL of 10% sterile glycerol at room temperature.

7. Spin down the suspended bacteria for 10 min at 10,000 × *g* at room temperature.
8. Resuspend the bacterial cells in 1.5 mL of 10% glycerol. After resuspension 50 μL of the cells' suspension can be used for each electroporation.
9. In a 1.5 mL tube, combine 50 μL of electrocompetent cells with 1 μL of plasmid DNA (1 ng to 100 ng/μL). Gently mix by flicking the side of the tube.
10. Transfer the cell/DNA mixture to a 1 mm gap electroporation cuvette. To remove air bubbles, gently tap the bottom of the cuvette.
11. Carry out electroporation under the following parameters: 1700 V, 100 Ω, 25 μF. With these parameters, the time constant should be ~5 ms (*see* **Notes** 5 and **6**).
12. After pulsing the cells, immediately add 950 μL of SOC media pre-warmed to 37 °C to the cuvette. Mix the cells and media thoroughly, and then transfer the media to a round-bottomed 14 mL conical tube. Incubate the cells at 37 °C for 1 h with shaking at 250 rpm.
13. After recovering, plate 100 μL of the cells onto LB agar plates containing appropriate antibiotics. These plates are pre-warmed to 37 °C, and ~5–10 sterile glass beads are added to each plate. Shake the plates vigorously to spread the cells, and then pour off the glass beads (*see* **Note** 7).
14. Incubate the agar plates overnight at 37 °C (*see* **Notes 8** and **9**).
15. To confirm transformation of DNA intended to integrate into the chromosome, perform colony PCR using primers specific to the integrated DNA and the host chromosome. To confirm transformation of plasmid DNA, isolate plasmid DNA using a commercial plasmid isolation kit, and perform a diagnostic restriction enzyme fragment analysis.

4 Notes

1. LB formulations with higher concentrations of salt (10 g/L) may inhibit the action of certain antibiotics such as hygromycin B.
2. Cuvettes are stored at room temperature (~22 °C) and used immediately.
3. Many strains of *A. baumannii* are multidrug resistant and thus are resistant to common antibiotics used in molecular biology. It is imperative to identify to natural resistances of any isolate of *A. baumannii* to choose an appropriate selectable marker. It

may be necessary to empirically determine the appropriate concentration of any given antibiotic to use as a selection.

4. Depending on the strain, 24-h growth would give OD_{600} of 5.5 or 6.0. Construct growth curve of the strain(s) to be transformed. Although stationary phase of the bacterial growth appears to be optimum for some clinical strains [*see* ref. 11], keep in mind that late-exponential phase or early-stationary phase might still be optimal for some other clinical strains.
5. If the sample cuvette arcs, then prepare a new batch of competent cells.
6. The presence of divalent ions (magnesium) in the transformation medium tends to reduce the efficiency (*see* ref. 11).
7. It may be necessary to perform serial tenfold dilutions of the recovered cells to obtain isolated transformant colonies on the selection plate. As a control, plate untransformed bacterial cells in the presence of selection antibiotic. If one or two colonies grow, it might be due to random mutations at very low frequency; therefore, include more than one plate as negative control.
8. Transformed colonies should be visible after ~16 h of incubation.
9. If electroporation reactions frequently arc and/or generate very low transformation efficiency, suspect impurities contaminating DNA sample used in the transformation (e.g., salt or organic solvents from previous purification procedures). If so, consider adding dry centrifugation step during DNA purification to remove contaminating solvents or dialysis of the purified DNA on membrane against sterile DNA-grade water to the protocol.

References

1. Biswas I (2015) Genetic tools for manipulating Acinetobacter baumannii genome: an overview. J Med Microbiol 64:657–669. https://doi.org/10.1099/jmm.0.000081
2. Jacobs AC, Thompson MG, Gebhardt M, Corey BW, Yildirim S, Shuman HA, Zurawski DV (2014) Genetic Manipulation of Acinetobacter baumannii. Curr Protoc Microbiol 35:6G.2.1–6G.211. https://doi.org/10.1002/9780471729259.mc06g02s35
3. Chan AP, Sutton G, DePew J, Krishnakumar R, Choi Y, Huang XZ, Beck E, Harkins DM, Kim M, Lesho EP, Nikolich MP, Fouts DE (2015) A novel method of consensus pan-chromosome assembly and large-scale comparative analysis reveal the highly flexible pan-genome of Acinetobacter baumannii. Genome Biol 16:143. https://doi.org/10.1186/s13059-015-0701-6
4. Aune TE, Aachmann FL (2010) Methodologies to increase the transformation efficiencies and the range of bacteria that can be transformed. Appl Microbiol Biotechnol 85:1301–1313. https://doi.org/10.1007/s00253-009-2349-1
5. Wirth R, Friesenegger A, Fiedler S (1989) Transformation of various species of gram-negative bacteria belonging to 11 different genera by electroporation. Mol Gen Genet 216:175–177
6. Tarek M (2005) Membrane electroporation: a molecular dynamics simulation. Biophys J 88:4045–4053. https://doi.org/10.1529/biophysj.104.050617

7. Weaver JC (1995) Electroporation theory. Concepts and mechanisms. Methods Mol Biol 47:1–26. https://doi.org/10.1385/0-89603-310-4:1
8. van Uitert I, Le Gac S, van den Berg A (2010) The influence of different membrane components on the electrical stability of bilayer lipid membranes. Biochim Biophys Acta 1798:21–31. https://doi.org/10.1016/j.bbamem.2009.10.003
9. Bockmann RA, de Groot BL, Kakorin S, Neumann E, Grubmuller H (2008) Kinetics, statistics, and energetics of lipid membrane electroporation studied by molecular dynamics simulations. Biophys J 95:1837–1850. https://doi.org/10.1529/biophysj.108.129437
10. Polak A, Bonhenry D, Dehez F, Kramar P, Miklavcic D, Tarek M (2013) On the electroporation thresholds of lipid bilayers: molecular dynamics simulation investigations. J Membr Biol 246:843–850. https://doi.org/10.1007/s00232-013-9570-7
11. Yildirim S, Thompson MG, Jacobs AC, Zurawski DV, Kirkup BC (2016) Evaluation of parameters for high efficiency transformation of Acinetobacter baumannii. Sci Rep 6:22110. https://doi.org/10.1038/srep22110
12. Zhang H, Li Y, Chen X, Sheng H, An L (2011) Optimization of electroporation conditions for Arthrobacter with plasmid PART2. J Microbiol Methods 84:114–120. https://doi.org/10.1016/j.mimet.2010.11.002
13. Dallaire-Dufresne S, Emond-Rheault JG, Attere SA, Tanaka KH, Trudel MV, Frenette M, Charette SJ (2014) Optimization of a plasmid electroporation protocol for Aeromonas salmonicida subsp. salmonicida. J Microbiol Methods 98:44–49. https://doi.org/10.1016/j.mimet.2013.12.019
14. Sato'o Y, Aiba Y, Kiga K, Watanabe S, Sasahara T, Hayakawa Y, Cui L (2018) Optimized universal protocol for electroporation of both coagulase-positive and -negative Staphylococci. J Microbiol Methods. https://doi.org/10.1016/j.mimet.2018.01.006
15. Pyne ME, Moo-Young M, Chung DA, Chou CP (2013) Development of an electrotransformation protocol for genetic manipulation of Clostridium pasteurianum. Biotechnol Biofuels 6:50. https://doi.org/10.1186/1754-6834-6-50

Chapter 8

Methods for Natural Transformation in *Acinetobacter baumannii*

Gottfried Wilharm and Evelyn Skiebe

Abstract

The genomes of *Acinetobacter baumannii* tell us stories about horizontal gene transfer (HGT) events that steadily drive the evolution of this nosocomial pathogen toward multidrug resistance. Natural transformation competence constitutes one of the several possible pathways that mediate HGT in *A. baumannii*. Here, we describe and discuss the methods for studying DNA uptake in *A. baumannii* via natural transformation.

Key words *Acinetobacter baumannii*, Nosocomial pathogen, Natural transformation competence, DNA uptake, Antibiotic resistance, Motility

1 Introduction

Acquisition of novel resistance genes is a key driver of multidrug resistance in the nosocomial pathogen *Acinetobacter baumannii* [1–3]. It was the hallmark publication of Fournier et al. [2] that first brought to our attention the massive acquisition of resistance-associated genes into *A. baumannii* genomes. These authors illustrated the mosaic structure of a large resistance island found in the genome of a multidrug-resistant strain, highlighting multiple horizontal gene transfer (HGT) events presumably from many different donor taxa including *Enterobacteriaceae* and *Pseudomonas* spp. [2]. Again and again, subsequent studies confirmed extensive HGT into the *A. baumannii* pan-genome from various sources [4–6] and with a specific emphasis on the acquisition of novel resistance genes [1, 7]. However, it remains elusive how HGT is mediated in *A. baumannii* and which mechanistic pathways contribute the most to these acquisition events. Next to the classical pathways of conjugation, transduction and natural transformation, outer membrane vesicles (OMV) [8, 9] and nanotubes [10, 11] might contribute to HGT in *A. baumannii*.

Indranil Biswas and Philip N. Rather (eds.), *Acinetobacter baumannii: Methods and Protocols*, Methods in Molecular Biology, vol. 1946, https://doi.org/10.1007/978-1-4939-9118-1_8, © Springer Science+Business Media, LLC, part of Springer Nature 2019

A single strain belonging to the genus *Acinetobacter*, strain BD4 and its derivatives, later assigned to the species *A. baylyi* [12], has been studied for its natural transformation competence for decades [13–18]. Although it had long been suspected that also *A. baumannii* is naturally competent, it took up to the year 2010 until the first naturally competent strain of *A. baumannii* was described [19]. It is now clear that competence is a widespread trait among *A. baumannii* lineages, found at rates between 36% and 44% of the strains investigated [20, 21]. Interestingly, while the first *A. baumannii* strain found to be competent was transformable during planktonic growth [19], the vast majority of strains proved competent while moving along wet surfaces [20–23].

Despite their lack of flagella, many representatives of various *Acinetobacter* spp. are motile on surfaces [24–28]. However, at least two forms of motility can be distinguished. Type IV pili (T4P)-dependent so-called twitching motility can be observed after stab-inoculation of bacteria into solid or semisolid media, where cells move alongside the interface between the medium and the bottom of the Petri dish [28–30]. By contrast, surface-associated motility occurs at the upper surface of solid or semisolid media after deposition of bacteria and is independent of T4P [20, 26, 28]. Although the two forms of motility are mechanistically distinct, there is an overlap of factors involved in both forms, e.g., the polyamine 1,3-diaminopropane (DAP) is required for twitching and surface-associated motility [30]. Moreover, T4P are not only expressed under conditions used to assess twitching motility but also when surface-associated motility is facilitated [20].

The reason why many strains of *A. baumannii* show competence while they move along wet surfaces is unresolved. Natural competence and twitching motility both are abolished following inactivation of the T4P retraction ATPase gene *pilT* [20, 28], the prepilin gene *pilA*, or the prepilin peptidase gene *pilD* [28], but also after inactivation of *comEC* encoding the putative DNA uptake channel [20]. The linkage of natural transformation competence and twitching motility is probably due to dual roles of several competence components in both piliation and DNA uptake [31].

Here we describe and discuss the protocol for natural transformation of *A. baumannii* under conditions permitting surface-associated motility.

2 Materials

2.1 Motility Medium for Transformation

1. Motility medium: To produce 500 mL of motility medium (for approximately 25 Petri dishes), weigh 2.5 g tryptone and 1.25 g NaCl into a beaker, add deionized water to a volume of 500 mL, and dissolve it with the help of a magnetic stirrer. Check the pH of the solution to be in the range of 6–7. Adjust

with HCl or NaOH if necessary. Weigh 2.5 g agarose (DNA gel quality) (*see* **Note 1**) into an autoclavable glass bottle, add the dissolved tryptone/salt solution, and autoclave the mixture.

2. Petri dishes from any supplier.

2.2 Modified Luria-Bertani (LB) Broth

Weigh 10 g tryptone, 5 g yeast extract, and 5 g NaCl into a beaker, add deionized water to a volume of 800 mL, and dissolve it with the help of a magnetic stirrer. Check the pH of the solution to be around 7. Adjust with HCl or NaOH if necessary, and fill up to a final volume of 1 L before autoclaving.

2.3 A. baumannii Strains

1. Although *A. baumannii* is mostly a nosocomial pathogen, it can cause disease in seemingly healthy individuals. It is a risk group 2 biological agent and has to be treated under biosafety level 2 conditions in most countries.
2. Cultivate *A. baumannii* using standard media such as modified LB broth (*see* Subheading 2.2). *A. baumannii* is non-fastidious and many standard media are suitable.
3. Widely used *A. baumannii* strains that are naturally competent are DSM 30011 [20] and AB5075 [23] (*see* **Note 2**).

2.4 Transforming DNA

1. To purify transforming DNA, use any commercial purification kit according to the manufacturer's recommendations.
2. Sterile ultrapure water or sterile Tris–HCl (10 mM, pH 7.5) to elute or dissolve DNA.

2.5 Sterile Buffer Solution to Flush Off Bacteria from Motility Plates

1. Prepare a sterile phosphate-buffered saline (PBS) as follows. Add 8 g NaCl, 0.2 g KCl, 1.44 g Na_2HPO_4, and 0.24 g KH_2PO_4 into a beaker, add 800 mL of deionized water to dissolve the salts, and adjust the pH to 7.4 with HCl. Fill up to a final volume of 1 L with deionized water and autoclave the solution.
2. Alternatively, use sterile filters with 0.2 μm pore size to filter appropriate volumes of an unsterile PBS. The volume required to flush off bacteria from motility plates is 1 mL per Petri dish.
3. Sterile 1 mL filter tips for flushing off bacteria.

2.6 Medium for Selection of Transformants

1. To select for transformants, selective plates with appropriate antibiotics are to be prepared. To this end, modified LB medium as above (*see* Subheading 2.2) can be supplemented with 15 g/L agar.
2. After autoclaving and cooling to approximately 50 °C, supplement with selective antibiotic at appropriate concentration.
3. Triangle shape cell spreader or equivalent.

2.7 Determination of Transformation Rates

1. Sterile PBS (*see* Subheading 2.5).
2. Spectrophotometer (wavelength 600 nm).
3. Disposable cuvettes (semimicro).
4. 1.5 mL reaction tubes ("low binding" quality).
5. Pipette tips ("low binding" quality).
6. Nonselective LB agar plates (*see* Subheading 2.6, **item 1**).

2.8 PCR Analysis of Transformants

1. PCR machine.
2. PCR reaction components including suitable primer pairs. Primer sequences corresponding to example illustrated in Fig. 2 are as follows:

   P1: 5′-TCAGCAGCTGATTAATCAACGAG-3′

   P2: 5′-ATCGGTAATAAAGCCGATATGCG-3′

   P3: 5′-GAGTTGAAGGATCAGATCACGC-3′
3. Agarose gel electrophoresis following standard procedures.

3 Methods

3.1 Transformation on Motility Plates

1. Select strains for transformation (*see* **Note 2**), and select donor DNA (*see* **Note 3**) according to the scientific questions to be addressed.
2. Purify DNA to be used for transformation. When finally dissolving or eluting the purified DNA, adjust to a minimum concentration of 100 ng/μL DNA (ideally 400 ng/μL), and dissolve or elute in sterile ultrapure water or sterile Tris–HCl (10 mM, pH 7.5) (*see* **Note 4**). Adjust to a DNA concentration of 400 ng/μL to yield maximum transformation rates irrespective of whether plasmid or chromosomal DNA is transformed.
3. Produce overnight cultures of the strains to be transformed. Select single colonies (*see* **Note 5**) for inoculation into 3 mL of modified LB broth (*see* Subheading 2.2), and incubate the cultures at 37 °C overnight.
4. Transfer 200 μL overnight culture into a 1.5 mL reaction tube, and centrifuge it for 5 min at 6000 × *g*. Remove the supernatant, wash the pellet once in 1 mL sterile PBS (*see* Subheading 2.5, **item 1**), again resuspend the pellet in 1 mL PBS, and adjust the suspension to 0.1 OD at 600 nm (*see* **Note 6**). Store bacterial suspensions on ice until use.
5. Produce proper amount of motility plates (*see* Subheading 2.1, **item 1**). After autoclaving of the motility medium and after its cooling to approximately 50 °C, Petri dishes are filled with approximately 20 mL of motility medium. After gelation, the plates are dried under the safety cabinet with the lid halfway

open until the condensed water under the lid has evaporated. The motility plates are used the same day (*see* **Note 7**).

6. For every transformation plate, mix 7 μL of transforming DNA with 7 μL of the bacterial suspension.
7. Take 2 μL of this mixture with a pipette and stab-inoculate it into a motility plate. Repeat this six times for every plate (*see* **Note 8**) (*see* Fig. 1).
8. Do include controls: (1) Bacterial suspensions without transforming DNA. (2) Transforming DNA without bacteria.
9. Seal every plate with Parafilm before putting it into the incubator, or alternatively put all plates into a plastic bag to avoid drying-up during incubation. Incubate the plates upside down at 37 °C overnight (~16 h).

3.2 Flushing Off Bacteria from Motility Medium to Plate on Selective Medium

1. Pipette 1 mL of sterile PBS onto each plate, flush off the bacteria from the surface, and collect them in 1.5 mL tubes. Use filter tips to avoid contamination of the pipette shaft while floating off the motility biofilm from the plate. Avoid producing aerosols by gently pipetting, and use a safety cabinet to protect yourself and others (*see* **Note 9**).
2. Thoroughly vortex the collected suspensions.
3. If no transformation rates are to be determined, 100 μL of each suspension can be directly plated on appropriate selective plates.
4. If transformation rates are to be determined, the total number of bacteria plated on the selective media needs to be known. To this end, it is advisable to determine the optical density of the bacterial suspensions flushed off the motility medium using a spectrophotometer at a wavelength of 600 nm. Adjust the bacterial suspensions to an $OD_{600\ nm}$ of 1.0 (*see* **Note 10**) before serially diluting the OD-adjusted samples (steps of 1:10 in sterile PBS; *see* Subheading 2.5, **item 1**) (*see* **Note 11**) and subsequently plating on nonselective LB agar plates. At a rough estimate, an $OD_{600\ nm}$ of 1.0 corresponds to approximately 2×10^8 cells of *A. baumannii* per mL although there can be significant differences between strains especially due to chain formation. Typically, plating 100 μL each of the 10^{-4} and 10^{-5} dilution on nonselective media is sufficient to determine the colony-forming units (CFU) of the adjusted sample.
5. Plate 100 μL of the OD-adjusted bacterial suspension on appropriate selective medium, and incubate overnight at 37 °C.
6. Determine the number of CFU on selective and nonselective medium, correct for dilution, and calculate the ratio (= transformation rate) (*see* **Note 12**).

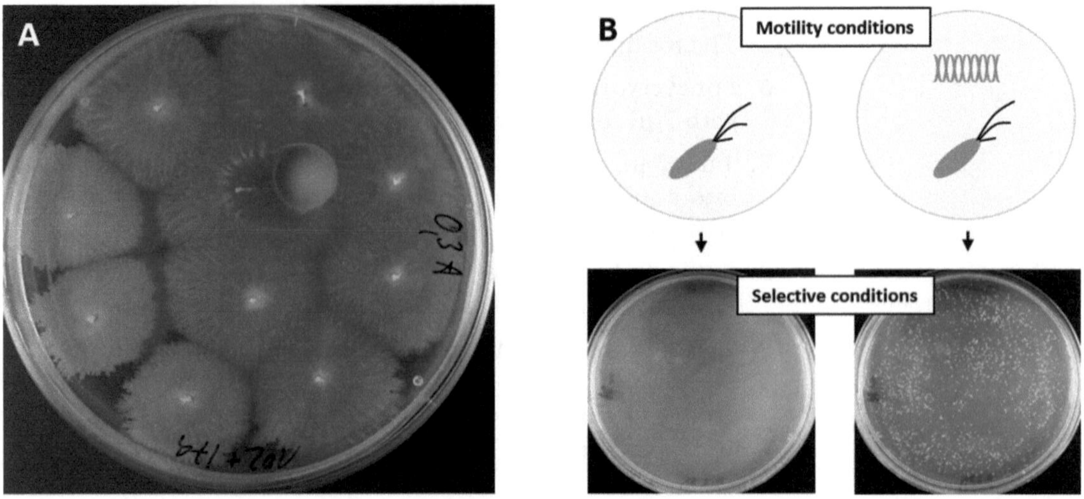

Fig. 1 The motility-dependent transformation assay. (**a**) Illustration of surface-associated motility. *Acinetobacter* strain 07-102 was stab-inoculated ten times into a semisolid medium (0.3% agarose supplemented with 5 g/L tryptone and 2.5 g/L NaCl) suitable for transformation and incubated for 16 h at 37 °C. (**b**) Scheme of the transformation procedure. Bacteria alone (top left) or bacteria mixed with transforming DNA (top right) are stab-inoculated on motility medium and incubated at 37 °C overnight. The next day motility biofilms are floated off the motility plates and spread on appropriate selective media. Only after effective transformation colonies appear (down right)

3.3 Confirmation of Transformation Events

1. Uptake of plasmid DNA can be confirmed by plasmid purification following the standard procedures.
2. Integration of transformed DNA into the chromosome can be confirmed by PCR (*see* example in Fig. 2) (*see* **Note 13**).

4 Notes

1. We recommend you to test different solidifying agents (agarose, agar) for your motility medium. This appears to be the most crucial point for optimization of transformation rates. Initially, when we established the protocol, we had used agarose of electrophoresis grade to produce motility medium simply because we found the motility phenotype more reproducible on agarose plates compared to agar plates. However, after the original agarose stock was used up, the next agarose that we ordered yielded very poor transformation rates so that we compared different products. A much cheaper standard agar (Difco) proved superior to any of the agarose products tested. Transformation rates have been elevated by up to two orders of magnitude compared to our previous study [20], reaching now 1% for the most competent strains.
2. We recommend to establish the transformation protocol with a strain that is already known to be transformable at high rates.

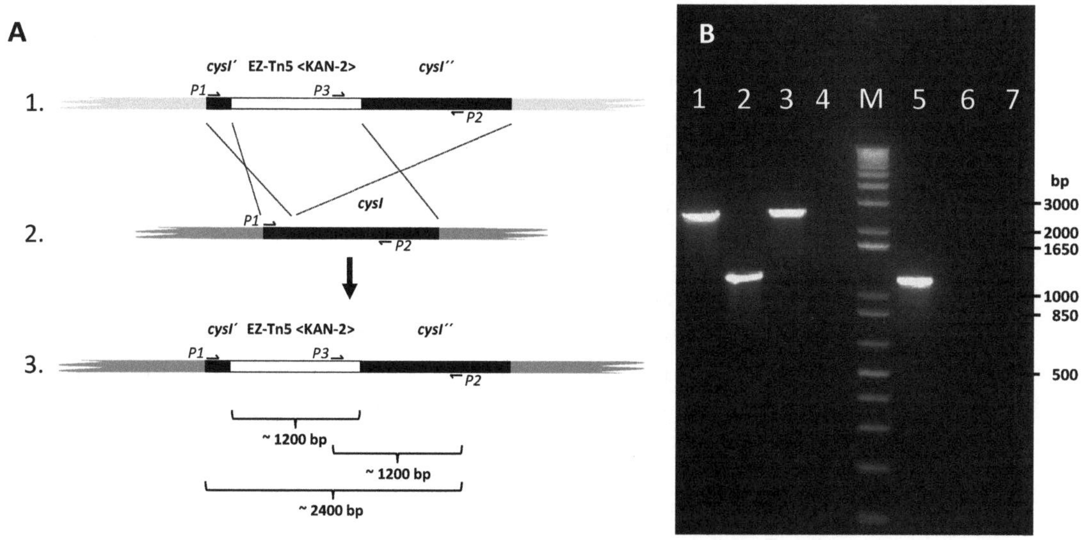

Fig. 2 Homologous recombination after DNA uptake and PCR analysis to confirm a transformation event. (**a**) Scheme of homologous recombination between a donor DNA (1) and the corresponding locus on the chromosome of an acceptor strain (2). The resulting DNA of such a recombination event shows the same overall arrangement as the donor DNA (3). The presentation of recombining regions in the scheme (dotted lines) is arbitrary and experimentally not well studied. The donor DNA exemplified here harbors a transposon EZ-Tn5 <KAN-2> insertion in gene A1S_2846 (*cysI*) of *A. baumannii* ATCC 17978 (mutant 179, [20]). (**b**) PCR analysis after natural transformation of acceptor strain *A. baumannii* DSM 30011 [41] treated with DNA of ATCC 17978 mutant 179. Lane 1: PCR with primers P1 and P2 using DNA from mutant 179 donor as template (product length approx. 2.4 kbp). Lane 2: PCR with primers P1 and P2 using DNA from acceptor strain *A. baumannii* DSM 30011 (not transformed) as template (product length approx. 1.2 kbp). Lane 3: PCR with primers P1 and P2 using DNA from transformant *A. baumannii* DSM 30011 x ATCC 17978 mutant 179 (product length approx. 2.4 kbp). Lane 4: PCR control with primers P1 and P2 without DNA template. Lane 5: PCR with primers P2 and P3 using DNA from transformant *A. baumannii* DSM 30011 x ATCC 17978 mutant 179 (product length approx. 1.2 kbp). Lane 6: PCR with primers P2 and P3 using DNA from acceptor *A. baumannii* DSM 30011 (not transformed). Lane 7: PCR control with primers P2 and P3 without DNA template. Lane M: DNA standard with length of fragments in base pairs (bp) as indicated

A. baumannii strain AB5075 has the advantage that it is competent at rates up to 1%, it is fully sequenced, and there is a comprehensive transposon mutant library available [23]. Most of these mutants harbor a tetracycline resistance cassette within mini-transposon T26. The DNA of such mutants can be used as a donor DNA to transform the parental strain AB5075. In this way you can easily establish and optimize the transformation procedure before proceeding with your original scientific question. We regularly use this library and our own EZ-Tn5 <KAN-2>-based library to introduce any gene inactivation of interest into our competent strains [20, 30, 32]. In case you make use of the tetracycline cassette for selection, check the susceptibility of the donor strain for tetracycline. Some of the AB5075 mutant strains can only withstand 5–10 μg/mL of tetracycline depending on the locus of insertion.

3. Alternative to exploiting the transposon mutant library as described above, transforming DNA suitable for establishing the protocol can be also a plasmid stably replicating in *A. baumannii* (e.g., pWH1266 and derivatives [20, 33], pBAV1K-T5 and derivatives [34]) or any selection marker integrated into a conserved genome region (e.g., *cysI* [20]; *see* Fig. 2).
4. Verify that the DNA you use for transformation is really sterile.
5. Be aware of phase variation [35, 36]. Opaque and translucent colonies may differ in their transformation efficiency.
6. If it is not meant to yield a maximum transformation efficiency or to determine a reproducible transformation rate, a fresh overnight culture is not required. A colony from a plate stored in the fridge can be resuspended in 50 μL sterile PBS and used without OD adjustment.
7. A recent report describes the miniaturization of the procedure using 2 mL reaction tubes filled with 1 mL of motility medium [37].
8. Inoculation of up to ten times per plate may increase the total number of transformants if desired (*see* Fig. 1).
9. Some strains tend to form very sticky biofilms within a few hours following the motility phase, which hinders flushing off and precludes proper OD measurement and serial dilution. In this case a reduction of the incubation time on the motility plate is advisable.
10. Be aware of the limitations of the Beer-Lambert law. The suspensions can reach several ODs and need to be diluted for proper OD determination. Typically, 1:10 dilutions of these suspensions yield $OD_{600\ nm}$ values between 0.1 and 1.
11. We recommend to use "low binding" tubes and "low binding" pipette tips to produce dilution series since adherence of some strains to plastic ware can dramatically distort CFU (*c*olony-*f*orming *u*nits) results.
12. Transformation rates unsatisfactory? Test different solidifying agents (*see* **Note 1**). Make sure that the protocol works well with an established strain (*see* **Note 2**). If your attempts to transform a specific strain should fail, try protocols based on planktonic growth [19, 38, 39]. Supplementing the medium with 1 mM $CaCl_2$ may improve the transformation efficiency [37, 39]. Test parameters known to affect transformation competence in *Acinetobacter baylyi* [40]. Control of pH (not above pH 7) during transformation seems to be crucial [37].
13. Note that the bacterial suspension floated off the motility plate and spread on the selective medium also contains the remains of the DNA used for transformation. Direct colony PCR from

the selective medium may thus yield false-positive results. Therefore, either design unambiguous PCR, e.g., one primer binding inside the putatively transferred DNA and one outside in a region specific to the recipient. Alternatively, transfer colonies to be checked from the selective plate to a fresh selective plate, and incubate transferred bacteria overnight before testing by colony PCR. Limiting the number of PCR cycles to 25 also reduces the probability of false-positive PCR results.

Acknowledgment

G.W. acknowledges financial support from the Deutsche Forschungsgemeinschaft (DFG-FOR 2251, WI 3272/3-1).

References

1. Da Silva GJ, Domingues S (2016) Insights on the horizontal gene transfer of carbapenemas e determinants in the opportunistic pathogen *Acinetobacter baumannii*. Microorganisms 4 (3). https://doi.org/10.3390/microorganisms 4030029
2. Fournier PE, Vallenet D, Barbe V, Audic S, Ogata H, Poirel L, Richet H, Robert C, Mangenot S, Abergel C, Nordmann P, Weissenbach J, Raoult D, Claverie JM (2006) Comparative genomics of multidrug resistance in *Acinetobacter baumannii*. PLoS Genet 2(1): e7. https://doi.org/10.1371/journal.pgen.0020007
3. Liu F, Zhu Y, Yi Y, Lu N, Zhu B, Hu Y (2014) Comparative genomic analysis of *Acinetobacter baumannii* clinical isolates reveals extensive genomic variation and diverse antibiotic resistance determinants. BMC Genomics 15:1163. https://doi.org/10.1186/1471-2164-15-1163
4. Wright MS, Haft DH, Harkins DM, Perez F, Hujer KM, Bajaksouzian S, Benard MF, Jacobs MR, Bonomo RA, Adams MD (2014) New insights into dissemination and variation of the health care-associated pathogen *Acinetobacter baumannii* from genomic analysis. MBio 5(1):e00963–e00913. https://doi.org/10.1128/mBio.00963-13
5. Jung J, Madsen EL, Jeon CO, Park W (2011) Comparative genomic analysis of *Acinetobacter oleivorans* DR1 to determine strain-specific genomic regions and gentisate biodegradation. Appl Environ Microbiol 77(20):7418–7424. https://doi.org/10.1128/aem.05231-11
6. Hassan A, Naz A, Obaid A, Paracha RZ, Naz K, Awan FM, Muhmmad SA, Janjua HA, Ahmad J, Ali A (2016) Pangenome and immuno-proteomics analysis of *Acinetobacter baumannii* strains revealed the core peptide vaccine targets. BMC Genomics 17(1):732. https://doi.org/10.1186/s12864-016-2951-4
7. Zarrilli R, Pournaras S, Giannouli M, Tsakris A (2013) Global evolution of multidrug-resistant *Acinetobacter baumannii* clonal lineages. Int J Antimicrob Agents 41(1):11–19. https://doi.org/10.1016/j.ijantimicag.2012.09.008
8. Chatterjee S, Mondal A, Mitra S, Basu S (2017) *Acinetobacter baumannii* transfers the blaNDM-1 gene via outer membrane vesicles. J Antimicrob Chemother 72(8):2201–2207. https://doi.org/10.1093/jac/dkx131
9. Rumbo C, Fernandez-Moreira E, Merino M, Poza M, Mendez JA, Soares NC, Mosquera A, Chaves F, Bou G (2011) Horizontal transfer of the OXA-24 carbapenemase gene via outer membrane vesicles: a new mechanism of dissemination of carbapenem resistance genes in *Acinetobacter baumannii*. Antimicrob Agents Chemother 55(7):3084–3090. https://doi.org/10.1128/aac.00929-10
10. Pande S, Shitut S, Freund L, Westermann M, Bertels F, Colesie C, Bischofs IB, Kost C (2015) Metabolic cross-feeding via intercellular nanotubes among bacteria. Nat Commun 6:6238. https://doi.org/10.1038/ncomms7238
11. Dubey GP, Ben-Yehuda S (2011) Intercellular nanotubes mediate bacterial communication. Cell 144(4):590–600. https://doi.org/10.1016/j.cell.2011.01.015
12. Vaneechoutte M, Young DM, Ornston LN, De Baere T, Nemec A, Van Der Reijden T, Carr E, Tjernberg I, Dijkshoorn L (2006) Naturally transformable *Acinetobacter* sp. strain ADP1

belongs to the newly described species *Acinetobacter baylyi*. Appl Environ Microbiol 72 (1):932–936. https://doi.org/10.1128/aem.72.1.932-936.2006

13. Juni E, Janik A (1969) Transformation of *Acinetobacter calcoaceticus* (Bacterium anitratum). J Bacteriol 98(1):281–288
14. Overballe-Petersen S, Harms K, Orlando LA, Mayar JV, Rasmussen S, Dahl TW, Rosing MT, Poole AM, Sicheritz-Ponten T, Brunak S, Inselmann S, de Vries J, Wackernagel W, Pybus OG, Nielsen R, Johnsen PJ, Nielsen KM, Willerslev E (2013) Bacterial natural transformation by highly fragmented and damaged DNA. Proc Natl Acad Sci U S A 110 (49):19860–19865. https://doi.org/10.1073/pnas.1315278110
15. Averhoff B, Friedrich A (2003) Type IV pili-related natural transformation systems: DNA transport in mesophilic and thermophilic bacteria. Arch Microbiol 180(6):385–393. https://doi.org/10.1007/s00203-003-0616-6
16. Friedrich A, Hartsch T, Averhoff B (2001) Natural transformation in mesophilic and thermophilic bacteria: identification and characterization of novel, closely related competence genes in *Acinetobacter* sp. strain BD413 and *Thermus thermophilus* HB27. Appl Environ Microbiol 67(7):3140–3148. https://doi.org/10.1128/aem.67.7.3140-3148.2001
17. Metzgar D, Bacher JM, Pezo V, Reader J, Doring V, Schimmel P, Marliere P, de Crecy-Lagard V (2004) *Acinetobacter* sp. ADP1: an ideal model organism for genetic analysis and genome engineering. Nucleic Acids Res 32 (19):5780–5790. https://doi.org/10.1093/nar/gkh881
18. Leong CG, Bloomfield RA, Boyd CA, Dornbusch AJ, Lieber L, Liu F, Owen A, Slay E, Lang KM, Lostroh CP (2017) The role of core and accessory type IV pilus genes in natural transformation and twitching motility in the bacterium *Acinetobacter baylyi*. PLoS One 12 (8):e0182139. https://doi.org/10.1371/journal.pone.0182139
19. Ramirez MS, Don M, Merkier AK, Bistue AJ, Zorreguieta A, Centron D, Tolmasky ME (2010) Naturally competent *Acinetobacter baumannii* clinical isolate as a convenient model for genetic studies. J Clin Microbiol 48 (4):1488–1490. https://doi.org/10.1128/jcm.01264-09
20. Wilharm G, Piesker J, Laue M, Skiebe E (2013) DNA uptake by the nosocomial pathogen *Acinetobacter baumannii* occurs during movement along wet surfaces. J Bacteriol 195 (18):4146–4153. https://doi.org/10.1128/jb.00754-13
21. Wilharm G, Skiebe E, Higgins PG, Poppel MT, Blaschke U, Leser S, Heider C, Heindorf M, Brauner P, Jackel U, Bohland K, Cuny C, Lopinska A, Kaminski P, Kasprzak M, Bochenski M, Ciebiera O, Tobolka M, Zolnierowicz KM, Siekiera J, Seifert H, Gagne S, Salcedo SP, Kaatz M, Layer F, Bender JK, Fuchs S, Semmler T, Pfeifer Y, Jerzak L (2017) Relatedness of wildlife and livestock avian isolates of the nosocomial pathogen *Acinetobacter baumannii* to lineages spread in hospitals worldwide. Environ Microbiol 19(10):4349–4364. https://doi.org/10.1111/1462-2920.13931
22. Yoon EJ, Chabane YN, Goussard S, Snesrud E, Courvalin P, De E, Grillot-Courvalin C (2015) Contribution of resistance-nodulation-cell division efflux systems to antibiotic resistance and biofilm formation in *Acinetobacter baumannii*. MBio 6(2). https://doi.org/10.1128/mBio.00309-15
23. Gallagher LA, Ramage E, Weiss EJ, Radey M, Hayden HS, Held KG, Huse HK, Zurawski DV, Brittnacher MJ, Manoil C (2015) Resources for genetic and genomic analysis of emerging pathogen *Acinetobacter baumannii*. J Bacteriol 197(12):2027–2035. https://doi.org/10.1128/jb.00131-15
24. Barker J, Maxted H (1975) Observations on the growth and movement of *Acinetobacter* on semi-solid media. J Med Microbiol 8 (3):443–446. https://doi.org/10.1099/00222615-8-3-443
25. LAUTROP H (1961) *Bacterium anitratum* transferred to the genus Cytophaga. Int J Syst Evol Microbiol 11(3):107–108
26. Clemmer KM, Bonomo RA, Rather PN (2011) Genetic analysis of surface motility in *Acinetobacter baumannii*. Microbiology 157 (Pt 9):2534–2544. https://doi.org/10.1099/mic.0.049791-0
27. Eijkelkamp BA, Stroeher UH, Hassan KA, Papadimitrious MS, Paulsen IT, Brown MH (2011) Adherence and motility characteristics of clinical *Acinetobacter baumannii* isolates. FEMS Microbiol Lett 323(1):44–51. https://doi.org/10.1111/j.1574-6968.2011.02362.x
28. Harding CM, Tracy EN, Carruthers MD, Rather PN, Actis LA, Munson RS Jr (2013) *Acinetobacter baumannii* strain M2 produces type IV pili which play a role in natural transformation and twitching motility but not surface-associated motility. MBio 4 4(4). https://doi.org/10.1128/mBio.00360-13
29. Antunes LC, Imperi F, Carattoli A, Visca P (2011) Deciphering the multifactorial nature of *Acinetobacter baumannii* pathogenicity.

PLoS One 6(8):e22674. https://doi.org/10.1371/journal.pone.0022674

30. Skiebe E, de Berardinis V, Morczinek P, Kerrinnes T, Faber F, Lepka D, Hammer B, Zimmermann O, Ziesing S, Wichelhaus TA, Hunfeld KP, Borgmann S, Grobner S, Higgins PG, Seifert H, Busse HJ, Witte W, Pfeifer Y, Wilharm G (2012) Surface-associated motility, a common trait of clinical isolates of *Acinetobacter baumannii*, depends on 1,3-diaminopropane. Int J Med Microbiol 302(3):117–128. https://doi.org/10.1016/j.ijmm.2012.03.003
31. Salzer R, Kern T, Joos F, Averhoff B (2016) The *Thermus thermophilus comEA/comEC* operon is associated with DNA binding and regulation of the DNA translocator and type IV pili. Environ Microbiol 18(1):65–74. https://doi.org/10.1111/1462-2920.12820
32. Heindorf M, Kadari M, Heider C, Skiebe E, Wilharm G (2014) Impact of Acinetobacter baumannii superoxide dismutase on motility, virulence, oxidative stress resistance and susceptibility to antibiotics. PLoS One 9(7): e101033. https://doi.org/10.1371/journal.pone.0101033
33. Hunger M, Schmucker R, Kishan V, Hillen W (1990) Analysis and nucleotide sequence of an origin of DNA replication in *Acinetobacter calcoaceticus* and its use for *Escherichia coli* shuttle plasmids. Gene 87(1):45–51
34. Bryksin AV, Matsumura I (2010) Rational design of a plasmid origin that replicates efficiently in both gram-positive and gram-negative bacteria. PLoS One 5(10):e13244. https://doi.org/10.1371/journal.pone.0013244
35. Tipton KA, Dimitrova D, Rather PN (2015) Phase-variable control of multiple phenotypes in *Acinetobacter baumannii* strain AB5075. J Bacteriol 197(15):2593–2599. https://doi.org/10.1128/jb.00188-15
36. Tipton KA, Rather PN (2016) An ompR/envZ two-component system ortholog regulates phase variation, osmotic tolerance, motility, and virulence in *Acinetobacter baumannii* strain AB5075. J Bacteriol. https://doi.org/10.1128/jb.00705-16
37. Godeux A-S, Lupo A, Haenni M, Guette-Marquet S, Wilharm G, Laaberki M-H, Charpentier X (2018) Fluorescence-based detection of natural transformation in drug resistant *Acinetobacter baumannii*. bioRxiv. https://doi.org/10.1101/262311
38. Quinn B, Traglia GM, Nguyen M, Martinez J, Liu C, Fernandez JS, Ramirez MS (2018) Effect of host human products on natural transformation in *Acinetobacter baumannii*. Curr Microbiol. https://doi.org/10.1007/s00284-017-1417-5
39. Traglia GM, Quinn B, Schramm ST, Soler-Bistue A, Ramirez MS (2016) Serum albumin and Ca2+ are natural competence inducers in the human pathogen *Acinetobacter baumannii*. Antimicrob Agents Chemother 60 (8):4920–4929. https://doi.org/10.1128/aac.00529-16
40. Leong CG, Boyd CM, Roush KS, Tenente R, Lang KM, Lostroh CP (2017) Succinate, iron chelation, and monovalent cations affect the transformation efficiency of *Acinetobacter baylyi* ATCC 33305 during growth in complex media. Can J Microbiol 63(10):851–856. https://doi.org/10.1139/cjm-2017-0393
41. Repizo GD, Viale AM, Borges V, Cameranesi MM, Taib N, Espariz M, Brochier-Armanet C, Gomes JP, Salcedo SP (2017) The environmental *Acinetobacter baumannii* isolate DSM30011 reveals clues into the preantibiotic era genome diversity, virulence potential, and niche range of a predominant nosocomial pathogen. Genome Biol Evol 9(9):2292–2307. https://doi.org/10.1093/gbe/evx162

Chapter 9

Vesicle-Mediated Gene Transfer in *Acinetobacter baumannii*

Shweta Fulsundar, Sara Domingues, and Kaare Magne Nielsen

Abstract

The role of vesicle-mediated gene transfer in *Acinetobacter baumannii* populations has been investigated in the last decade. Importantly, outer membrane vesicles (OMVs) secreted from *A. baumannii* cells have proven to be efficient agents of transfer of antimicrobial resistance genes to other bacterial species. However, the measurement of vesicle-mediated transfer depends on many experimental parameters. Here, we describe an experimental method useful to study transfer of DNA via membrane vesicles of *A. baumannii* in various bacterial populations.

Key words Outer membrane vesicle, *Acinetobacter baumannii*, Antimicrobial resistance, Horizontal gene transfer, Transfer frequency

1 Introduction

Living cells belonging to all domains of life secret membrane vesicles [1]. Outer membrane vesicles (OMVs) are produced by a great variety of Gram-negative bacteria and are formed by budding of the outer membrane (OM) of these cells [2]. OMVs can pack a wide variety of cargos, including lipids, nucleic acids, proteins, and polysaccharides, and can have numerous functions associated with bacterial physiology and pathogenesis [2–4]. Production of OMVs by *Acinetobacter baumannii* cells was demonstrated for the first time in 2009 [5]. Later, it was demonstrated that OMVs released by this species can have different sizes depending on the growth phase [6]. OMV secretion seems to be common among species of the *Acinetobacter* genus and has been detected in *A. baylyi* [7], *A. nosocomialis* [8, 9], and *A. radioresistens* [10].

So far, proteins and nucleic acids related to antimicrobial resistance [11–14] and virulence effects [14–17] have been detected in OMVs released by *A. baumannii*. DNA carried in OMVs has been horizontally transferred between *A. baumannii* cells [11, 13] and

Indranil Biswas and Philip N. Rather (eds.), *Acinetobacter baumannii: Methods and Protocols*, Methods in Molecular Biology, vol. 1946, https://doi.org/10.1007/978-1-4939-9118-1_9, © Springer Science+Business Media, LLC, part of Springer Nature 2019

from *A. baumannii* to *Escherichia coli* cells [13], leading to expression of resistance genes and changes to the phenotype. The release of *A. baumannii* OMVs containing high levels of the beta-lactamase OXA-58 was shown to have a sheltering effect on neighbor carbapenem-susceptible *A. baumannii* cells [12]. OMVs can also interact with human host cells; they can deliver OM components or other cargos, activating the human immune system [17, 18] as well as causing cytotoxic effects [14, 17].

There are several factors that limit the extent of studies involving OMVs [19]. OMV production involves different steps, namely, cultivation of bacteria, separation between intact cells and OMVs, concentration of OMVs, and purification [20]. The production and isolation in sufficient amount and purification of the OMVs are critical steps and some of the main difficulties of OMV studies [20, 21]. The cultivation time and the growth media are essential for an efficient vesiculation [20]. Often, ultrafiltration and centrifugation are applied to isolate and purify OMVs, and these may lead to random inclusion of non-natural cargo or components [6, 20, 21].

OMV-mediated transfer has the advantage of protecting the cargo from extracellular degradation, as well as delivering molecules in an active and concentrated form [21]. OMV-mediated gene transfer seems to occur at a low frequency when compared to other laboratory-optimized routes of horizontal gene transfer [7]. Several factors such as the secretion rate, the concentration of the genetic cargo, and the membrane properties of the donor and recipient strain, will influence the transfer rate [22]. The amount of DNA present or associated with vesicles is considered the main determinant of the transfer frequency [7, 23]. The OMV-recipient cell ratio, the incubation periods, and the growth medium are also some of the factors that need to be taken into account during setting up vesicle-mediated gene transfer experiments.

In this chapter, we present an optimized and detailed protocol to test and confirm OMV-mediated transfer in *A. baumannii*.

2 Materials

2.1 Purification of Cell-Free Vesicles

1. HEPES buffer (50 mM, pH 6.8).
2. 0.22 μm pore size syringe filters (Pall Corporation, New York, USA).
3. Sterile Eppendorf tubes.
4. Proteinase K (100 μg/mL).
5. DNase I (100 ng/mL).

2.2 Vesicle Quantification

Bradford assay kit (Bio-Rad, USA) (*see* **Note 1**).

2.3 Vesicle DNA Isolation

1. Zyppy Plasmid Miniprep Kit, Zymo Research, USA (*see* **Note 2**).
2. TE buffer (pH 8.0).
3. NanoDrop instrument.
4. PicoGreen reagent (Molecular Probes, Invitrogen, USA).
5. 96-well black fluorescence assay plates (lid covered with aluminum foil).
6. Hybrid multimode microplate reader.

2.4 Transfer/Incubation Media

Prepare the media using ultrapure water, and autoclave/filter sterilize to ensure sterility.

1. SOC (super optimal broth with catabolite repression).

   2% bacto-tryptone, 0.5% bacterial yeast extract, 10 mM NaCl, 2.5 mM KCl, 10 mM $MgCl_2$, 10 mM $MgSO_4$, and 20 mM glucose (pH 7.0) (*see* **Note 3**).
2. LB (Luria-Bertani) (*see* **Note 4**).
   1% tryptone, 1% NaCl, and 0.5% yeast extract (pH 7.5).

2.5 Recipient Cells

1. Shaker-Incubator with temperature setting.
2. Spectrophotometer.
3. Sterile saline.
4. Sterile Eppendorf tubes (2.5 mL).
5. Sterile culture tubes with caps.

2.6 Confirmation of Gene Transfer

1. Specific primers for amplification of marker gene.
2. PCR thermocycler.
3. PCR reagents.
4. Gel electrophoresis.
5. Gel documentation system.

2.7 Agar Plates With Antibiotic Supplements

1. LB agar.
2. Antibiotics (*see* **Note 5**).

3 Methods

Carry out all procedures at room temperature unless otherwise specified.

3.1 Pre-treatment of Vesicles

1. Filter the solubilized vesicles (in HEPES buffer) through 0.22 μm pore size syringe filters (*see* **Note 6**), and make number of aliquots of vesicles (500–600 μL) in the Eppendorf tubes.
2. Store the filtered purified form of vesicles at −20 °C until further use.
3. Treat the vesicles with proteinase K (100 μg/mL) (*see* **Note 7**) and DNase I (100 ng/mL) (*see* **Notes 8–10**).

3.2 Quantification of Vesicles

1. Quantify the purified vesicles (in triplicate) to determine their protein concentration using Bradford assay [24] or Bradford assay kit.
2. Use the aliquoted vesicles that are quantified for the gene transfer assays (*see* **Note 11**).

3.3 Isolation and Quantification of Vesicular DNA (V-DNA)

1. Use the DNA isolation kit containing purification columns to isolate intravesicular or associated DNA.
2. Perform the isolation procedure in triplicate from three independent batches of vesicle samples with a protein concentration equivalent to at least 100 μg/mL.
3. Resuspend the extracted DNA in 15 μL of TE buffer.
4. Measure the V-DNA from three independent batches of vesicles using the Quant-iT PicoGreen dsDNA assay [7], or it can also be measured using NanoDrop [11] or spectrophotometer [13].

3.4 Recipient Cell Preparation

1. Determine the growth curve of the recipient strains using a spectrophotometer for $OD_{600\ nm}$ measurements to identify the number of hours required for the cells to enter into log phase.
2. Culture the recipient strains in SOC broth until log phase.
3. Harvest the cells by centrifugation (5000 × *g* for 10 min); wash and dilute with saline to produce a bacterial suspension of 2×10^4 CFU/mL (*see* **Note 12**).

3.5 Vesicle-Mediated Gene Transfer (Fig. 1)

Perform the gene transfer experiments in triplicate and at three independent times.

1. To prepare the gene transfer incubation mix, add 50 μL of recipient cells to 500 μL of SOC medium in the Eppendorf tubes.
2. Add 500 μL of purified OMVs with a known protein concentration (mg/mL) and DNA content (*see* **Note 13**).
3. Add 1 μL of 100 μg/μL DNase (final concentration 100 ng/mL) followed by incubation at 37 °C for 1 h without shaking (*see* **Note 14**). The static incubation can be increased to 4 h as per the recent report [13].

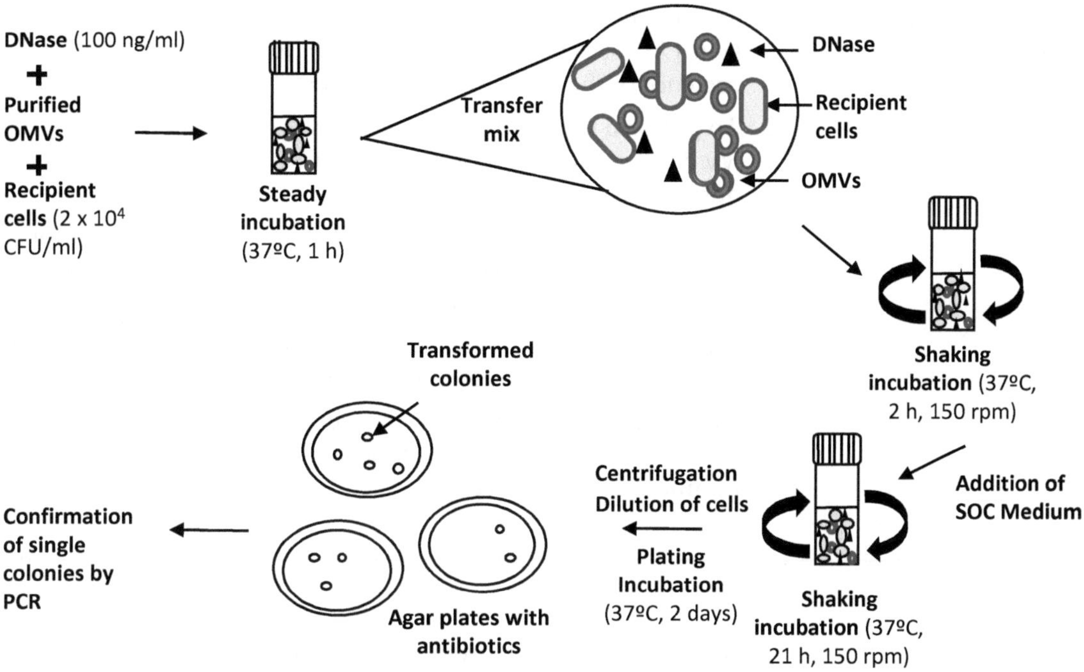

Fig. 1 Schematic presentation of the steps involved in vesicle-mediated transfer in *A. baumannii*

4. After 1 h of static incubation, transfer the mixture aseptically to culture tubes with cap, and incubate further for 2 h with shaking at 150 rpm.
5. Next, add 2 mL of SOC medium per assay, and continue the incubation with shaking for additional 21 h.
6. Prepare control tubes-A as above without adding vesicles in the mixture. Make another set of control tubes-B by replacing vesicles with free DNA/plasmid in purified form (DNA concentration same as the vesicle DNA concentration). Additionally, prepare another set of control tubes-C containing only recipient cells and DNase (final concentration 100 ng/mL) (*see* **Note 15**).
7. Pellet the cells by centrifugation after 24 h of incubation, and resuspend in 1 mL SOC medium.
8. Determine the total number of recipient titers by plating on LB agar plates and the proportion of bacterial cells that had acquired antibiotic resistance due to gene transfer via vesicles (vesiculants) [7] on LB agar supplemented with specific antibiotic concentration.
9. Incubate the plates at 37 °C for 2 days, and evaluate by colony-forming unit (CFU) counting. Determine the gene transfer frequency by dividing the number of vesiculants (CFU/mL) in the total cell count (CFU/mL).

3.6 Confirmation of Gene Transfer by PCR

1. To screen for acquired antibiotic resistance (newly formed vesiculants [7]), select a minimum of 10 CFUs per assay, and check the presence of the transferred marker gene by PCR (*see* **Note 16**).
2. To perform the colony PCR, use 2 μL of the supernatant of the lysed cell suspension (after boiling at 80 °C for 5 min) from the pure-streaked and restreaked CFUs.
3. Use donor bacteria containing DNA/plasmid with antibiotic-resistant gene as a marker gene (positive control) and bacteria without resistance or water (negative control).
4. Analyze the PCR products by gel electrophoresis on 1% agarose, and visualize with a gel documentation system.
5. Calculate the gene transfer frequency (from ≥3 independent experiments) as the number of gene transfer events over the number of recipient cells.

4 Notes

1. A commercial kit is preferable to determine the protein concentration of vesicles as a lower quantity of vesicles (microliters) is required for the microtiter plate measurements.
2. Isolation of DNA from vesicles is necessary to quantify the amount of V-DNA present relative to protein concentrations of vesicles, as well as for the PCR confirmation of the marker gene. Another advantage of using a DNA isolation kit is that V-DNA can be isolated in a purified form even from a small amount of vesicle sample.
3. The SOC medium components should not be mixed together and autoclaved in order to avoid the quality to be compromised. Each component should be autoclaved separately and later mixed, followed by a final step of filter sterilization.
4. LB medium can be used in the transfer mix instead SOC medium, as per the recent report [13].
5. Addition of antibiotics at high temperature could deteriorate the antibiotics. The plates can be stored at room temperature for 1–2 days in the dark and later at 4 °C. It is also advisable not to use plates containing antibiotic supplement that are more than 1 week old.
6. Filtration through 0.22 μm pore size syringe filters will make the vesicles free from bacterial cells, if present. This also blocks vesicles larger than 220 nm in size.
7. Treatment with proteinase K is necessary to digest any phage coats, if present.

8. Vesicle suspensions are incubated at 37 °C after addition of DNase I for 20 min followed by deactivation of the DNase at 80 °C for 10 min. DNase I treatment will digest free DNA. Thus, it will not interfere with vesicular DNA measurements.
9. The vesicle suspensions can be stored at −20 °C in aliquotes, but avoid freezing and thawing multiple times.
10. To confirm the absence of bacterial cells, 100 microliters of the vesicle suspension can be spread on LB agar and also inoculated in fresh LB medium followed by culturing overnight at 37 °C.
11. For comparisons, it is important that the quantified vesicles correspond to the same protein concentration before using them for gene transfer assays. It can be adjusted by diluting vesicle samples with HEPES buffer.
12. The recipient cells should not be overcrowded in the gene transfer mixture as compare to vesicles. Once they are diluted and suspended in saline, use immediately for the gene transfer assay.
13. Make sure that the purified vesicle preparations are available in large quantity (protein concentration). It can be achieved by pooling several batches of OMVs as not every vesicle contains DNA [7].
14. Static incubation increases the close contact/attachment/adhesion of vesicles with recipient cells [7].
15. Control tubes-A will be useful to detect the mutant colonies on the agar plates supplemented with antibiotics. Control tubes-B will further confirm that the DNA transfer was vesicle mediated. Control tubes-C will be necessary to determine the total cell count (CFU/mL) for calculating the gene transfer frequency.
16. If PCR-negative isolates are encountered, adjust the gene transfer frequencies accordingly.

References

1. Domingues S, Nielsen KM (2017) Membrane vesicles and horizontal gene transfer in prokaryotes. Curr Opin Microbiol 38:16–21. https://doi.org/10.1016/j.mib.2017.03.012
2. Jan AT (2017) Outer membrane vesicles (OMVs) of Gram-negative bacteria: a perspective update. Front Microbiol 8:1053. https://doi.org/10.3389/fmicb.2017.01053
3. Schwechheimer C, Kuehn MJ (2015) Outer-membrane vesicles from Gram-negative bacteria: biogenesis and functions. Nat Rev Microbiol 13:605–619. https://doi.org/10.1038/nrmicro3525
4. Kulkarni HM, Jagannadham MV (2014) Biogenesis and multifaceted roles of outer membrane vesicles from Gram-negative bacteria. Microbiology 160:2109–2121. https://doi.org/10.1099/mic.0.079400-0
5. Kwon SO, Gho YS, Lee JC, Kim SI (2009) Proteome analysis of outer membrane vesicles from a clinical *Acinetobacter baumannii* isolate. FEMS Microbiol Lett 297:150–156. https://doi.org/10.1111/j.1574-6968.2009.01669.x
6. Koning RI, de Breij A, Oostergetel GT, Nibbering PH, Koster AJ, Dijkshoorn L (2013) Cryo-electron tomography analysis of

membrane vesicles from *Acinetobacter baumannii* ATCC19606 T. Res Microbiol 164:397–405. https://doi.org/10.1016/j.resmic.2013.02.007

7. Fulsundar S, Harms K, Flaten GE, Johnsen PJ, Chopade BA, Nielsen KM (2014) Gene transfer potential of outer membrane vesicles of *Acinetobacter baylyi* and effects of stress on vesiculation. Appl Environ Microbiol 80:3469–3483. https://doi.org/10.1128/AEM.04248-13
8. Kim SW, Oh MH, Jun SH, Jeon H, Kim SI, Kim K, Lee YC, Lee JC (2016) Outer membrane protein A plays a role in pathogenesis of *Acinetobacter nosocomialis.* Virulence 7:413–426. https://doi.org/10.1080/21505594.2016.1140298
9. Nho JS, Jun SH, Oh MH, Park TI, Choi CW, Kim SI, Choi CH, Lee JC (2015) *Acinetobacter nosocomialis* secretes outer membrane vesicles that induce epithelial cell death and host inflammatory responses. Microb Pathog 81:39–45. https://doi.org/10.1016/j.micpath.2015.03.012
10. Fulsundar S, Kulkarni HM, Jagannadham MV, Nair R, Keerthi S, Sant P, Pardesi K, Bellare J, Chopade BA (2015) Molecular characterization of outer membrane vesicles released from *Acinetobacter radioresistens* and their potential roles in pathogenesis. Microb Pathog 83-84:12–22. https://doi.org/10.1016/j.micpath.2015.04.005
11. Rumbo C, Fernandez-Moreira E, Merino M, Poza M, Mendez JA, Soares NC, Mosquera A, Chaves F, Bou G (2011) Horizontal transfer of the OXA-24 carbapenemase gene via outer membrane vesicles: a new mechanism of dissemination of carbapenem resistance genes in *Acinetobacter baumannii.* Antimicrob Agents Chemother 55:3084–3090. https://doi.org/10.1128/AAC.00929-10
12. Liao YT, Kuo SC, Chiang MH, Lee YT, Sung WC, Chen YH, Chen TL, Fung CP (2015) *Acinetobacter baumannii* extracellular OXA-58 is primarily and selectively released via outer membrane vesicles after Sec-dependent periplasmic translocation. Antimicrob Agents Chemother 59:7346–7354. https://doi.org/10.1128/AAC.01343-15
13. Chatterjee S, Mondal A, Mitra S, Basu S (2017) *Acinetobacter baumannii* transfers the *bla*$_{\mathrm{NDM\text{-}1}}$ gene via outer membrane vesicles. J Antimicrob Chemother 72:2201–2207. https://doi.org/10.1093/jac/dkx131
14. Jin JS, Kwon SO, Moon DC, Gurung M, Lee JH, Kim SI, Lee JC (2011) *Acinetobacter baumannii* secretes cytotoxic outer membrane protein A via outer membrane vesicles. PLoS One 6:e17027. https://doi.org/10.1371/journal.pone.0017027
15. Mendez JA, Soares NC, Mateos J, Gayoso C, Rumbo C, Aranda J, Tomas M, Bou G (2012) Extracellular proteome of a highly invasive multidrug-resistant clinical strain of *Acinetobacter baumannii.* J Proteome Res 11:5678–5694. https://doi.org/10.1021/pr300496c
16. Jha C, Ghosh S, Gautam V, Malhotra P, Ray P (2017) In vitro study of virulence potential of *Acinetobacter baumannii* outer membrane vesicles. Microb Pathog 111:218–224. https://doi.org/10.1016/j.micpath.2017.08.048
17. Li ZT, Zhang RL, Bi XG, Xu L, Fan M, Xie D, Xian Y, Wang Y, Li XJ, Wu ZD, Zhang KX (2015) Outer membrane vesicles isolated from two clinical *Acinetobacter baumannii* strains exhibit different toxicity and proteome characteristics. Microb Pathog 81:46–52. https://doi.org/10.1016/j.micpath.2015.03.009
18. Jun SH, Lee JH, Kim BR, Kim SI, Park TI, Lee JC, Lee YC (2013) *Acinetobacter baumannii* outer membrane vesicles elicit a potent innate immune response via membrane proteins. PLoS One 8:e71751. https://doi.org/10.1371/journal.pone.0071751
19. Orench-Rivera N, Kuehn MJ (2016) Environmentally controlled bacterial vesicle-mediated export. Cell Microbiol 18:1525–1536. https://doi.org/10.1111/cmi.12676
20. Klimentova J, Stulik J (2015) Methods of isolation and purification of outer membrane vesicles from Gram-negative bacteria. Microbiol Res 170:1–9. https://doi.org/10.1016/j.micres.2014.09.006
21. Kulp A, Kuehn MJ (2010) Biological functions and biogenesis of secreted bacterial outer membrane vesicles. Annu Rev Microbiol 64:163–184. https://doi.org/10.1146/annurev.micro.091208.073413
22. Tran F, Boedicker JQ (2017) Genetic cargo and bacterial species set the rate of vesicle-mediated horizontal gene transfer. Sci Rep 7:8813. https://doi.org/10.1038/s41598-017-07447-7
23. Renelli M, Matias V, Lo RY, Beveridge TJ (2004) DNA-containing membrane vesicles of *Pseudomonas aeruginosa* PAO1 and their genetic transformation potential. Microbiology 150:2161–2169. https://doi.org/10.1099/mic.0.26841-0
24. Bradford MM (1976) A rapid and sensitive method for the quantitation of microgram quantities of protein utilizing the principle of protein-dye binding. Anal Biochem 72:248–254

Chapter 10

Targeted Gene Replacement in *Acinetobacter baumannii*

Indranil Biswas and Joshua Mettlach

Abstract

Acinetobacter baumannii is characterized as a nosocomial, gram-negative, multidrug-resistant bacterium, which has gained increasing attention due to its prevalence in hospital settings and high mortality rates upon infection. Currently, a number of different protocols have been developed in attempts to genetically alter *A. baumannii*, including multidrug-resistant strains. Although the bacterium has an unusual ability to uptake exogenous DNA in the natural environments, within the laboratory setting, gene manipulation to study virulence properties can be challenging. In this chapter we describe a general protocol for modification of specific genes using homologous recombination and a counterselectable marker.

Key words *Acinetobacter*, Multidrug resistance, Homologous recombination, Counterselection, Gene deletion, Gene replacement, Allelic exchange

1 Introduction

Acinetobacter baumannii is a nosocomial, gram-negative, multidrug-resistant bacterium that has gained increasing attention due to its prevalence within hospital settings. Recent studies by the CDC have placed this pathogen under a "serious threat" level due to its ability to thrive in harsh environments, high mortality rate upon infection, and limited availability of antibiotics to combat *A. baumannii* infections.

To better understand the virulence mechanisms of a pathogen, it is necessary to carry out targeted modifications (such as deletion, substitution, and others) of selected genes. While gene modification techniques are well established for many gram-negative bacteria, it is still challenging for *A. baumannii* despite the fact that more than 30% of clinical isolates are naturally competent [1, 2]. Nevertheless, a few techniques are available that rely on single-crossover integration mediated by RecA or site-specific recombination mediated by RecET [3–5].

Allelic exchange is a commonly used technique for gene replacement that is based on integration and excision of a circular

Indranil Biswas and Philip N. Rather (eds.), *Acinetobacter baumannii: Methods and Protocols*, Methods in Molecular Biology, vol. 1946, https://doi.org/10.1007/978-1-4939-9118-1_10, © Springer Science+Business Media, LLC, part of Springer Nature 2019

plasmid DNA via homologous recombination. The plasmid used for gene replacement is unable to replicate in *A. baumannii* and therefore called a suicide plasmid. Thus, under appropriate antibiotic selection, the plasmid integrates into the chromosome using a gene segment (homologous region) that is cloned in the plasmid. This gene segment carries the desired modifications (deletion, insertion, or mutation) and the up and down flanking regions that are homologous to the host chromosome (Fig. 1). The plasmid can integrate using any up or down homologous segment (Fig. 2). In the next step, using a second recombination vent, the vector backbone is excised out of the chromosome using one of the flanking homologous segments. If the second event occurs via the segment that was not used during initial integration, it will generate the desired modification on the host chromosome (Fig. 2). The second recombination event depends on both RecA and RecBCD enzymes and occurs very infrequently. To screen for this rare allelic exchange event, a counterselection method is applied that preferentially selects for cells that have lost the vector backbone.

The commonly used plasmids for allelic replacement contain the ColE1-type replication origin, which only functions in *Escherichia coli* but not in *A. baumannii* [6]. Many of these plasmids often contain the origin of transfer (*oriT*) from RP4, a broad-host-range plasmid. The presence of *oriT* provides opportunity for transferring ("mobilizing") these plasmids from *E. coli* to *A. baumannii* by conjugation instead of solely relying on electroporation. The choice of antibiotic resistance markers on these plasmids is often limited. Therefore, one should verify that the recipient strains are sensitive to the available markers before starting gene replacement experiments. In case of multidrug-resistant strains, a tellurite resistance marker is available for selection [5].

The efficiency of overall gene replacement strategy depends on the length of homologous segments. Homology less than 0.2 kb in length is not desirable since it generates very few recombination events. The optimum length is between 0.5 kb and 1.0 kb. Homology length greater than 1.0 kb does not add further benefit. The other factor that influences the efficiency is the counterselection marker. A few counterselection markers are available for use in *A. baumannii* [1]. But the most frequently used marker is the *sacB* gene from *Bacillus subtilis*. The *sacB* gene encodes levansucrase, which hydrolyzes sucrose to produce levan polysaccharides that are toxic to gram-negative bacteria. The method described here is based on *sacB* counterselection and tested in many *A. baumannii* strains including clinical isolates.

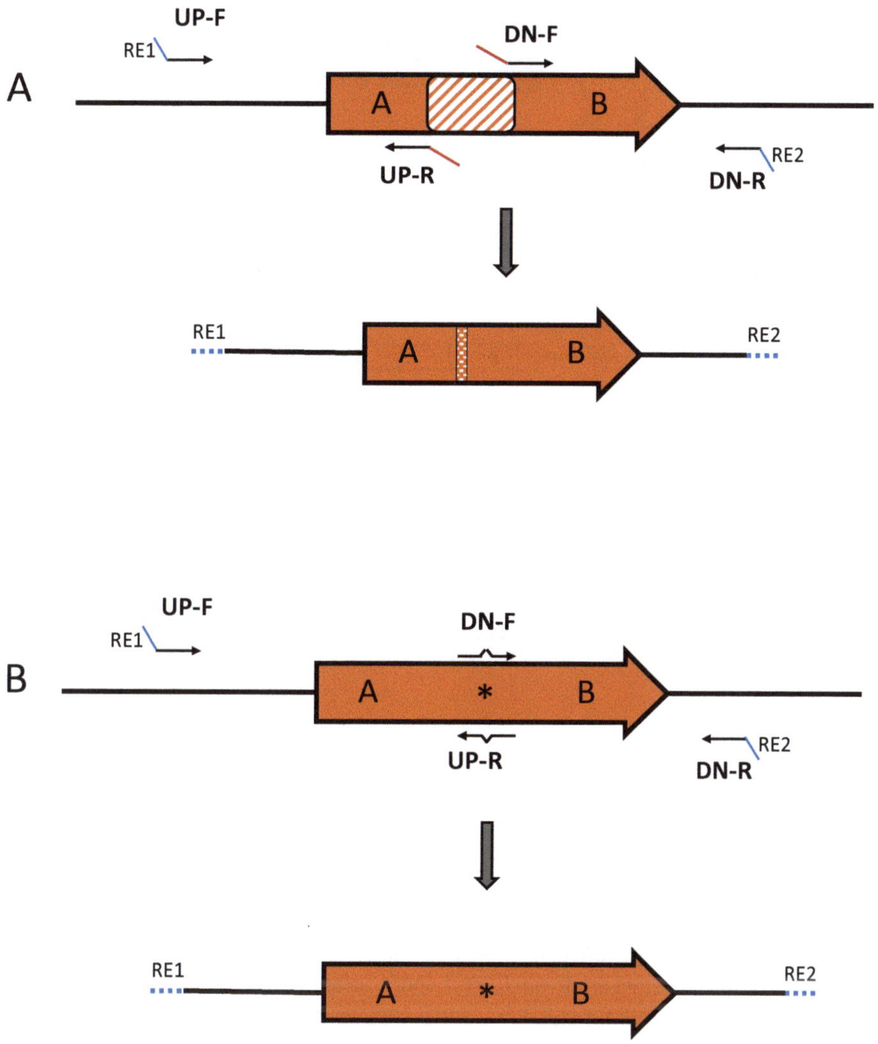

Fig. 1 Schematic diagram of generation of mutant allele by PCR for cloning into an integrative allelic exchange plasmid. (**a**) Two homologous fragments corresponding to the upstream and the downstream regions of interest are first amplified using UP-F/UP-R and DN-F/DN-R primer sets. During primer designing, restriction endonuclease sites are incorporated into UP-F and DN-R primers (RE1 and RE2, respectively). A short overlapping region of ~25 nt is also incorporated into both UP-R and DN-F primers. After the first round of PCR, equimolar amounts of the two fragments are mixed and used as template for a second round of PCR using only UP-F and DN-R primers. During the second PCR, both the fragments will be fused to generate a mutant allele. The locations of the UP-R and DN-F primers on the target site dictate the length of the deletion. (**b**) For incorporation of a point mutation, the DN-F primer is designed in such a way that it contains the desired base change, preferably in the center of the primer. The UP-R primer is the reverse-complement of DN-F. After the second fusion PCR, the desired point mutation will be inserted into the fragment. Using RE1 and RE2, these mutant alleles are cloned into pEX18Tc or similar allelic exchange vector

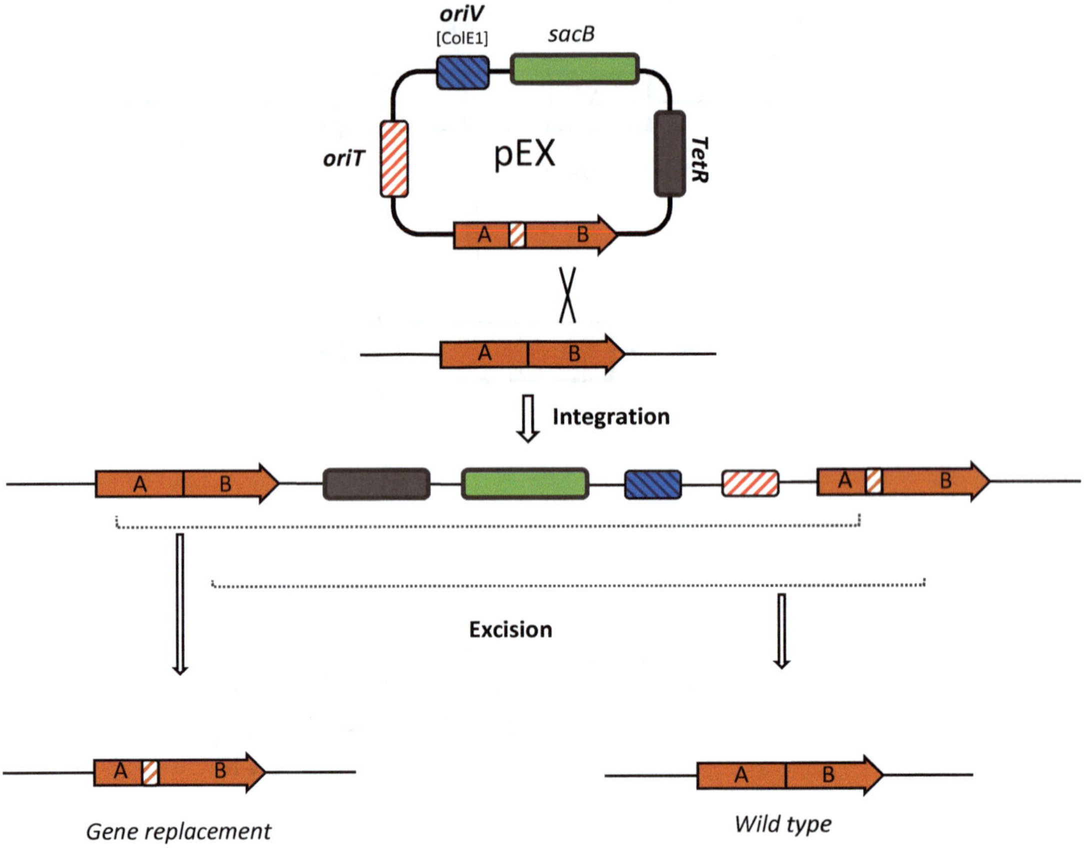

Fig. 2 Schematic diagram of a two-step gene replacement event with a pEX-based exchange vector and applying counterselection. The allelic exchange vector contains four essential parts: a narrow host range origin of replication that does not function in *A. baumannii* (*oriV*), an origin of transfer that is used for conjugative transfer of the plasmid from an *E. coli* donor strain to the recipient *A. baumannii* strain, an antibiotic resistance gene (e.g., TetR for pEX18Tc), and a *sacB* gene for sucrose counterselection. The plasmid also contains multicloning sites for the insertion of mutated alleles. Upon introduction of the vector into the recipient host, selection for tetracycline resistance produces integrants via a single-crossover recombination using either upstream or downstream homologies. Subsequent selection for the sucrose-resistant colonies (due to loss of *sacB* gene) leads to a second recombination event that promotes excision of the integrated plasmid from the chromosome. If the second recombination event occurs through the homology that was not used for initial integration, it will create the desired gene replacement onto the chromosome

2 Materials

All reagents should be analytical grade and can be prepared at room temperature unless otherwise noted. After preparation of solutions/materials, autoclave at 121 °C for 30 min prior to use to ensure no contamination. All waste should be considered a biohazard and disposed of in appropriate containers.

1. Lysogeny broth (LB: 2.5 g NaCl, 1.25 g yeast extract, 2.5 g tryptone, 1 tablet NaOH, 250 mL water).

2. LBA without NaCl (LBANS) containing 10% or 15% (w/v) sucrose.
3. Phosphate-buffered saline (PBS: dissolve 8 g NaCl, 0.2 g KCl, 1.44 g Na_2HPO_4, 0.24 g KH_2PO_4 in 800 mL water, adjust pH to 7.4 with HCl, and bring the final volume to 1.0 L with water). Autoclave to sterilize.
4. *A. baumannii* strains (clinical or laboratory standard) (*see* **Note 1**).
5. Genomic DNA isolated from *A. baumannii* strain for PCR template.
6. *E. coli* DH5α chemical competent cells (for cloning).
7. *E. coli* SM10 (λ*pir*) or S17-1 (λ*pir*) (*see* **Note 2**).
8. Bacterial strains streaked and grown overnight at 37 °C on LB agar plate (LBA: 1.5% agar; *see* **Note 1**).
9. Integration plasmid: pEX18Tc or similar plasmid (*see* **Note 3**).
10. Water, Milli-Q (ultrapure).
11. 10× PCR reaction buffer (without $MgCl_2$).
12. dNTP mixture (10 mM each).
13. 25 mM $MgCl_2$.
14. 10 μM oligonucleotide primers.
15. Taq DNA polymerase (5 U/μL).
16. Restriction enzymes (as required by integration vector).
17. T4 ligase.
18. Sucrose, molecular biology grade.
19. 20 mg/mL tetracycline.
20. 50 mg/mL ampicillin.
21. Agarose gel electrophoresis apparatus and power supply.
22. Plasmid isolation kit (miniprep).
23. Sterile disposable pipette tips: 1 mL, 200 μL, 10 μL.
24. Sterile disposable plastic inoculation loop.
25. Sterile disposable culture tubes (10 mL).
26. 50 mL conical tube.
27. 100 mm × 15 mm petri dishes.
28. DNase-free PCR tubes.
29. Filter holder (Swinnex, SX00 025, Millipore); autoclave.
30. Membrane filters (sterile), 0.45 μm PVDF (Durapore HVLP 02500, Millipore).
31. Disposable sterile plastic syringe, with Luer Lock (10 mL).
32. Laminar hood.

33. Thermocycler.
34. NanoDrop or a regular spectrophotometer.
35. Shaker incubator.
36. Incubator.
37. Water bath (42 °C).

3 Methods

The gene replacement method described here can be divided into three basic steps. These steps are (1) generation and cloning of a mutant allele into a suitable allelic exchange vector, (2) generation of an integrant after introduction of the mutant allele by either electroporation or via conjugation. and (3) sucrose counterselection and verification of the mutants by PCR and sequencing. To obtain a mutant clone by this gene replacement method, it generally requires 2–3 weeks provided all the steps are successful at the first trial.

3.1 Construction and Cloning of a Mutant Allele into a pEX18Tc Vector

1. Design two sets of primers to amplify upstream (UP) and downstream (DN) regions of the target gene using an online primer design software. Mark the primers for UP region as UP-F and UP-R. Similarly, mark the DN primers as DN-F and DN-R. While designing the primers, insert a synthetic restriction site at the 5′ end of UP-F primer and another restriction site at the 5′ end of DN-R primer. These restriction sites should be unique sites present on the MCS of the plasmid. While designing the UP-R and DN-F primers, make sure to insert a region of overlapping sequence for the fusion PCR (*see* **Notes 4–6**).
2. Using UP-F and UP-R primers and *A. baumannii* genomic DNA as the template (50 ng), set up a PCR reaction for the upstream region (*see* Table 1).
3. Similarly, set up a PCR reaction for the downstream region using DN-F and DN-R primers (*see* Table 1).
4. Carry out PCR amplification for 30 cycles. Adjust annealing temperature as needed for the primer sets.
5. After the PCR amplifications, clean up the PCR-amplified fragments using a purification kit, and elute with 40 μL of warm ultrapure H_2O.
6. Using NanoDrop (or a regular spectrophotometer), confirm that the two desired fragments were amplified. Run an agarose gel electrophoresis to ensure the amplified fragments were of correct sizes.

Table 1
Reagents and amounts used for a typical PCR reaction

10× PCR reaction buffer	5 μL
25 mM $MgCl_2$	6 μL
dNTPs (10 mM each)	1 μL
10 μM forward primer	1 μL
10 μM reverse primer	1 μL
Template DNA	1 μL
Water	34 μL
5 U/μL Taq DNA pol	1 μL
Final volume	50 μL

7. Now set up a second PCR reaction to generate a fusion product of the upstream and downstream fragments. Prepare template DNA by mixing equimolar amounts of both the fragments. Using only the UP-F and DN-R primers and ~20–50 ng of the template DNA mix, carry out the fusion PCR amplification for 30 cycles.
8. After the second PCR amplification, clean up the amplified fragment using the PCR purification kit, and elute the DNA with 50 μL of warm ultrapure H_2O.
9. Run an agarose gel electrophoresis to ensure the fusion fragment was of correct sizes (*see* **Note 7**).
10. Now to clone the fusion fragment into pEX18Tc, set up a restriction digestion reaction using the restriction sites that were incorporated into the UP-F and DN-R primers. Take about 2 μg of the plasmid DNA and the fragment DNA, and separately restrict them with the appropriate restriction enzymes following the supplier's protocol. After restriction digestion, purify the samples using a column purification kit.
11. Carry out an agarose gel electrophoresis to ensure restriction digestions were complete.
12. Set up a ligation reaction that consists of 1:6 molar ratio of digested plasmid and digested insert. By following the vendor's protocol, ligate these two fragments using T4 DNA ligase (typically by incubating for 1 h at 22 °C).
13. Add ~100 ng of the ligation mixture to 50 μL of chemically competent *E. coli* cells. Incubate for 30 min on ice, and heat shock at 42 °C for 45 s before further incubating for 5 min on ice.
14. Add 450 μL of prewarmed LB broth to the transformed cells, and incubate at 37 °C for 1 h shaking at 220 rpm.

15. In the meantime, prepare LB agar plates containing 20 μg/mL tetracycline, and dry them underneath a laminar hood.
16. Plate 100 μL of the transformed cells per plate, and incubate the plates at 37 °C for 12–18 h.
17. Next day, inoculate ~10–15 colonies into LB with 20 μg/mL tetracycline, and incubate at 37 °C shaking 220 rpm for 12–18 h.
18. Prepare plasmids from these cultures, and run an agarose gel to identify the potential clones.
19. Since pEX18-based vectors also contain M13-For and M13-Rev primers binding sites, verify the clones by PCR amplification using these two primers.
20. Select one correct clone (mutant allele) and proceed to the next step.

3.2 Introduction of the Mutant Allele into A. baumannii for Integrant Generation

The following protocol is one of the several methods to transfer the plasmid carrying the mutant allele into the target recipient *A. baumannii* strain. This method relies on interspecies conjugation via filter mating using *E. coli* as the donor strain (*see* **Note 8**).

1. Using a regular chemical transformation method, move the allelic exchange plasmid to the donor *E. coli* SM10 (λ*pir*) or S17-1 (λ*pir*) strain (*see* Subheading 3.1).
2. In the meantime, streak out the recipient *A. baumannii* strain on LBA.
3. At this point, make sure the recipient strain is intrinsically resistant to some antibiotics, while the donor *E. coli* strain is sensitive. This will allow the exconjugant cells to be selected. As an example, we will use ampicillin to counter select donor *E. coli* cells (*see* **Note 9**).
4. Inoculate a single isolated colony from the transformed plate (*E. coli* donor strain) in 5 mL LB with appropriate antibiotic, and grow at 37 °C with shaking at 220 rpm for 12–18 h (from **step 1**).
5. Also, inoculate the recipient *A. baumannii* strain from a single isolated colony in 5 mL LB, and incubate at 37 °C with shaking at 220 rpm for 12–18 h (from **step 2**).
6. Next day, inoculate by diluting the overnight cultures in LB with appropriate antibiotic to prepare fresh cultures for both the donor *E. coli* and the recipient *A. baumannii*.
7. Incubate at 37 °C with shaking at 220 rpm to an OD_{600} approximately between 0.6 and 0.8.
8. While waiting for the cultures to reach the desired density, assemble a mating filter device by inserting the filter

(0.45 μm) into the filter holder, and attach the 10 mL syringe (without the plunger) to the filter holder.

9. Open the cap of a 50 mL tube, and place the entire filter setup on the top (the tube will serve as a liquid collector).
10. Now add 3.0 mL of the fresh *E. coli* donor cell culture and 1.0 mL of the fresh *A. baumannii* recipient cell culture to the open syringe, and pass the bacterial cultures through the filter by inserting and pressing the plunger. At this point, the bacterial culture will form a thin film on the filter.
11. Aseptically transfer the filter on LBA in such a way that the bacteria on the surface of the filter are facing upward. Incubate the plate with the filter at 37 °C overnight (*see* **Note 10**).
12. Using a sterile plastic loop, collect the bacterial lawn from the filter, and resuspend the cell mass in 1.0 mL PBS by either vortexing or with a 1 mL pipette tip. Now plate different volumes (10 μL, 50 μL, and 100 μL) of the bacterial suspension on LBA with 20 μg/mL tetracycline and 40 μg/mL ampicillin (our recipient *A. baumannii* is resistant to up to 40 μg/mL ampicillin).
13. Incubate the plates at 37 °C for up to 2 days (*see* **Note 11**).
14. Once the exconjugants start appearing on the plates, streak the colonies again on LBA containing 20 μg/mL tetracycline and 40 μg/mL ampicillin. Select several isolated colonies and proceed to the next step.

3.3 Counterselection and Verification of the Gene Replacement Event

1. Using sterile inoculation loops, streak about 10–20 isolated colonies (from the last step) onto LBANS plates containing 10% or 15% sucrose. Use one plate for each isolated colony. Incubate the plates at 37 °C for up to 2 days (*see* **Note 12**).
2. Select about 20 isolated colonies from the LBANS/sucrose plate for colony PCR.
3. Resuspend each colony into 40 μL ultrapure water. Ensure that the final resuspension in not very cloudy.
4. Take a fresh LBA plate and spot 5 μL of the resuspension, and mark the spots for future reference. Also, spot 5 μL of the resuspension onto LBA containing tetracycline. If too many colonies grow on LBA tetracycline plate (merodiploid), it will indicate that the sucrose counterselection might be leaky.
5. Prepare a PCR master mix by increasing the volumes to ten times of each reagent listed in Table 1, except the template DNA. Use either UP-F and DN-R primer pair or SQ-F and SQ-R primer pair (*see* **Note 7**).
6. Distribute 25 μL master mix solution into each PCR tube (20 tubes).

7. Take 10 μL of the bacterial resuspension from the remaining, and add 100 μL ultrapure water. Boil the samples at 100 °C for 10 min using a thermocycler to disrupt the cells so that the chromosomal DNA is released.
8. Add 1.0 μL of the sample to each PCR tube, and run a PCR amplification using a suitable annealing temperature.
9. Confirm gene replacement events by running the PCR-amplified products in an agarose gel and comparing the size of the product with the wild-type gene (*see* **Note 13**).
10. Once the correct gene placement clone is identified by PCR, recover the strain from **step** 7 by inoculating into a fresh LB medium, and store appropriately.

4 Notes

1. *A. baumannii* strains can be fresh clinical isolates or can be procured from ATCC or a similar microbial resources facility. When freshly isolated unknown clinical isolates are used, make sure these strains are sensitive to the antibiotic whose resistance gene is present on the allelic exchange plasmid.
2. The SM10 strain is commercially available from BCCM (http://bccm.belspo.be) or from Life Science Market (www.lifescience-market.com).
3. pEX18Tc and its cousin pEX18Gm are commercially available from Life Science Market. Other integration vectors that offer sucrose counterselection are pMo130 and pMo130-Tel; these two plasmids are available from Addgene (www.addgene.org). The latter plasmid is useful when *A. baumannii* MRD strains are under study since pMo130-Tel offers tellurite resistance.
4. These primers are designed in such a way that the entire open reading frame (ORF) is removed. They can also be designed such that a part of the internal region or the 3′-end of the target gene is deleted. Moreover, the primers can be designed in such a way that they can introduce a single base pair mutation and even create a restriction site (*see* Fig. 1).
5. For initial experiments, plan to amplify ~500 bp upstream and downstream of the regions for single nucleotide mutation or typical single gene deletion experiments. If the deletion segment is larger than ~1.0 kb (e.g., deletion of an operon), consider using longer homologous regions. Homology lengths greater than 1.0 kb do not offer additional benefit. Numerous online tools are available for primer design. You can find a few of them from this website (https://molbiol-tools.ca/PCR.htm).

6. The overlapping regions of UP-R and DN-F primers should be between 20 and 30 nt. The theoretical Tm value should be close to the other two primers. If the aim of the experiment is to insert a single point mutation, both the UP-R and DN-F primers should contain the desired mutation and share significant overlaps. A useful online toll for designing the overlapping regions can be found here: http://www.ohalloranlab.net/STITCHER_2_0/index.html#about.
7. It is a good idea to verify that the sequence of fusion fragment is correct. For this, design two extra primers (SQ-F and SQ-R) that flank the desired deletion or the mutation site. These primers should be able to amplify a fragment of ~ 500 bp in length so that the junction region is well within the sequencing range.
8. An alternative to filter mating is puddle mating. In this case, both the donor and recipient strains are grown separately and then mixed together. The mixed cells are then harvested by low-speed centrifugation and spotted onto the LBA plate as "puddle." Conjugation will take place inside the "puddle" during incubation. One can also transfer the allelic exchange vector by electroporation. However, electroporation is >100-fold less efficient than conjugation.
9. Many *A. baumannii* strains are intrinsically resistant to several antibiotics. One needs to find a suitable antibiotic to which the *A. baumannii* under study is resistant but not the *E. coli* donor strain. If none are available, one can simply generate spontaneous resistant mutants by spreading *A. baumannii* cultures onto LBA containing 100 μg/mL streptomycin or rifampicin. The resistant colonies will appear within 2 days.
10. Sometimes incubation at 30 °C or even lower temperature increases the efficiency of conjugation. If a lower temperature is used, incubate for a longer time to obtain the desired conjugation events.
11. Depending on the nature of the antibiotic selection, the time of appearance of the exconjugants varies. If one does not see any pinpointed colonies after 3 days of incubation, it will mean that the conjugation experiment did not work. If one observes a bacterial lawn after overnight incubation, it means that the antibiotic selection did not work properly. In this case, one needs to increase the antibiotic concentration so that the donor *E. coli* strain does grow or consider using a different antibiotic for restricting growth of the donor cells.
12. Sucrose counterselection is leaky and often produces false positives. If one observes numerous false positives, incubate the plates at 30 °C or even at room temperature. Also consider increasing the sucrose concentration up to 20%.

13. A simple PCR amplification and analyzing on agarose gel will only identify gene replacement events that include a large deletion or insertion. If one needs to identify a single base mutation, purify the PCR products, and subject them for DNA sequencing. If the single base mutation generates a restriction endonuclease site, which can be done during primer design (*see* **Note 4**), the PCR products can be restricted with the appropriate restriction enzymes and verified by agarose gel electrophoresis.

Acknowledgment

Work in this publication was supported in part by the National Institute of General Medical Sciences of the National Institutes of Health under Award Number P20 GM113117.

References

1. Biswas I (2015) Genetic tools for manipulating Acinetobacter baumannii genome: an overview. J Med Microbiol 64(7):657–669
2. Wilharm G et al (2013) DNA uptake by the nosocomial pathogen Acinetobacter baumannii occurs during movement along wet surfaces. J Bacteriol 195(18):4146–4153
3. Tucker AT et al (2014) Defining gene-phenotype relationships in Acinetobacter baumannii through one-step chromosomal gene inactivation. MBio 5(4):e01313–e01314
4. Aranda J et al (2010) A rapid and simple method for constructing stable mutants of Acinetobacter baumannii. BMC Microbiol 10:279
5. Amin IM et al (2013) A method for generating marker-less gene deletions in multidrug-resistant Acinetobacter baumannii. BMC Microbiol 13:158
6. Jacobs AC et al (2014) Genetic manipulation of Acinetobacter baumannii. Curr Protoc Microbiol 35:6G.2.1–6G.211

Chapter 11

RecET-Mediated Recombineering in *Acinetobacter baumannii*

Ashley T. Tucker, Matthew J. Powers, M. Stephen Trent, and Bryan W. Davies

Abstract

Acinetobacter baumannii rapidly acquires antibiotic resistance, and its genome encodes mechanisms to tolerate biocides and desiccation, enhancing its persistence in hospital settings. Tools to rapidly dissect the *A. baumannii* genome are needed to understand cellular factors that contribute to its resiliency at a genetic and mechanistic level. While a substantial amount of clinical data has documented the global rise of *A. baumannii* as an antibiotic-resistant pathogen, genetic tools to dissect its molecular details have been limited. This procedure describes a recombination-mediated genetic engineering (recombineering) system for targeted genome editing of *A. baumannii*. This system can perform directed mutagenesis on wide-ranging genes and operons and has broad application in various strains of *A. baumannii*.

Key words *Acinetobacter baumannii*, Recombineering, Genetic engineering, Gene knockout, Mutagenesis, Recombinase

1 Introduction

Recombination-mediated genetic engineering (recombineering) is a quick and efficient way to perform rapid genome editing in vivo. Recombineering is mediated by phage proteins that belong to either the Red proteins of phage λ (Gam, Beta, Exo) or the RecET proteins from the Rac prophage, which catalyze homologous recombination between short DNA segments. Both of these systems are functionally equivalent with Exo/RecE exonucleases exposing 3′ single-stranded DNA that Beta/RecT bind and pair with the homologous genomic target region [1, 2]. The Red system also encodes a Gam protein that inhibits host nucleases and is absent from RecET systems. The *Escherichia coli* λ Red system is the most well-studied recombineering system [3, 4]. The development of recombineering revolutionized *E. coli* genetics

Indranil Biswas and Philip N. Rather (eds.), *Acinetobacter baumannii: Methods and Protocols*, Methods in Molecular Biology, vol. 1946, https://doi.org/10.1007/978-1-4939-9118-1_11, © Springer Science+Business Media, LLC, part of Springer Nature 2019

allowing for rapid gene substitutions and deletions in a single step without the need for restriction enzymes and building editing constructs in vitro [3, 4]. Recombineering rapidly accelerated our genetic analysis of *E. coli*, solidifying *E. coli* as a model bacterium and facilitating the prominence of *E. coli* for use in synthetic biology and biotechnology. The development of recombineering for *A. baumannii* helps to advance our ability to genetically interrogate this emerging pathogen. This protocol outlines the utility of this simple, one-step system, for rapidly creating marked or unmarked deletion of genes and operons in *A. baumannii* using a RecET system we call Rec_{Ab}. This method uses a PCR-generated template for recombination. In addition, this system is portable between *A. baumannii* strains with the appropriate choice of selection cassettes.

2 Materials

Strains and plasmids required for recombineering in *A. baumannii* are listed in Table 1. Request materials from Dr. Bryan Davies: bwdavies@austin.utexas.edu.

2.1 Generation of PCR Template for Recombination

1. TOPO TA 2.1 plasmid containing kanamycin resistance cassette. Alternatively, plasmid pKD4 for generating kanamycin cassette flanked by FRT sites can be used if the resistance cassette must be removed later (*see* **Note 1**).
2. High-fidelity DNA polymerase master mix for PCR cloning.
3. Taq polymerase master mix for recombinant checking.
4. Oligonucleotides (IDT ultramers) to amplify the kanamycin resistance cassette with homologous sequences containing 125 bases flanking the target sequence. Dissolve lyophilized ultramers with sterile H_2O to 100 μM.

Table 1
Strains and plasmids for RecET recombineering in *A. baumannii*

Designation	Relevant characteristics	Source
AT02	*A. baumannii* 17978 carrying pAT02	[5]
TOPO TA 2.1	Kan^R	Invitrogen
pKD4	Kan^R flanked by FRT sites	[3]
pAT02	pMMB67EH with Rec_{Ab} system	[5]
pAT03	pMMB67EH with FLP recombinase	[5]
pAT04	pMMB67EH with Rec_{Ab} system, Tet^R	[5]

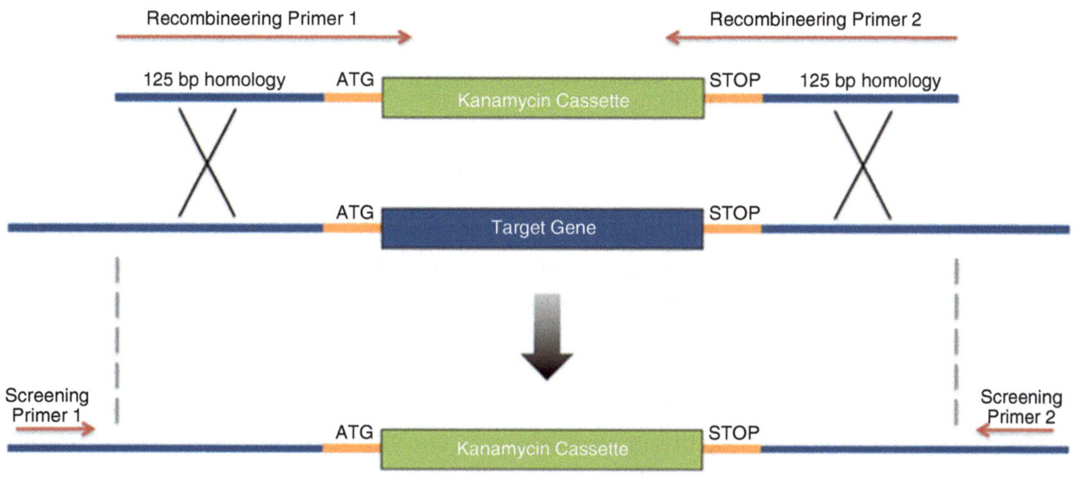

Fig. 1 Schematic diagram of the one-step recombineering method. Primers containing 125 bp homology flanking the gene of interest are used to amplify the PCR product. The insert is then recombined into the genome using the RecAb system. Screening primers outside the region of homology are used to confirm insertion of the kanamycin cassette [5]

5. Design additional screening primers outside of the junction sites to confirm recombination (*see* Fig. 1). Dissolve lyophilized oligos in sterile water to a final concentration of 100 μM.
6. PCR spin column kit to clean and concentrate PCR product.
7. Sterile H_2O.
8. Thermocycler.

2.2 Rec$_{Ab}$ Competent Cells and Transformation

1. 75 mg/mL carbenicillin (1000×)
2. 25 mg/mL and 7.5 mg/mL kanamycin (1000×).
3. 10% glycerol solution in H_2O (filter sterilize) and chilled to 4 °C.
4. Lysogeny broth (LB): 10 g tryptone, 5 g yeast extract, and 5 g NaCl dissolved in 1 L H_2O and autoclaved.
5. LB agar: 10 g tryptone, 5 g yeast extract, 5 g NaCl, and 15 g agar dissolved in 1 L H_2O and autoclaved.
6. *A. baumannii* strain with plasmid to express the Rec$_{Ab}$ recombination system (*see* **Note 1**).
7. 2 mM IPTG (isopropyl-beta-D-thiogalactoside, $C_9H_{18}O_5S$).
8. Centrifuge.
9. Speedvac (optional).
10. Electroporator.
11. Chilled 0.2 cm electroporation cuvettes.
12. Sterile H_2O.

2.3 Colony PCR and Recombinant Checking

1. 1% agarose gel.
2. DNA gel apparatus with power source.
3. UV imager.

3 Methods

3.1 Primer Design

1. Design two primers containing sequences (approximately 18–25 nucleotides) to amplify the kanamycin cassette followed by an additional 125 bp sequence homologous to the flanking regions of the target sequence (*see* Fig. 1 and **Note 2**).
2. Design two additional primers which will be used to confirm recombination. Primers should begin outside of the homologous regions used for recombination.

3.2 Preparation of Substrate PCR Product

1. It is recommended to run at least eight reactions to obtain sufficient amount of DNA for recombination (>1–5 μg) (*see* **Note 3**). For example, to obtain a master mix for eight PCR reactions, combine a high-fidelity polymerase, 0.2 μM each primer, ~50 ng plasmid containing kanamycin cassette (TOPO TA 2.1), and H_2O per manufacturer's instructions. Aliquot reaction mixture into eight PCR tubes.
2. Incubate PCR reactions: 98 °C, 2 min; 40 cycles of (98 °C, 10 s; 60 °C, 15 s; 72 °C, 60 s); 72 °C, 5 min; 10 °C hold.
3. Purify the PCR product using guidelines provided by the manufacture for PCR clean and concentration.
4. Pool all reactions, and concentrate PCR product in Vacufuge speedvac under the aqueous setting at 30 °C until final volume reaches ~5–15 μL.

3.3 Preparation of Electrocompetent *A. baumannii* Expressing Rec_{Ab}

1. Grow *A. baumannii* containing Rec_{Ab} plasmid on LB agar medium supplemented with 75 μg/mL carbenicillin overnight at 37 °C.
2. Scrape a loop of bacteria from the plate, and resuspend in 5 mL LB supplemented with 75 μg/mL carbenicillin. Measure OD_{600}.
3. Dilute bacteria into 500 mL LB supplemented with 75 μg/mL carbenicillin to a final OD_{600} of 0.005.
4. Grow shaking at 220 rpm and 37 °C for 45 min.
5. Add 2 mM IPTG and continue growing the culture to an $OD_{600} = 0.4$.
6. Split the culture into half and pellet bacteria at 8600 × *g* for 6 min at 4 °C. Pour off supernatant and keep pellet chilled on ice.

7. Wash each pellet in 250 mL of chilled 10% glycerol.
8. Repeat wash step with 125 mL of 10% glycerol. Pellet bacteria.
9. Combine pellets and wash in 100 mL of 10% glycerol. Pellet bacteria.
10. Resuspend pellet in 500 μL of 10% glycerol. Chill on ice.
11. Aliquot 100 μL of competent bacteria into a chilled microcentrifuge tube for electroporation (*see* **Note 4**).

3.4 Transform *A. baumannii* Expressing Rec$_{Ab}$ with PCR Substrate

1. Add 5 μL (>5 μg) of recombineering PCR product to 100 μL aliquot of *A. baumannii* electrocompetent cells.
2. Immediately transfer to a chilled 0.2 cm electroporation cuvette, and electroporate at 1800 V.
3. Add 1 mL of LB directly to the cuvette, and gently pipette up and down and transfer to test tube with an additional 3 mL LB (*see* **Note 5**).
4. Add IPTG to a final concentration of 2 mM.
5. Grow induced culture for 4 h at 37 °C with shaking at 200 rpm.
6. Pellet cells at 1800 × *g* for 5 min and resuspend in 300 μL LB.
7. Plate equal amounts (100 μL) of electroporation outgrowth culture on 3 LB agar plates supplemented with 7.5 μg/mL kanamycin.
8. Incubate plates for 16–24 h at 37 °C (*see* **Note 6**).

3.5 Genotyping by Colony PCR

1. Dilute an *A. baumannii* colony by lightly touching one colony with a pipette tip and mixing into 100 μL sterile H_2O.
2. Set up colony PCR reaction for each colony by mixing Taq polymerase master mix per manufacturer's instructions with 0.6 μM each screening primer, 2 μL of the colony/H_2O mixture, and sterile H_2O to volume. 30 ng of genomic DNA may be used in place of colony/H_2O mixture as a control.
3. Run colony PCR by incubating at 95 °C, 2 min; 30 cycles (95 °C, 10 s; 57 °C, 15 s; 68 °C, 1 min/kb product); 68 °C, 5 min.
4. Check products on a standard 1% agarose gel (*see* **Note 7**). An example of the expected outcomes from the confirmation colony PCR is shown in Fig. 2.

3.6 Generating an Unmarked Mutant

1. Successful recombinants can be patched to ensure removal of the Rec$_{Ab}$ recombineering plasmid from successful clones. Streak PCR confirmed single colonies onto kanamycin agar plates as well as carbenicillin agar plates. Clones that no longer contain the plasmid will only grow on kanamycin plates.

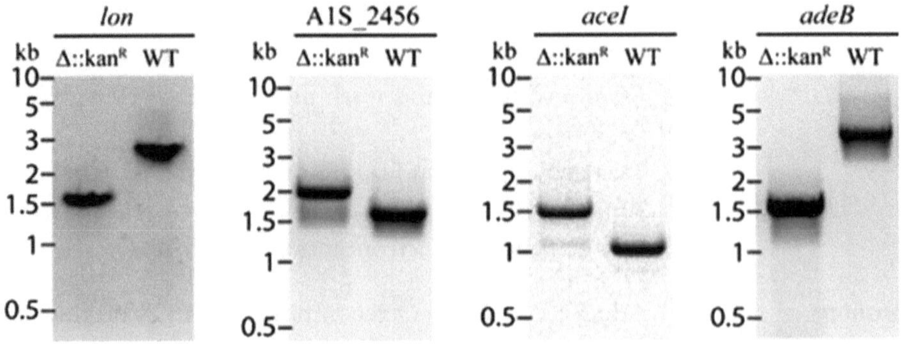

Fig. 2 PCR verification of constructed mutants. Following recombineering, PCR was used to verify the presence of the wild-type allele or the replacement of that allele with a kanamycin resistance cassette to create the mutant strain. The wild-type control is marked "WT." Shown are gene replacements with a kanamycin resistance cassette for targets lon, A1S_2456, aceI, and adeB [5]

2. Post recombination, the resistance cassette can be excised from the bacterial chromosome using the FRT sites that flank the cassette as described previously [3].
3. Once the resistant marker has been eliminated, additional mutations can be performed in the genome by repeating the procedure as described above.

4 Notes

1. Rec_{Ab} recombineering in *A. baumannii* has been performed successfully in the 17978 strain as outlined in this protocol, as well as Ab 19606 using a Rec_{Ab} plasmid conferring tetracycline resistance. Although kanamycin is our preferred marker, any resistance or nutritional marker could be used.
2. We found that 125 bp of homology provided the highest recombination efficiency; however, any length greater than 100 bp is sufficient.
3. Recombineering efficiency has been tested with 100 ng, 500 ng, 1 μg, 5 μg, and 10 μg of PCR product. We did not observe any recombinants using less than 1 μg of PCR product. Using 10 μg of PCR product did not increase the efficiency of recombinant production compared with that of 5 μg.
4. *A. baumannii* competent cells expressing Rec_{Ab} must be made fresh for each recombination and cannot be stored for later use.
5. Using less than 3 mL of LB did not result in identification of positive colonies. More than 4 mL of LB can also enhance recovery of recombinants.
6. Following our optimized protocol, on average, we obtained 100 transformants. At the kanamycin concentration of 7.5 μg/

mL, pinpoint colony background is commonly observed. Larger colonies representing potential mutants can be restreaked onto higher concentrations of kanamycin (25 μg/mL) prior to colony PCR screening.

7. The expected outcome of the DNA bands is dependent upon the size of the gene that was deleted. The length of the kanamycin insertion cassette between primer annealing sequences should yield a band at approximately 1500 bp.

Acknowledgment

This work was supported by NIH grant AI125337 to B.W.D, NIH F32 GM116523 to A.T.T., NIH grants AI064184, AI76322, AI119879 and AI129940 to M.S.T., and NSF 2017238867 M. J. P.

References

1. Cassuto E, Radding CM (1971) Mechanism for the action of lambda exonuclease in genetic recombination. Nat New Biol 229(1):13–16
2. Karakousis G, Ye N, Li Z, Chiu SK, Reddy G, Radding CM (1998) The beta protein of phage lambda binds preferentially to an intermediate in DNA renaturation. J Mol Biol 276(4):721–731. https://doi.org/10.1006/jmbi.1997.1573
3. Datsenko KA, Wanner BL (2000) One-step inactivation of chromosomal genes in Escherichia coli K-12 using PCR products. Proc Natl Acad Sci U S A 97(12):6640–6645. https://doi.org/10.1073/pnas.120163297
4. Yu D, Ellis HM, Lee EC, Jenkins NA, Copeland NG, Court DL (2000) An efficient recombination system for chromosome engineering in Escherichia coli. Proc Natl Acad Sci U S A 97 (11):5978–5983. https://doi.org/10.1073/pnas.100127597
5. Tucker AT, Nowicki EM, Boll JM, Knauf GA, Burdis NC, Trent MS, Davies BW (2014) Defining gene-phenotype relationships in Acinetobacter baumannii through one-step chromosomal gene inactivation. MBio 5(4):e01313–e01314. https://doi.org/10.1128/mBio.01313-14

Chapter 12

Methods for Tn-Seq Analysis in *Acinetobacter baumannii*

Larry A. Gallagher

Abstract

Transposon insertion sequencing (Tn-seq) is a powerful method for identifying genes required for virtually any growth or survival trait in bacteria. The technology employs next-generation DNA sequencing to identify and quantify the relative abundances of individual transposon mutants within complex pools of such mutants. When applied to pools of thousands to millions of random transposon mutants grown under selective pressure, the technique can rapidly identify, at genome scale, the mutants and corresponding genes negatively or positively selected. This chapter presents core protocols for Tn-seq analysis of *Acinetobacter baumannii*: generation of a high-saturation random transposon mutant pool, and isolation and sequencing of transposon-genome junctions from such a pool for identifying and quantifying the individual mutants. With these tools, the researcher can address diverse biological questions by carrying out selective growth of a mutant pool followed by Tn-seq analysis to identify genotype-phenotype associations.

Key words Tn-seq, T26, Tn5, AB5075, C-tailing

1 Introduction

Transposon insertion sequencing (Tn-seq) is a powerful genome-scale method for identifying bacterial genes and genetic elements needed for normal growth or survival under particular conditions (for reviews, *see* [1, 2]). Typically, a complex pool of random transposon mutants, representing multiple insertions in virtually every nonessential gene, is generated and then grown under a selective condition of interest, such as exposure to a stressor or survival within an infected host. Next-generation DNA sequencing is then used to identify the individual mutants in the pool and quantify their relative abundances. Genes with mutant fitness phenotypes are identified by depletion (or amplification) of the corresponding mutants relative to the starting pool or to the pool grown under a control condition. The technique can also be applied to smaller, rationally assembled pools representing specific gene sets of interest [1, 3].

With *Acinetobacter baumannii*, Tn-seq has been used to catalog the genes essential for growth on nutrient media [4], identify

Indranil Biswas and Philip N. Rather (eds.), *Acinetobacter baumannii: Methods and Protocols*, Methods in Molecular Biology, vol. 1946, https://doi.org/10.1007/978-1-4939-9118-1_12,

genes required for survival in an insect-based infection model [5], and analyze the relative importance of core and accessory genes for extreme antibiotic resistance [6].

A variety of molecular methodologies have been devised for identifying and quantifying individual transposon mutants within a mutant pool for Tn-seq [2]. The most common fundamental strategy is to isolate and sequence (usually by Illumina sequencing) DNA fragments representing the junctions between the transposon and the genome at transposon insertion locations. The portion of each sequence read representing genomic sequence is then mapped to the genome sequence to determine the insertion location, and reads per location are counted to estimate the relative abundances of the corresponding mutants.

This chapter presents detailed protocols for generating saturation-level transposon mutant pools in *Acinetobacter* spp. and for isolating and sequencing transposon junctions from such pools. Numerous biological questions can be addressed through carefully designed selective growth experiments of such mutant pools [2] followed by analysis of their composition using the junction-sequencing protocol.

When generating a transposon mutant pool for Tn-seq analysis, a key consideration is the choice of transposon [1]. Most studies use either mariner- or Tn*5*-based transposons. Mariner transposes exclusively into TA dinucleotide sites, limiting the degree of genome saturation possible. Tn*5* displays less target site preference and can provide extremely high genome saturation. With either transposon, statistically robust fitness analyses can be carried out by comparing relative mutant abundance in pools before and after selective growth, or after growth under both test and control conditions [1, 7].

The method presented here utilizes Tn*5*-based mutagenesis to analyze *Acinetobacter baumannii*. Stable transposon-transposase complexes ("transposomes") are generated by incubating a Tn*5*-based transposon with hyperactive Tn*5* transposase in the absence of Mg^{2+} [8]. The transposomes are then introduced into the bacterium by electroporation, where cytoplasmic Mg^{2+} enables completion of the transposition reaction, resulting in random insertions into the host genome. The insertion mutants are selectively grown by plating on selective media, pooled, and then used for biological selections and Tn-seq analysis.

The methodology presented here for isolating and sequencing transposon junctions, known as the "C-tailing" method, represents an adaptation of a method first employed with *Porphyromonas gingivalis* [9]. The method is relatively easy to carry out, produces results with minimal bias [2, 4], and can be used with virtually any transposon. The C-tailing method has been employed in *A. baumannii* to analyze virulence and antibiotic resistance [5, 6].

The C-tailing method starts by adding poly(C) tails to the 3′ ends of fragmented DNA using terminal deoxynucleotidyl transferase (TdT) and dCTPs. The average length of the tails is controlled by including a defined proportion of dideoxy CTPs (ddCTP) in the reaction (ddCTP incorporation terminates tail elongation). Following C-tailing, transposon junctions are selectively amplified by PCR using one primer that anneals within the transposon near one of its ends and another that contains a 3′ poly (G) tract to anneal to the C-tails. A second round of PCR amplification then incorporates the necessary adaptor sequences for Illumina sequencing, using one primer that anneals within the transposon (but closer to the junction than the first transposon-specific primer) and another that corresponds to the pre-defined 5′ end of the poly(G) primer. Finally, sequencing is carried out using a transposon-specific primer that anneals just a few bases from the junction. The sequence read thus covers a few bases of the transposon end, allowing verification that the amplicon represents a true transposon junction, and enough bases of flanking genomic sequence for unambiguous mapping. Figure 1 depicts the primer alignments to a typical transposon junction following C-tailing. Table 1 lists the primer sequences.

The protocols presented are based on a particular Tn*5* transposon (T26) [4], but can be adapted easily to other transposons. The pool generation protocol will work with any transposon bearing hyperactive 19-bp Tn*5* mosaic end sequences [10], and the junction-sequencing protocol can be used with virtually any transposon by modifying the transposon-specific primers (*see* protocols and Notes).

Bioinformatic processing and analysis of Tn-seq sequence data can entail sophisticated statistical approaches [7, 11–15], but at minimum requires mapping and counting of individual sequence reads. The junction-sequencing protocol presented below provides a link to a rudimentary bioinformatic tool for carrying out such basic data processing.

2 Materials

2.1 Strains, Media, and Solutions

1. *Acinetobacter baumannii* strain of choice (e.g., AB5075) [16].
2. LB liquid media: 1% tryptone, 0.5% yeast extract, 0.8% NaCl.
3. No-salt LB: 1% tryptone, 0.5% yeast extract.
4. LB agar media: 1% tryptone, 0.5% yeast extract, 0.8% NaCl, 1.5% agar.
5. SOC: 0.5% yeast extract, 2% tryptone, 10 mM NaCl, 2.5 mM KCl, 10 mM $MgCl_2$, 10 mM $MgSO_4$, 20 mM glucose.
6. TE buffer: 10 mM Tris, 1 mM EDTA, pH 8.0.
7. 10 mM Tris-Cl, pH 8.5.
8. 5 M NaCl.

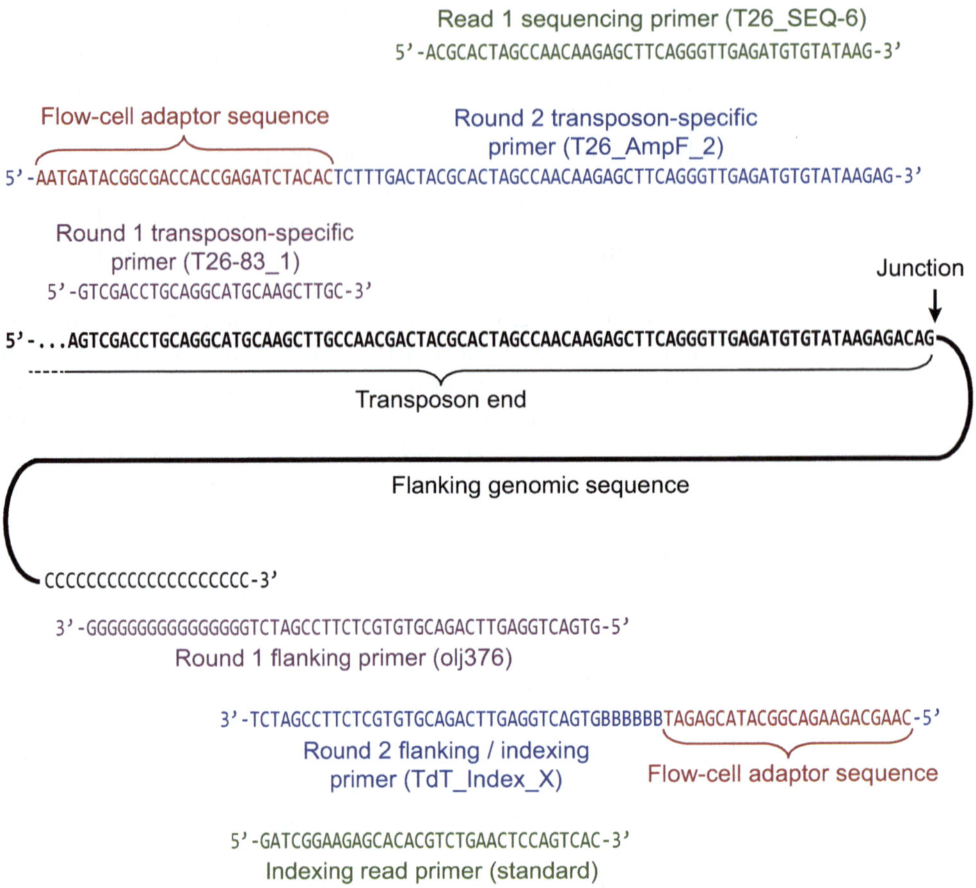

Fig. 1 Primer alignments to a transposon junction following C-tailing. The central fragment (black) represents one strand of a transposon T26 junction fragment following C-tailing. The fragment includes one end of the transposon, flanking genomic sequence, and a C-tail of average length. The primers above and below the fragment are aligned to show their annealing positions. They include the PCR round 1 primers (purple), the PCR round 2 primers (which incorporate flow cell adaptor sequences) (blue/red), and the sequencing and indexing primers (green). The names of the specific primers used with transposon T26 are displayed in parentheses. The "B" bases in the round 2 flanking/indexing primer represent the barcode portion

9. 30% PEG 8000/1.5 M NaCl.
10. 40% glycerol.
11. 10% glycerol.

2.2 Reagents, Kits, and Consumables

1. Transposon template DNA (*see* **Note 1**).
2. Oligonucleotide primers (*see* Table 1 and **Note 24**).
3. Tetracycline or other appropriate antibiotics (*see* **Note 9**).
4. EZ-Tn5™ Transposase (Lucigen).
5. Terminal deoxynucleotidyl transferase with reaction buffer (Promega).
6. dCTP and ddCTP (Sigma-Aldrich).

Table 1
Oligonucleotide primer sequences

Name	Sequence (5′ to 3′)
Transposon amplification primers	
T26_ME_F (5PH)	<5′phos>CTGTCTCTTATACACATCTCAACCATCA
T26_ME_R (5PH)	<5′phos>CTGTCTCTTATACACATCTCAACCCTGA
Round 1 PCR primers	
T26-83_1	GTCGACCTGCAGGCATGCAAGCTTGC
olj376	GTGACTGGAGTTCAGACGTGTGCTCTTCCGATC TGGGGGGGGGGGGGGGG
Round 2 PCR primers	
T26_AmpF_2	AATGATACGGCGACCACCGAGATCTACACTCTTTGACTACGCAC TAGCCAACAAGAGCTTCAGGGTTGAGATGTGTATAAGAG
TdT_Index_1	CAAGCAGAAGACGGCATACGAGATCGTGATGTGACTGGAGTTCAGACGTG TGCTCTTCCGATCT
TdT_Index_2	CAAGCAGAAGACGGCATACGAGATACATCGGTGACTGGAGTTCAGACGTG TGCTCTTCCGATCT
TdT_Index_3	CAAGCAGAAGACGGCATACGAGATGCCTAAGTGACTGGAGTTCAGACGTG TGCTCTTCCGATCT
TdT_Index_4	CAAGCAGAAGACGGCATACGAGATTGGTCAGTGACTGGAGTTCAGACGTG TGCTCTTCCGATCT
TdT_Index_5	CAAGCAGAAGACGGCATACGAGATTTCAGAGTGACTGGAGTTCAGACGTG TGCTCTTCCGATCT
TdT_Index_6	CAAGCAGAAGACGGCATACGAGATGAACTCGTGACTGGAGTTCAGACGTG TGCTCTTCCGATCT
TdT_Index_7	CAAGCAGAAGACGGCATACGAGATATCTCGGTGACTGGAGTTCAGACGTG TGCTCTTCCGATCT
TdT_Index_8	CAAGCAGAAGACGGCATACGAGATTCGAGAGTGACTGGAGTTCAGACGTG TGCTCTTCCGATCT
TdT_Index_9	CAAGCAGAAGACGGCATACGAGATCTGCTAGTGACTGGAGTTCAGACGTG TGCTCTTCCGATCT
TdT_Index_10	CAAGCAGAAGACGGCATACGAGATTCTGCTGTGACTGGAGTTCAGACGTG TGCTCTTCCGATCT
Sequencing primers	
T26_SEQ-6	ACGCACTAGCCAACAAGAGCTTCAGGGTTGAGATGTGTATAAG
Read1_SEQ	ACACTCTTTCCCTACACGACGCTCTTCCGATCT

For transposons other than T26, replace the transposon amplification primers and the transposon-specific junction amplification and sequencing primers (T26-83_1, T26_AmF_2 and T26_SEQ-6) accordingly (*see* **Note 24**). The barcode portions of the round 2 flanking/indexing primers are underlined. Additional corresponding flanking/indexing primers with unique barcodes may be created if multiplexing more than ten samples (*see* **Note 25**). <5′phos>, primer is 5′ phosphorylated

7. 2X KAPA HiFi HotStart ReadyMix (Kapa Biosystems).
8. SYBR Green I (Invitrogen).
9. NEBNext End Repair Module (New England Biolabs).
10. DNeasy Blood and Tissue Kit (Qiagen).

11. MinElute PCR Purification Kit (Qiagen).
12. Illumina sequencing flow cell and reagent cartridge (Illumina).
13. PhiX v3 control library (Illumina).
14. 1-mm electroporation cuvettes.
15. Vented QTrays (Molecular Devices) or other large-format petri plates.
16. 5-mm sterile glass beads.
17. L-shaped sterile plastic spreaders.
18. Covaris microTUBEs with AFA fiber (Covaris).
19. Performa® DTR Gel Filtration Cartridges (Edge Biosystems).
20. Nanosep MF 0.2-μM columns (Andwin Scientific).

2.3 Equipment

1. Electroporator.
2. Real-time PCR thermocycler.
3. Covaris focused-ultrasonicator.
4. Illumina sequencing instrument capable of utilizing custom sequencing primers (e.g., MiSeq).
5. Fluorometric DNA quantification instrument (e.g., Invitrogen Qubit).
6. Spectrophotometer for measuring optical density at 600 nm.
7. Agarose gel electrophoresis equipment.
8. Microcentrifuge.
9. Centrifuge and tubes for centrifugation of volumes up to 25 mL.
10. Tube roller at 37 °C.
11. 37 °C incubator.

3 Methods

Use molecular biology grade water to bring DNA reactions to the specified volumes. Prepare and keep all DNA samples and reaction mixtures on ice unless otherwise indicated. For procedures with bacterial cells, use sterile media, reagents, and tubes.

3.1 Generating a Random Transposon Mutant Pool

3.1.1 Making Tn5-Based Transposon-Transposase Complexes (Transposomes)

1. Amplify the Tn*5*-based transposon by PCR (*see* **Note 1**) using primers that correspond to the exact ends of the transposon and whose 5′ ends are phosphorylated. For example, for transposon T26, use primers T26_ME_F(5PH) and T26_ME_R (5PH) (Table 1). In a 50-μL reaction, combine 0.1 ng of template DNA (purified transposon or plasmid bearing the transposon), 25 μL of 2X HiFi HotStart ReadyMix (*see* **Note**

2), and 1.5 μL each of 10-μM stocks of the forward and reverse primers. Thermocycle as follows: 2 min at 95 °C, followed by 25 cycles of 20 s at 98 °C, 20 s at 64 °C, and 90 s at 72 °C, followed by 5 min at 72 °C. Perform at least four parallel 50-μL amplifications to generate ample product, and combine the reactions after amplification.

2. Use gel electrophoresis to verify that the amplification produced a single, pure band of the expected size (e.g., T26 is 1786 bp).
3. Purify the product by PEG precipitation. Bring the total volume of the combined mixture to 500 μL by addition of TE. Add 250 μL of 5 M NaCl and 250 μL of 30% PEG 8000/1.5 M NaCl. Mix well and incubate at 4 °C for at least 30 min. Centrifuge at 4 °C for 10 min at ≥10,000 × *g*. Carefully remove and discard the supernatant. Centrifuge again briefly and remove any remaining supernatant. Dissolve the pellet in 40-μL TE (or 10 μL × the number of combined reactions). It is essential to use TE to dissolve the pellet, or transposome assembly (**step 5**) will fail (Tris buffer without EDTA is insufficient).
4. Quantify the DNA by fluorometry (*see* **Note 3**). A minimum concentration of at least 100 ng/μL is required for transposome assembly. We recommend 400 ng/μL. The concentration after PEG precipitation is often greater than 400 ng/μL; dilute with additional TE to achieve 400 ng/μL.
5. Assemble transposomes. Prepare the reaction mixture by adding in the following order: 1 μL purified transposon DNA (≥100 ng/μL in TE buffer, ideally 400 ng/μL), 2 μL EZ-Tn5™ Transposase, and 1 μL 100% glycerol. The reaction can be scaled up as needed. Mix by vortexing, and incubate for 45 min at room temperature and then overnight or longer at 4 °C (*see* **Note 4**). Store at −20 °C.

3.1.2 Transposon Mutagenesis of *Acinetobacter baumannii*

1. Prepare electrocompetent *A. baumannii* (*see* **Note 5**). From a freshly grown colony on agar media, inoculate 5 mL of LB in an 18-mm test tube, and grow overnight at 37 °C on a roller. Use 1 μL of the overnight culture to inoculate 50 mL of no-salt LB, mix well, and distribute 5-mL aliquots into ten 18-mm test tubes. Grow at 37 °C on a roller until the cultures reach an OD_{600} of at least 6.0 (requires at least 17 h for strain AB5075), and then recombine the cultures. Perform the following centrifugations and resuspensions at room temperature. Centrifuge the culture for 10 min at 15,000 × *g* in large sterile centrifuge tubes, discard the supernatant, and resuspend the pellet in 20-mL sterile 10% glycerol by vortexing. Centrifuge again, discard the supernatant, and resuspend the pellet in

10-mL 10% glycerol. Centrifuge again, discard the supernatant, and resuspend the pellet in 1.0-mL 10% glycerol. Transfer the cell suspension to a microcentrifuge tube, centrifuge in a microcentrifuge at maximum speed for 5 min, discard the supernatant, and resuspend the pellet in a minimal volume of 10% glycerol. Use only enough glycerol for the cells to be pipette-able (for AB5075, the total volume of resuspended cells is usually between 0.5 and 0.7 mL). Place the electrocompetent cells on ice until used for electroporation.

2. Transform the cells with transposomes by electroporation. For each transformation, mix 50-μL electrocompetent cells with 0.4-μL prepared transposomes (Subheading 3.1.1) (*see* **Note 6**). In a 0.1-mm cuvette, pulse at 1.6 kV, 100 Ω, and 25 μF (*see* **Note 7**). Immediately after pulsing, add 1-mL SOC to the cells in the cuvette, mix by pipetting, transfer to a 13-mm test tube, and incubate at 37 °C on a roller for 1 h. We suggest performing multiple identical transformations in parallel. Also carry out a control transformation that excludes the transposome.
3. Combine all the sample transformations that didn't arc (except the control transformation) into a single mixture, and place on ice. Add 1/3 volume of sterile 40% glycerol, mix by vortexing, and then divide into multiple 1-mL aliquots and at least one 0.1-mL test aliquot (see below) in microcentrifuge tubes. Flash-freeze the aliquots by placing the tubes in a dry ice/ethanol bath. Freeze the control transformation similarly (with glycerol). Store the frozen aliquots at −80 °C.
4. Determine the titer of recoverable transposon insertion mutants in the frozen mixtures. Thaw the test aliquot from **step 3** by placing the tube on ice for 30 min then at room temperature for 15 min, and then mix by brief vortexing. Make tenfold serial dilutions of the cells in LB (or other appropriate media), and plate 100 μL of each dilution on selective media (*see* **Notes 8** and **9**) (for T26 mutagenesis of AB5075 plated on LB agar, use 5- to 7.5-μg/mL tetracycline, and expect 10^4–10^5 tetracycline-resistant colonies per mL of frozen transformation mixture). Incubate the plates at 37 °C until colonies are apparent. Count the colonies (*see* **Note 10**), and back-calculate the titer and total quantity of recoverable transposon insertion mutants in the remaining frozen aliquots of combined transformation mixture. Also similarly thaw, dilute, and plate the test aliquot of the control transformation to verify the absence of background growth at a plating density similar to what will be used for the mutant pool assembly (*see* Subheading 3.1.3).

5. Repeat the above steps as necessary to obtain enough frozen recoverable transposon insertion mutants for the intended mutant pool (*see* **Note 11**).

3.1.3 Mutant Pool Assembly

1. Prepare large solid media plates for selective plating of transformation mixes (*see* **Notes 9** and **12**). We recommend using vented QTrays with 250-mL selective agar media per tray (*see* **Note 13**). Prepare enough QTrays for all of the transposon mutants intended for the mutant pool (*see* **Note 11**), assuming a plating density of 20,000–50,000 transposon mutants per QTray (35–90 mutants per cm^2). For example, to make a pool of 350,000 mutants, prepare 10 QTrays. Also pour several regular 10-cm plates with the same media. After pouring, allow the QTrays and plates to dry at room temperature for 2 full days with lids on. Store in bags at 4 °C if not used immediately.
2. Based on the calculated titer(s) of recoverable transposon insertion mutants (*see* Subheading 3.1.2), thaw enough aliquots of the frozen transformation mixture(s) for the entire mutant pool (*see* **Note 11**). To thaw, place the frozen tubes on ice for 30 min and then at room temperature for 15 min. Combine all of the aliquots in a single large tube, and add additional LB or other appropriate media to bring the total volume to approximately 1.2 mL × the number of prepared QTrays (*see* **Note 13**). Mix by vortexing and place on ice.
3. Use a small quantity of the cell mixture to make tenfold serial dilutions to 10^{-3} in LB or other appropriate media. Plate 100 μL of each dilution on one of the pre-prepared 10-cm selective media plates.
4. Plate 1.2 mL of the cell mixture on each pre-prepared selective media QTray, taking care to spread the mixture thoroughly and evenly over the entire agar surface (*see* **Note 14**). After spreading, keep the plates level on the bench with lids off until any remaining surface liquid has fully evaporated (10 min is usually sufficient).
5. Incubate the QTrays at 37 °C until robust colonies have formed and cover most of the surface at high density, but not so long that no spaces between colonies are visible. For strain AB5075 with T26 mutagenesis selected on LB agar with 5-μg/mL tetracycline, 14 h is appropriate. Incubate all the QTrays for the same amount of time. The total amount of growth is less important than the consistency of growth between QTrays. Incubate the 10-cm plates similarly.
6. Harvest the colonies from each QTray as follows. Flood the agar surface with 10 mL of LB (or other appropriate liquid media or buffer) (*see* **Note 13**), scrape the cells into the media using a sterile L-shaped cell spreader (taking care not to gauge

the agar surface), and pipette the media containing the cells into a tube kept on ice. Repeat the process with another 10 mL of liquid media to recover the remaining cells (by flooding and scraping twice, >99% of the cells will be recovered). The total volume of recovered media is usually substantially less than 20 mL due to absorption of liquid into the agar. The cell density is usually very high (>10^{10} cfu/mL).

7. Combine the harvests from all the QTrays and measure the total volume. Add 1/3 volume of 40% glycerol (*see* **Note 15**) and mix thoroughly by vortexing. Place 1 mL aliquots of the cell mixture (or other volumes of choice) into microcentrifuge tubes, and flash-freeze by placing the tubes in a dry ice/ethanol bath. Store at −80 °C. These tubes represent your random transposon mutant pool.
8. Count colonies from the 10-cm dilution series plates (**step 3**) (*see* **Note 10**). From these counts and from the total volume of transformation mixture plated on the QTrays (**step 4**), calculate the approximate number of unique transposon mutants in the mutant pool (the pool complexity).
9. Thaw one aliquot of the mutant pool by placing the tube on ice for 30 min and then at room temperature for 15 min. Make tenfold serial dilutions in LB (or other appropriate media or buffers) to 10^{-8}, and plate 100 μL of each dilution on both selective (containing the antibiotic used to select the transposon mutants) and nonselective media. After growth, count colonies (*see* **Note 10**), and calculate the titer of mutant and nonmutant colony-forming units in the mutant pool (*see* **Note 16**).
10. Use a portion of the same thawed aliquot for isolation and sequencing of the transposon junctions (Subheading 3.2) to identify and quantify the relative abundances of the mutants in the pool. This Tn-seq analysis will serve as a useful comparison for analyses done after subsequent biological selections. The number of insertion locations identified should correspond approximately to the number of mutants estimated in **step 8**. The analysis can also define the set of genes essential under the conditions used to generate the pool [4].

3.2 Isolation and Sequencing of Transposon Junctions from a Mutant Pool

1. Prepare genomic DNA from the mutant pool sample to be analyzed. Use the Qiagen DNeasy Blood and Tissue Kit, following the manufacturer's instructions for isolation of genomic DNA from Gram-negative bacteria, or use another method of choice. Quantify the DNA by fluorometry (*see* **Note 3**).
2. Shear 1.5 μg of sample genomic DNA (*see* **Note 17**) to an average size of approximately 300 bp (*see* **Note 18**). Dilute the DNA to 130 μL total volume in 10 mM Tris-Cl, pH 8.5, place it in a Covaris microTUBE with AFA Fiber, briefly centrifuge to remove bubbles, and shear with a Covaris focused-

ultrasonicator (*see* **Note 19**) using the instrument's recommended settings for achieving 300-bp average fragment size (for a Covaris M220, shear for 65 s using peak incident power of 50, duty factor of 20%, and 200 cycles per burst).

3. Recommended (*see* **Note 20**): End-repair the sheared DNA using the NEBNext® End Repair Module. Mix the sheared DNA with the supplied 10× reaction buffer (1× final concentration) and enzyme mix (1/20th of the total reaction volume). Incubate in a thermocycler for 30 min at 20 °C. Purify the end-repaired DNA over a single column using the MinElute PCR Purification Kit, following the recommended protocol but eluting with 10 μL buffer EB twice into one tube. Optionally, use 1 μL of the eluate to quantify the DNA by fluorometry (*see* **Note 3**).

4. C-tail the DNA (*see* **Note 21**). Make a fresh 10× dCTP/ddCTP solution composed of 9.5 mM dCTP and 0.5 mM ddCTP in water (keep on ice). In a 30-μL reaction, combine the end-repaired DNA, 6 μL of 5× TdT Reaction Buffer (1× final concentration), 3 μL of the dCTP/ddCTP solution (1× final concentration), and 1 μL (30 units) of terminal deoxynucleotidyl transferase (TdT). Incubate in a thermocycler for 1 h at 37 °C and 20 min at 75 °C. Purify the C-tailed DNA using a Performa® DTR Gel filtration cartridge as follows: centrifuge the gel filtration cartridge for 2 min at 750 × *g*; remove the column and transfer to a new 1.5 mL microcentrifuge tube; add the DNA sample to the packed column, making sure that the fluid runs into the gel; and centrifuge for 2 min at 750 × *g* (*see* **Note 22**). Optionally, use 1 μL of the eluate to quantify the DNA by fluorometry (*see* **Note 3**).

5. Perform PCR round 1 (*see* **Note 21**). In a 50-μL reaction, combine 7.5 μL of C-tailed DNA (*see* **Note 23**), 25 μL of 2× HiFi HotStart ReadyMix (*see* **Note 2**), 3 μL of a 10-μM stock of the round 1 flanking primer (olj376), 1 μL of a 10-μM stock of the transposon-specific round 1 primer (e.g., T26-83_1) (*see* **Note 24**), and, optionally, 0.25 μL of 100× SYBR Green I. Carry out thermocycling as follows: incubate for 2 min at 95 °C, followed by 22 cycles of 30 s at 98 °C, 30 s at 63 °C, 90 s at 72 °C, and, optionally, a read of fluorescence, followed by 5 min at 72 °C. If tracking amplification by fluorescence, it is typical to see detectable amplification after 22 cycles, but even in the apparent absence of strong amplification, or apparent linear amplification during the early cycles, proceed to **step 6**.

6. Perform PCR round 2 to determine the optimal cycle number for **step 7** (*see* **Note 21**). In a 50-μL reaction, combine 1 μL of the round 1 PCR reaction, 25 μL of 2× HiFi HotStart

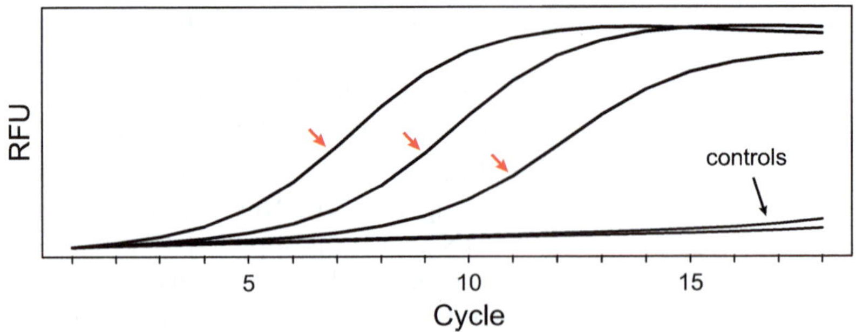

Fig. 2 Recommended cycles at which to stop round 2 PCR reactions for sequencing. Shown are real-time PCR curves for PCR round 2 amplifications of three typical samples (**step 6** of the junction-sequencing protocol). The inflection points of the curves (red arrows) represent appropriate cycles for stopping amplification (**step 7** of the junction-sequencing protocol). These points correspond to cycles where ample product has been produced but where amplification remains exponential. *RFU* relative fluorescence units

ReadyMix (*see* **Note 2**), 1 μL of a 10-μM stock of a round 2 flanking/indexing primer (e.g., TdT_Index_1) (*see* **Notes 24** and **25**), 1 μL of a 10-μM stock of the transposon-specific round 2 primer (e.g., T26_AmpF_2) (*see* **Note 24**), and 0.25 μL of 100× SYBR Green I. Carry out thermocycling as follows: incubate for 2 min at 95 °C, followed by 16 cycles of 30 s at 98 °C, 30 s at 63 °C, and 90 s at 72 °C with a read of fluorescence, followed by 5 min at 72 °C. From the real-time fluorescence data, identify the cycle number at which to stop amplification in **step 7**. This is typically at the inflection point of the real-time curve, where sufficient amplification has been achieved (at least 20% of maximum) but before the rate of exponential amplification begins to noticeably decline (Fig. 2). The ideal cycle number may differ between samples.

7. Repeat PCR round 2, stopping each sample after the corresponding cycle number identified in **step 6**. Purify the amplified DNA using the MinElute PCR Purification Kit, following the recommended protocol. Quantify the DNA by fluorometry (*see* **Note 3**). Calculate the sample molarity from the fluorometric quantification and the average fragment length. The average fragment length can be estimated by agarose or acrylamide gel, by Agilent TapeStation or Bioanalyzer, or by adjusting the predicted average shearing size (**step 2**) to account for additional bases incorporated during C-tailing and PCR (Fig. 1).
8. If multiplexing (*see* **Note 25**), mix together equal molar amounts of each sample's purified round 2 PCR product.
9. Optional: Size-select the DNA to narrow the range of fragment sizes to be sequenced (*see* **Note 26**). Run the DNA sample on a

6% precast TBE acrylamide gel or other suitable gels. Cut out sizes between approximately 150 and 500 bp (*see* **Note 18**). Break the gel slice into small fragments (e.g., by centrifugation through a small hole), and then incubate in 200-μL TE for a least 1 h at 50–60 °C with periodic vortexing. Separate the buffer from the gel material using a Nanosep 0.2-μM column, following the manufacturer's protocol. Purify the amplified DNA from the buffer using the MinElute PCR Purification Kit, following the recommended protocol. Quantify the DNA by fluorometry (*see* **Note 3**), and calculate the sample molarity as in **step 7**.

10. Sequence the DNA by Illumina sequencing. Follow standard Illumina protocols for library dilution and denaturation. Spike in PhiX Control Library at an amount appropriate for the sequencing platform used (for a MiSeq, spike in at 5%) (*see* **Note 27**). Use a custom Read 1 primer mix that includes both the transposon-specific sequencing primer (e.g., T26_SEQ-6) (*see* **Note 24**) and Illumina's standard Read 1 sequencing primer (Read1_SEQ) for sequencing the PhiX fragments. If multiplexing, specify the barcode associated with each sample when setting up the sequencing run (see Illumina documentation). The barcode corresponds to the reverse complement of the barcode portion of the round 2 flanking/indexing primer (e.g., for primer TdT_Index_1 (Table 1), the barcode is ATCACG). The standard Illumina index sequencing primer (Fig. 1) is used to read the barcodes. Retrieve the output as FASTQ files.
11. Bioinformatically process the data. Discard reads whose initial bases don't display transposon end sequence (usually less than 5%). Map the remaining portion of each read to the genome to determine the position and orientation of the corresponding transposon insertion. Tabulate and count the reads per unique insertion. Compare the insertion positions to the genome annotation to tabulate the hits and reads per genome feature (e.g., gene). Normalize read counts for comparison between samples. The following link provides a bioinformatic tool for carrying out the above steps: https://github.com/lg9/Tn-seq. For more sophisticated normalization, statistical analysis, and visualization tools, consult appropriate literature [7, 11–15].

4 Notes

1. The required features of the transposon are 19-bp Tn*5* mosaic end (ME) sequences on each end (5′-CTGTCTCTTATACA-CATCT-3′) [10] and an antibiotic resistance gene (or other

selectable markers) with its own promoter. The resistance gene should be for an antibiotic to which the *A. baumannii* strain under study is sensitive, and the promoter should drive expression sufficiently for robust mutant selection independent of the genetic context at any given insertion location. If the promoter insufficiently drives expression, the mutant collection will be biased toward co-oriented insertions (where the resistance gene within the transposon is oriented in the same direction as the gene of insertion) in highly expressed genes (such that local transcription drives expression of the resistance gene), and poorly expressed genes will be underrepresented in the mutant pool. Transposon T26 carries a tetracycline-resistance gene with a promoter originally isolated from *Francisella* spp. [4]. In strain AB5075 (*see* **Note 5**), the promoter drives expression sufficiently for selection at low tetracycline levels (5–7.5 μg/mL in LB agar) but not for selection at higher levels (>10 μg/mL). An ideal transposon for *Acinetobacter* will independently drive expression of its resistance gene sufficiently for selection at drug concentrations substantially higher in the MIC of the recipient strain.

2. The protocols presented here were optimized using KAPA HiFi polymerase for PCR reactions, but any high-fidelity thermostable polymerase and PCR methodology can be used. The researcher will need to work out conditions for amplification with other polymerases.
3. For quantifying DNA in these protocols, fluorometry (e.g., Qubit) is recommended over spectrophotometry (e.g., Nanodrop), since fluorometric methods specifically quantify dsDNA, while spectrophotometric measurements can be influenced by other solvents.
4. Experience suggests that longer incubation at 4 °C can improve transposome assembly.
5. This protocol for electrocompetent cell preparation was developed for *A. baumannii* strain AB5075 [16] and with minor variation has been used with strain ATCC 17978. For other strains, the researcher may need to optimize conditions for electrocompetent cell preparation and high efficiency transformation.
6. The volume of transposome preparation can be increased to achieve more mutants per transformation, though doing so increases the potential for arcing during electroporation. Additionally, because of variability between transposome preparations, it is suggested to start with small volumes to assess the mutagenesis efficiency possible with each preparation.
7. The electroporation conditions were optimized using a Bio-Rad Gene Pulser. Pulsing conditions may need to be

modified for other instruments and strains. Lowering the voltage may help prevent arcing.

8. The research goals may inform the choice of growth media. In most cases, rich-nutrient media are preferred so that the maximum number of genes will be represented in the mutant pool (for many catabolic function genes, mutants do not grow on minimal media), though defined media may also be used. The same media type used for selection (but without antibiotic) should also be used for the serial dilutions.
9. Care should be taken in choosing the amount of antibiotic included in the selective media. For transposons with weaker independent expression of their selectable marker, using an excessively high level of antibiotic may result in a mutant collection biased toward co-oriented insertions in genes that are highly expressed under the plating conditions used (*see* **Note 1**).
10. When counting colonies from a plated dilution series for titer calculations, count the colonies from whichever plate corresponds to the least amount of dilution but has a countable number of colonies. For example, count the colonies from a plate that has at least 50 colonies. Counts from plates with fewer colonies are of lower precision due to randomness of sampling.
11. Carefully consider how many unique transposon mutants should make up the mutant pool (the pool complexity). In general, having more mutants (higher saturation) will provide greater power in downstream statistical analyses. Greater saturation can also facilitate analysis of smaller-scale genomic tracts (e.g., promoters, protein domains, and snRNA genes). However, making more mutants increases reagent and labor costs, and pools with large numbers of mutants require greater sequencing depth to detect the majority of mutants. Stochastic bottlenecks during biological selection and/or molecular processing may also reduce the proportion of hits detected for high-saturation pools. Detecting all the mutants in very high-saturation pools may not be critical, however, since detecting even a fraction of mutants from such pools may still be sufficient for robust statistical analysis of gene-level selection. With *A. baumannii*, a pool of 100,000 mutants will have, on average, one insertion every 35 bases or so within the nonessential genome, which is sufficient for gene-level analyses of many conditions that don't impose experimental bottlenecks (e.g., growth on selective laboratory media). In cases of known experimental bottlenecks (e.g., infection assays), the researcher may opt for lower complexity pools so that each unique mutant can be reliably tracked started from relatively small inocula.

12. Solid media is strongly recommended over liquid media for selective outgrowth of transposon mutants when making a mutant pool. Two potential problems are associated with growth in liquid media: fast-growing mutants quickly dominating the culture and prophage or other extracellular inhibitory factors produced by specific mutants inhibiting or killing other mutants. Both possibilities can drastically skew mutant representation and reduce the complexity of the mutant pool. Outgrowth on solid media largely circumvents these potential problems.
13. Other plate types can also be used. Adjust the inoculum volume to achieve a plating density similar to the recommendation for QTrays. Also proportionally adjust the volume of media used for harvesting.
14. For even and consistent spreading, we suggest pre-placing a set number of 5-mm sterile glass beads (e.g., 30 per QTray) on the agar surface. After adding the inoculum, shake the plate vigorously back and forth in numerous directions for at least 60 s. The shaking may be done with stacks of plates. Other spreading methods may also be used, provided the spreading is even and consistent between plates.
15. DMSO may also be used as a cryoprotectant. Add 1/19th volume DMSO to achieve a final concentration of 5%.
16. The mutant pool may contain nonmutant bacteria which survived selective plating (the high density of mutant bacteria plated on the selective media may have inactivated or sequestered much of the antibiotic, allowing survival and growth of some nonmutant cells). The presence of such nonmutant bacteria does not usually interfere with subsequent Tn-seq analysis (the nonmutant bacteria contain no transposon junctions), but knowing the proportion of mutant and nonmutant bacteria in the pool can help in the planning of biological selection experiments.
17. The amount of starting DNA required depends upon the proportion of bacterial mutant DNA in the sample, the complexity of the mutant pool, and the average multiplicity of mutant representation desired. If most of the DNA was derived from transposon mutants, such as in cases of pure mutant pools grown in laboratory culture, 1 μg is sufficient (1 μg of *A. baumannii* genomic DNA corresponds to more than 2×10^8 genome copies; for a pool of 10^5 unique insertion mutants, this represents an average multiplicity of 2000 genome copies per unique mutant). In contrast, if bacterial mutant DNA represents only a small percentage of the total DNA, such as in infection assays where most of the DNA is host eukaryotic DNA, more DNA may be needed to sufficiently represent the

full complexity of the mutant pool at adequate multiplicity. An average multiplicity of 50 implies that mutants represented at less than 1/50th of the average mutant concentration in the pool will be below the limit of detection, assuming 100% detection of individual junction molecules. Note that randomness of shearing, loss of sample between steps, and other methodological factors make levels of detection of individual junction molecules substantially less than 100%.

18. The target size for shearing (or size selection) should be adjusted based on the annealing position of the first transposon-specific primer relative to the end of the transposon (*see* **Note 24**). Calculate the minimum required fragment size for the transposon. For example, for T26, the first transposon-specific primer (T26-83_1) anneals 83 bases from the end of the transposon (Fig. 1). If 44 bases of flanking sequence are desired (with the first six bases of a 50-bp sequencing read representing transposon sequence), then the minimum fragment size needed is 127 bp (83 + 44). At an average shearing size off 300 bp, most fragments will be longer than this minimum, but short enough so that the majority will be within a narrow size range. Fragments of a relatively narrow size range sequence more efficiently on Illumina platforms.
19. Covaris instruments work well for shearing to a defined and relatively narrow size range, but other shearing methods and instruments can be used. If the shearing produces a broad range of fragment sizes, it may be beneficial to include a size-selection step either immediately after shearing or as in **step 9** (Subheading 3.2) (*see* **Note 18**).
20. The end-repair step was not part of the original C-tailing protocol [9], but improves the efficiency of detection of individual junction fragments. The step may not significantly improve results for samples which represent the insertion mutant population at high complexity (*see* **Note 17**).
21. Control reactions are recommended for key steps in the protocol if permitted by the amount of DNA available. Control reactions help identify and troubleshoot problems that may arise before the final sequencing step. For example, for the C-tailing step, include a control reaction that excludes the TdT enzyme. A lack of observable difference between the true sample and this control in the amplification steps (**steps 5** and **6**) is an indication that the C-tailing reaction may have failed. Table 2 provides a prioritized list of suggested control reactions.
22. For purification of the C-tailed DNA, alternatives to the Performa® DTR Gel filtration cartridge may work equally well, but have not been attempted by our laboratory.

Table 2
Suggested control reactions for junction amplification and sequencing. Controls are listed in order of suggested priority

Step	Control reaction	Control for	Expected result
C-tailing	No TdT enzyme	Successful C-tailing	Reduced amplification in PCR rounds 1 and 2 relative to true sample
Round 1 PCR	No Tn-specific primer	Specific amplification of junctions	Reduced amplification in PCR rounds 1 and 2 relative to true sample
Round 2 PCR	No Tn-specific primer	Specific amplification of junctions	Reduced PCR round 2 amplification relative to true sample
Round 1 PCR	No round 1 flanking primer (olj376)	Successful C-tailing, specific amplification of C-tailed junctions	Reduced amplification in PCR rounds 1 and 2 relative to true sample
Round 2 PCR	No round 2 flanking/ indexing primer	Specific amplification of junctions	Reduced PCR round 2 amplification relative to true sample

23. For starting DNA samples that represent the complexity of the mutant pool at high multiplicity (e.g., thousands of genome copies per unique mutant, on average) (*see* **Note 17**), a majority of the unique mutants can be identified and quantified by using only a portion (7.5 of 30 μL) of the C-tailed DNA for PCR round 1, and the additional DNA can be used for control reactions (*see* **Note 21**). Conversely, if the starting DNA sample represents limiting multiplicity, then the controls should be omitted and the entire C-tailed sample should be used for PCR round 1. The amount of multiplicity that is "limiting" depends on the researcher's tolerance for loss of detection of low-abundance mutants and may need to be assessed empirically.
24. For transposons other than T26, redesign the three transposon-specific primers. For the PCR round 1 transposon-specific primer (**step 5**), use a primer that anneals within a few hundred bases of the transposon end (*see* **Note 18**) and that has a Tm similar to that of T26-83_1. For the PCR round 2 transposon-specific primer (**steps 6** and 7), the 5′ end should represent the required flow cell adaptor sequence (Fig. 1) and the 3′ end should anneal closer to the junction than the round 1 transposon-specific primer and have a Tm similar to the transposon-specific portion of primer T26_AmpF_2 (Fig. 1). For the read 1 sequencing primer (**step 10**), use a primer that anneals within a few bases of the junction and that has a Tm

similar to that of T26_SEQ-6. The primers may correspond to either end of the transposon.

25. If multiplexing (sequencing multiple samples together on a single Illumina flow cell lane), use uniquely barcoded round 2 flanking/indexing primers for each sample (*see* Table 1 and "B" bases in Fig. 1). The set of flanking/indexing primers used for the samples to be sequenced together must include at least one red channel base (A or C) and one green channel base (G or T) at every position of the barcode. For example, if multiplexing two samples, primers TdT_Index_A and TdT_Index_B may be used together, but not primers TdT_Index_A and TdT_Index_C (Table 1). See Illumina documentation on maintaining index read color balance for additional information. If not multiplexing, the choice of barcode is irrelevant.
26. Size-selecting a narrower range of fragment sizes can improve efficiency of Illumina sequencing, since a broad fragment size range may cause a greater proportion of reads to fail Illumina sequence quality filtering. However, the loss of unique fragments resulting from size selection should be considered when processing samples of limiting multiplicity (*see* **Notes 17** and **23**). Size selection is not necessary if the shearing method used produces a relatively narrow range of fragment sizes (*see* **Notes 18** and **19**).
27. Since the vast majority of sample reads will display identical sequence for the first few bases (corresponding to transposon end sequence), spiking in PhiX Control Library is essential to prevent failure of the sequencing run due to lack of base diversity. See Illumina documentation on low diversity libraries for more information.

References

1. Chao MC, Abel S, Davis BM, Waldor MK (2016) The design and analysis of transposon insertion sequencing experiments. Nat Rev Microbiol 14(2):119–128. https://doi.org/10.1038/nrmicro.2015.7
2. Kwon YM, Ricke SC, Mandal RK (2016) Transposon sequencing: methods and expanding applications. Appl Microbiol Biotechnol 100(1):31–43. https://doi.org/10.1007/s00253-015-7037-8
3. Troy EB, Lin T, Gao L, Lazinski DW, Camilli A, Norris SJ, Hu LT (2013) Understanding barriers to *Borrelia burgdorferi* dissemination during infection using massively parallel sequencing. Infect Immun 81(7):2347–2357. https://doi.org/10.1128/IAI.00266-13
4. Gallagher LA, Ramage E, Weiss EJ, Radey M, Hayden HS, Held KG, Huse HK, Zurawski DV, Brittnacher MJ, Manoil C (2015) Resources for genetic and genomic analysis of emerging pathogen *Acinetobacter baumannii*. J Bacteriol 197(12):2027–2035. https://doi.org/10.1128/JB.00131-15
5. Gebhardt MJ, Gallagher LA, Jacobson RK, Usacheva EA, Peterson LR, Zurawski DV, Shuman HA (2015) Joint transcriptional control of virulence and resistance to antibiotic and environmental stress in *Acinetobacter baumannii*. MBio 6(6):e01660–e01615. https://doi.org/10.1128/mBio.01660-15
6. Gallagher LA, Lee SA, Manoil C (2017) Importance of core genome functions for an extreme antibiotic resistance trait. MBio 8(6). https://doi.org/10.1128/mBio.01655-17

7. Burby PE, Nye TM, Schroeder JW, Simmons LA (2017) Implementation and data analysis of Tn-seq, whole-genome resequencing, and single-molecule real-time sequencing for bacterial genetics. J Bacteriol 199(1). https://doi.org/10.1128/JB.00560-16
8. Goryshin IY, Jendrisak J, Hoffman LM, Meis R, Reznikoff WS (2000) Insertional transposon mutagenesis by electroporation of released Tn*5* transposition complexes. Nat Biotechnol 18(1):97–100. https://doi.org/10.1038/72017
9. Klein BA, Tenorio EL, Lazinski DW, Camilli A, Duncan MJ, Hu LT (2012) Identification of essential genes of the periodontal pathogen *Porphyromonas gingivalis*. BMC Genomics 13:578. https://doi.org/10.1186/1471-2164-13-578
10. Zhou M, Bhasin A, Reznikoff WS (1998) Molecular genetic analysis of transposase-end DNA sequence recognition: cooperativity of three adjacent base-pairs in specific interaction with a mutant Tn*5* transposase. J Mol Biol 276(5):913–925. https://doi.org/10.1006/jmbi.1997.1579
11. DeJesus MA, Ambadipudi C, Baker R, Sassetti C, Ioerger TR (2015) TRANSIT – a software tool for Himar1 TnSeq analysis. PLoS Comput Biol 11(10):e1004401. https://doi.org/10.1371/journal.pcbi.1004401
12. DeJesus MA, Nambi S, Smith CM, Baker RE, Sassetti CM, Ioerger TR (2017) Statistical analysis of genetic interactions in Tn-Seq data. Nucleic Acids Res 45(11):e93. https://doi.org/10.1093/nar/gkx128
13. Pritchard JR, Chao MC, Abel S, Davis BM, Baranowski C, Zhang YJ, Rubin EJ, Waldor MK (2014) ARTIST: high-resolution genome-wide assessment of fitness using transposon-insertion sequencing. PLoS Genet 10(11):e1004782. https://doi.org/10.1371/journal.pgen.1004782
14. Solaimanpour S, Sarmiento F, Mrazek J (2015) Tn-seq explorer: a tool for analysis of high-throughput sequencing data of transposon mutant libraries. PLoS One 10(5):e0126070. https://doi.org/10.1371/journal.pone.0126070
15. Zomer A, Burghout P, Bootsma HJ, Hermans PW, van Hijum SA (2012) ESSENTIALS: software for rapid analysis of high throughput transposon insertion sequencing data. PLoS One 7(8):e43012. https://doi.org/10.1371/journal.pone.0043012
16. Jacobs AC, Thompson MG, Black CC, Kessler JL, Clark LP, McQueary CN, Gancz HY, Corey BW, Moon JK, Si Y, Owen MT, Hallock JD, Kwak YI, Summers A, Li CZ, Rasko DA, Penwell WF, Honnold CL, Wise MC, Waterman PE, Lesho EP, Stewart RL, Actis LA, Palys TJ, Craft DW, Zurawski DV (2014) AB5075, a highly virulent isolate of acinetobacter baumannii, as a model strain for the evaluation of pathogenesis and antimicrobial treatments. MBio 5(3):e01076–e01014. https://doi.org/10.1128/mBio.01076-14

Chapter 13

Tn*7*-Based Single-Copy Insertion Vectors for *Acinetobacter baumannii*

Kaleigh Ducas-Mowchun, P. Malaka De Silva, Rakesh Patidar, Herbert P. Schweizer, and Ayush Kumar

Abstract

Acinetobacter baumannii is considered a problematic Gram-negative pathogen due to its widespread resistance to antibiotics. Understanding of resistance mechanisms in *A. baumannii* is critical for designing new and effective therapeutic options. However, this is hampered by the lack of tools to carry out genetic manipulations in *A. baumannii*. Here, we describe methods to use a chromosomal mini-Tn*7*-based single-copy gene expression system in *A. baumannii*. This system can be effectively used for performing genetic complementation studies, for tagging with fluorescent proteins, or for reporter fusion assays.

Key words Gene complementation, Gene expression, Electroporation, Four-parental mating

1 Introduction

Acinetobacter baumannii has recently emerged as a major opportunistic pathogen in healthcare facilities around the globe [1]. The rising threat of *Acinetobacter* infection reflects this species' ability to display multidrug-resistant and hyper-virulent phenotypes [1]. This has become a major concern for healthcare facilities around the world, and as a result, the World Health Organization has urged research and development to focus on *Acinetobacter* as a top priority [2]. Increased attention *Acinetobacter* has received in recent years has also consequently resulted in vast numbers of *A. baumannii* genomes becoming available. However, tools to carry out genetic manipulation in *A. baumannii* remain scarce [3].

Mini-Tn*7* vectors are a versatile set of tools that can be used for a comprehensive range of molecular genetic applications in Gram-negative bacteria including molecular cloning, gene complementation and reporter gene tagging, and other applications [4, 5]. These systems are particularly useful in gene complementation, where chromosomal integration of the mini-Tn*7* element overcomes

Indranil Biswas and Philip N. Rather (eds.), *Acinetobacter baumannii: Methods and Protocols*, Methods in Molecular Biology, vol. 1946, https://doi.org/10.1007/978-1-4939-9118-1_13,

many of the limitations associated with orthodox methods of gene complementation using plasmid vectors such as the multi-copy nature in vivo, requirement of an antibiotic dosage to maintain the plasmid, and limited host range [6]. Mini-Tn*7* transposon element delivery plasmids are capable of delivering a single copy of a particular gene that is cloned into the multiple cloning site of the element to a defined neutral location of the host genome [5–7]. Chromosomal integration occurs at the *att*Tn*7* attachment site situated within an intergenic region immediately downstream of the highly conserved *glmS* gene-encoding glucosamine-fructose-6-phosphate aminotransferase [5, 6]. The vast majority of Gram-negative bacteria contain a single *glmS*-linked *att*Tn*7* site for site-specific Tn*7* insertion. However, in some bacteria, e.g., most *Burkholderia* species, multiple *glmS* genes and *att*Tn*7* sites are present. *Burkholderia thailandensis* [5] and *B. mallei* [8] contain two *att*Tn*7* sites, and *B. pseudomallei* contains three *att*Tn*7* sites [9]. Rarely does mini-Tn*7* insertion occur at non-*glmS*-linked *att*Tn*7* sites, such as in *Proteus mirabilis* [10]. Insertion at *att*Tn*7* is aided by Tn*7* helper plasmids, which contain the Tn*7* *tnsABCD* transposase genes that encode the components of the site-specific Tn*7* transposition pathway [5, 11]. The mini-Tn*7* elements are engineered to allow excision of the resident antibiotic resistance selection marker using *Saccharomyces cerevisiae* Flp recombinase. This enzyme recognizes the Flp recombinase target (*FRT*) sites that flank the resistance marker [4]. The mini-Tn*7* system has been shown to be a useful tool for genetic manipulation of *Acinetobacter baumannii* (Fig. 1) [4]. The original mini-Tn*7* delivery method in this bacterium employed a four-parental conjugation insertion method [4]. In this study, we outline electroporation and four-parental conjugation procedures as two possible methods for introducing mini-Tn*7* vectors in *A. baumannii*.

2 Materials

2.1 Materials for Subheading 3.1

2.1.1 Reagents

1. Bacterial strains
   (a) Recipient strain (e.g., *A. baumannii* ATCC17978).
2. Sterile distilled water.
3. Antibiotic
   (a) Gentamicin (Gm): 50 mg/mL stock solution in water.
4. DNA
   (a) Mini-Tn*7* delivery vector (in this example we use pUC18T-mini-Tn*7*T-LAC-Gm (Fig. 2), GenBank accession no. EU240902 [9]).
   (b) pTNS2 helper plasmid (GenBank accession no. AY884833).

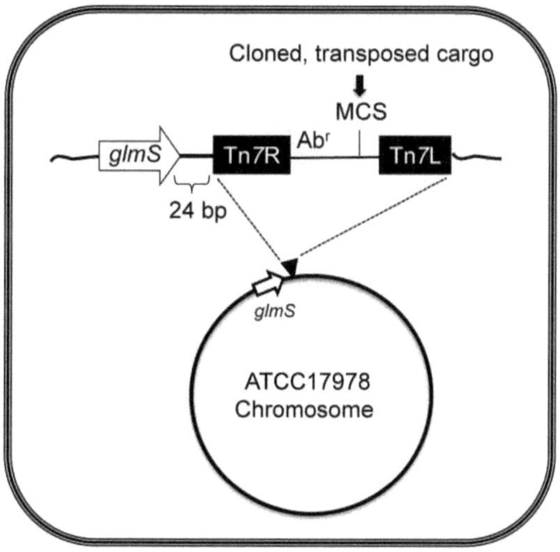

Fig. 1 Integration of mini-Tn*7* vectors into the *A. baumannii* chromosome. It was previously shown that mini-Tn*7* elements integrate into the *A. baumannii* ATCC19978 chromosome in a site- and orientation-specific manner at a single *glmS*-associated Tn*7* attachment (*att*Tn*7* site) located downstream of *glmS*. In strain ATCC17978, *glmS* (A1S_3396) is located between nucleotides 3,903,352 and 3,904,041 of the 3,976,747 bp chromosome (GenBank accession no. CP000521.1). The mini-Tn*7* elements contain a multiple cloning site (MCS) for cloning of genetic cargo and an antibiotic resistance marker (Ab^r) for selection of transformants. The boxed arrow marks the *glmS* gene and its transcriptional orientation. Black boxes indicate the Tn*7* left (Tn*7*L) and right (Tn*7*R) ends

5. LB Lennox broth: 10 g/L tryptone, 5 g/L yeast extract, 5 g/L NaCl.
6. LB agar Gm50 plates: 10 g/L tryptone, 5 g/L yeast extract, 5 g/L NaCl, 15 g/L agar, 50 μg/mL gentamicin (Gm50).

2.1.2 Equipment and Supplies

1. Sterile culture tubes (13 mL).
2. Sterile microcentrifuge tubes (1.7 mL).
3. Electroporation cuvettes (placed on ice prior to use).
4. Electroporator (e.g., Eppendorf Electroporator 2510).

2.2 Materials for Subheading 3.2

2.2.1 Reagents

1. Bacterial strains
   (a) Donor strain (*E. coli* DH5α/pUC18T-mini-Tn*7*T-LAC-Gm [9]).
   (b) Recipient strains (e.g., *A. baumannii* ATCC17978).
   (c) Helper strains.
   - *E. coli* DH5α(λ*pir*)/pTNS2.
   - *E. coli* HB101/pRK2013.

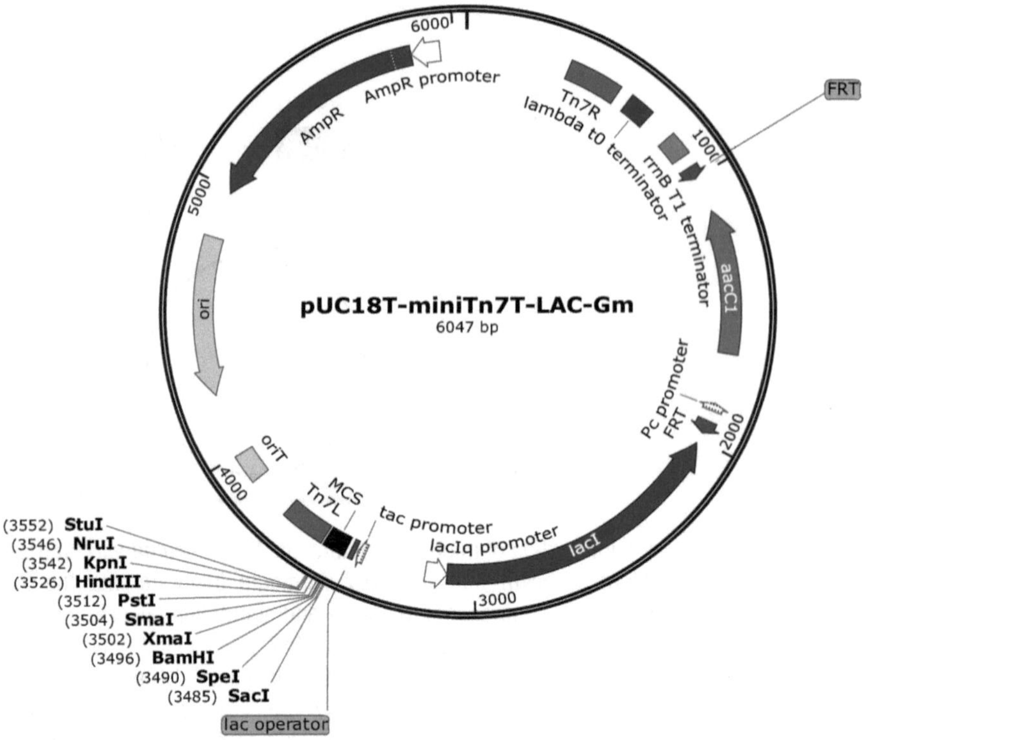

Fig. 2 Plasmid map and features of the pUC18T-mini-Tn*7*T-LAC-Gm plasmid used in this study. The extent of mini-Tn*7*T-LAC-Gm DNA ranges from Tn*7*L (Tn*7* left end) to Tn*7*R (Tn*7* right end). This segment includes the MCS (multiple cloning site), the *tac* promoter (P_{tac}), *lacI* (*lac* repressor), gentamicin resistance (Gm^R) genes flanked *FRT* (Flp recombinase target) sites, and the double lambda t_0 and *E. coli rrnB* T_1 transcription terminators. The suicide delivery vector portion contains *bla* (beta-lactamase-encoding gene), *oriT* (origin of conjugative transfer), and the narrow-host-range ColE1 origin of replication (*ori*). The map was created using SnapGene

2. Sterile distilled water.
3. Sterile 0.85% saline solution.
4. Antibiotics
   (a) Gentamicin (Gm): 50 mg/mL stock solution in water.
   (b) Kanamycin (Km): 50 mg/mL stock solution in water.
   (c) Ampicillin (Ap): 50 mg/mL stock solution in water.
5. LB Lennox broth: 10 g/L tryptone, 5 g/L yeast extract, 5 g/L NaCl.
6. LB Lennox agar plates: 10 g/L tryptone, 5 g/L yeast extract, 5 g/L NaCl, 15 g/L agar.
7. LB Lennox agar Gm50 plates: 10 g/L tryptone, 5 g/L yeast extract, 5 g/L NaCl, 15 g/L agar, 50 μg/mL gentamicin.

8. Simmons citrate agar Gm50 plates: 0.2 g/L $MgSO_4.7H_2O$, 0.2 g/L $NH_4H_2PO_4$, 0.8 g/L $Na_2NH_4PO_4$, 2 g/L $Na_3C_6H_5O_7$, 5 g/L NaCl, 0.08 g/L bromothymol blue, 15 g/L agar (final pH 7.0), 50 μg/mL gentamicin.

2.2.2 Equipment and Supplies

1. Sterile culture tubes (13 mL).
2. Sterile microcentrifuge tubes (1.7 mL).
3. Sterile nitrocellulose membrane.

2.3 Materials for Subheading 3.3

2.3.1 Reagents

1. *Taq* DNA polymerase and buffer.
2. dNTPs (10 mM working stock solution),
3. Primers
   (a) ABglmS_F: 5′-TATGGAAGAAGTTCAGGCTC.
   (b) Tn*7*R: 5′-CACAGCATAACTGGACTGATTTC.
4. Agarose.
5. TAE buffer: 242 g Tris-base, 57.1 mL glacial acetic acid, 100 mL 0.5 M EDTA (pH 8.0) in 1 L of dH_2O.
6. 1.8% agarose gel in TAE buffer
7. DNA molecular weight marker ladder.

2.3.2 Equipment and Supplies

1. Thermal cycler.
2. Sterile 0.2 mL PCR tubes.

2.4 Materials for Subheading 3.4

2.4.1 Reagents

1. Bacterial strains
   (a) Donor strain (*E. coli* SM10/pFLP2^{ori1266})* *pFLP2^{ori1266} was created by inserting the broad host origin of replication from the cyptic plasmid pWH1266 [12] into the *Bam*HI restriction site to allow for replication in *Acinetobacter* spp.
   (b) Recipient strain (e.g., ATCC 17978::mini-Tn*7*T-LAC-Gm).
2. Sterile distilled water.
3. Antibiotics
   (a) Ampicillin (Ap): 50 mg/mL stock solution in water.
   (b) Gentamicin (Gm): 50 mg/mL stock solution in water.
   (c) Carbenicillin (Cb): 50 mg/mL stock solution in water.
4. LB Lennox broth: 10 g/L tryptone, 5 g/L yeast extract, 5 g/L NaCl.
5. LB Lennox agar plates: 10 g/L tryptone, 5 g/L yeast extract, 5 g/L NaCl, 15 g/L agar.

6. LB Lennox agar plates with 10% w/v sucrose: 10 g/L tryptone, 5 g/L yeast extract, 5 g/L NaCl, 15 g/L agar, 10% (w/v) sucrose.
7. LB Lennox agar Gm50 plates: 10 g/L tryptone, 5 g/L yeast extract, 5 g/L NaCl, 15 g/L agar, 50 μg/mL gentamicin.
8. LB Lennox agar Cb200 plates: 10 g/L tryptone, 5 g/L yeast extract, 5 g/L NaCl, 15 g/L agar, 200 μg/mL carbenicillin.
9. Simmons citrate agar Cb200 plates: 0.2 g/L $MgSO_4.7H_2O$, 0.2 g/L $NH_4H_2PO_4$, 0.8 g/L $Na_2NH_4PO_4$, 2 g/L $Na_3C_6H_5O_7$, 5 g/L NaCl, 0.08 g/L bromothymol blue, 15 g/L agar (final pH 7.0), 200 μg/mL carbenicillin.

2.4.2 Equipment and Supplies

1. Sterile culture tubes (13 mL).
2. Sterile microcentrifuge tubes.
3. Sterile nitrocellulose membrane.

3 Methods

3.1 Electroporation Method (Time Required: 4 Days)

3.1.1 Day 1: Inoculation (Time Required: ~10 min)

1. Inoculate *A. baumannii* ATCC17978 in 4 mL of LB broth in a sterile 13-mL culture tube, and incubate overnight at 37 °C with shaking at 250 rpm (*see* **Note 1**).
2. Place a bottle of sterile distilled water in the 4 °C refrigerator overnight to be used on the following day.

3.1.2 Day 2: Preparation of A. baumannii Electrocompetent Cells Followed by DNA Transformation (Time Required: ~3 h) (See Note 2)

1. Transfer 1.5 mL of the overnight bacterial culture into a microcentrifuge tube kept on ice, and centrifuge at 13,000 × *g* for 2 min.
2. Discard the supernatant carefully using a P1000 pipette without disturbing the cell pellet.
3. Repeat **steps 1** and **2** for the remaining culture volume using the same microcentrifuge tube (*see* **Note 3**).
4. Add 1 mL of cold sterile distilled water that was stored in the 4 °C refrigerator the previous day into the microcentrifuge tube with the final cell pellet.
5. Resuspend the cell pellet using the P1000 pipette until the cell pellet is visibly removed from the bottom of the microcentrifuge tube (*see* **Note 4**).
6. Centrifuge the resuspended cells at 13,000 × *g* for 2 min, and return the tubes back onto ice.
7. Carefully remove the supernatant using a P1000 pipette making sure that the cell pellet remains undisturbed.

8. Repeat the washing **steps 4** through 7 three times, taking extra care to keep the cell pellets undisturbed when discarding the supernatant (*see* **Note 5**).
9. After the washing steps, resuspend the final pellet from 4 mL of initial culture in 200 μL of cold sterile distilled water.
10. Aliquot 100 μL of the cells from the above step into a separate sterile microcentrifuge tube sitting on ice to serve as the negative control (*see* **Note 6**).
11. To the microcentrifuge tube containing 100 μL of electrocompetent cells, add 100–200 ng each of pTNS2 and pUC18T-mini-Tn*7*T-LAC-Gm delivery vector, and mix by gentle fingertip tapping to ensure complete mixing of the plasmids with the electrocompetent cells (*see* **Note** 7).
12. Add the corresponding volume of sterile distilled water into the negative control, and incubate tubes on ice for 20 min.
13. After the incubation period, transfer the entire DNA-cell mixture into a cold electroporation cuvette and return the cuvette onto ice.
14. Electroporate using the following settings: 25 μF, 200 Ω, and 2.0 kV for both the mini-Tn*7* delivery vector carrying cuvette and the negative control.
15. After electroporation, immediately add 1 mL of LB broth (pre-warmed to 37 °C) to the cuvette containing the DNA-cell mixture, and transfer the entire volume into a fresh microcentrifuge tube (*see* **Notes 8** and **9**).
16. Incubate the electroporated cells at 37 °C with shaking at 250 rpm for 1 h to facilitate the recovery of the cells.
17. In the meantime, pre-warm three LB agar Gm50 plates for 30–45 min in a 37 °C static incubator.
18. After 1 h of incubation, spread plate 100 μL of the electroporated cells onto a pre-warmed LB agar Gm50 plate, and incubate for 16–18 h at 37 °C (*see* **Note 10**).
19. The remaining cells from the recovery tube should be centrifuged at 13,000 × *g* for 2 min to collect all cells as a pellet.
20. Carefully remove the supernatant and resuspend the cell pellet in 100 μL LB broth. Spread the entire 100 μL onto a fresh LB agar Gm50 plate.
21. Centrifuge the negative control tube at 13,000 × *g* for 2 min and discard the supernatant.
22. Resuspend the cell pellet in 100 μL LB broth, and plate the entire 100 μL on a fresh LB agar Gm50 plate and incubate for 16–18 h at 37 °C.

3.1.3 Day 3: Selecting Colonies for Screening (Time Required: ~15 min)

1. Pick single colonies from the LB agar Gm50 plates, and patch them onto a fresh LB agar Gm50 plate.
2. Incubate for 16–18 h at 37 °C.

3.1.4 Day 4: PCR Screening for Positive Colonies (Time Required: ~2 h)

Carry out the PCR screening for the selected colonies using the protocol described below.

3.2 Four-Parental Conjugation Method (Time Required: 6 Days)

3.2.1 Day 1: Inoculation (Time Required: ~15 min)

1. Inoculate the four strains listed below in 3 mL LB broth in a 13-mL culture tube (with appropriate antibiotic, if applicable, at concentrations indicated in parentheses), and incubate overnight at 37 °C with shaking at 250 rpm.
   (a) *A. baumannii* ATCC17978.
   (b) *E.coli* DH5α/pUC18T-mini-Tn7T- LAC-Gm (100 μg/mL ampicillin).
   (c) *E.coli* DH5α/pTNS2 (100 μg/mL ampicillin).
   (d) *E.coli* HB101/pRK2013 (35 μg/mL kanamycin).

3.2.2 Day 2: Setting Up the Conjugation (Time Required: ~40 min)

1. Place a sterile 2 × 2 cm nitrocellulose filter paper onto a LB agar plate, and place the LB agar plate in a 37 °C static incubator to pre-warm the plate before the following steps:
   (a) Remove the culture tubes from the incubator, and transfer 100 μL of each strain into a 1.5 mL microcentrifuge tube containing 600 μL of fresh LB broth (the total volume of mixture of the cultures with fresh LB is 1 mL).*

   *The antibiotics present in the overnight cultures do not affect the cells in the mixture as the antibiotics are diluted beyond their effective concentration. However, it is important to carry out the following washing steps immediately after to remove the antibiotics from the mixture completely.
   (b) Centrifuge at 13,000 × *g* for 2 min to pellet the cells to the bottom of the microcentrifuge tube.
   (c) Carefully remove the supernatant using a P1000 pipette making sure not to disturb the cell pellet.
   (d) After removing the supernatant, add 1 mL of fresh LB broth into the microcentrifuge tube and resuspend the pellet.
   (e) Centrifuge the tube at 13,000 × *g* for 2 min to pellet the cells.
   (f) Repeat the washing **steps 4–6** three times to remove remaining antibiotics from the overnight cultures.

(g) After the washing steps, resuspend the cell pellet in 10 μL of LB broth making sure not to introduce air bubbles.

(h) Retrieve the pre-warmed LB agar plate with the nitrocellulose filter paper from the incubator, and spot the 10 μL cell suspension onto the center of the nitrocellulose paper (*see* **Note 11**).

(i) Let the spot containing the conjugation mixture to air-dry at room temperature, and then place the inverted LB agar plate in a 30 °C static incubator.

(j) Incubate for 20–24 h at 30 °C.

3.2.3 Day 3: Plating the Conjugation Mixture (Time Required: ~30 min)

1. Place three Simmons citrate agar Gm50 plates in the 37 °C incubator to pre-warm the plates before proceeding with the following steps:

   (a) Remove the LB agar plate containing the conjugation mixture from the 30 °C incubator.

   (b) Carefully remove the nitrocellulose paper from the LB agar plate using sterile forceps.

   (c) Cut the extra nitrocellulose filter paper from the edges so that the filter paper containing the conjugation mixture fits into a 1.5 mL microcentrifuge tube.

   (d) Place the nitrocellulose filter paper into a clean 1.5 mL microcentrifuge tube, and add 300 μL of sterile 0.85% saline solution into the microcentrifuge tube.

   (e) Let stand for 5–7 min at room temperature to facilitate the suspension of cells into the saline.

   (f) After 5–7 min, gently resuspend the cells using a P200 pipette making sure there are no cells left attached to the nitrocellulose membrane. Alternatively, the tube can be gently vortexed to remove cells from the nitrocellulose membrane.

   (g) Remove the Simmons citrate Gm50 plates placed in the incubator earlier, and add 100 μL of the conjugation mixture suspended in saline to the center of each plate.

   (h) Spread the cells using a cell spreader and incubate for 42–48 h at 37 °C (*see* **Note 12**).

3.2.4 Day 5: Selecting Colonies for Screening (Time Required: ~15 min)

1. Pick single colonies from the Simmons citrate Gm50 plate, and patch them onto a fresh LB agar Gm50 plate.
2. Incubate for 16–18 h at 37 °C.

3.2.5 Day 6: PCR Screening for Positive Colonies (Time Required: ~3 h)

Carry out the PCR screening for the selected colonies using the protocol mentioned below.

3.3 Screening for Mini-Tn7 Insertion by Colony PCR (Time Required: ~3 h)

1. Remove the LB agar Gm50 plates containing the patched colonies from the incubator.
2. Select the colonies that are to be PCR screened, and using a sterile toothpick, remove a small portion of the patch.
3. Mix the removed portion of the patch in 20 μL of sterile distilled water in a 0.2 mL PCR tube (*see* **Note 13**).
4. Incubate the PCR tubes containing the bacterial suspension at 100 °C for 10 min using a thermal cycler.
   (a) This will serve as the template DNA for the PCR reaction as mentioned below.
5. Prepare a master mix following the recipe below based on the number of PCR reactions required. Each PCR reaction constitutes a 25 μL volume with the following components, and prepare a negative control reaction by using all the following components except template DNA which should be substituted with dH_2O:

   2.5 μL *Taq* 10X PCR buffer

   0.5 μL dNTPs (final concentration of 200 μM)

   0.5 μL forward primer (ABglmS_F) (final concentration of 0.18 μM).

   0.5 μL reverse primer (Tn*7*R) (final concentration of 0.18 μM).

   1 unit of *Taq* DNA polymerase.

   1 μL template DNA.

   dH_2O to a total of 25 μL reaction volume.

6. Run the PCR reaction using a thermal cycler using the following conditions:

   Initial denaturation at 95 °C for 2 min 1×

   Denaturation at 95 °C for 30 s
   Annealing at 49 °C for 30 s — 30×
   Extension at 72 °C for 30 s

   Final extension at 72 °C for 10 min 1×

   Hold at 12 °C

   (a) The expected band size is 368 bp. Therefore, use a 1.8% agarose gel to visualize the positive bands (Fig. 3).

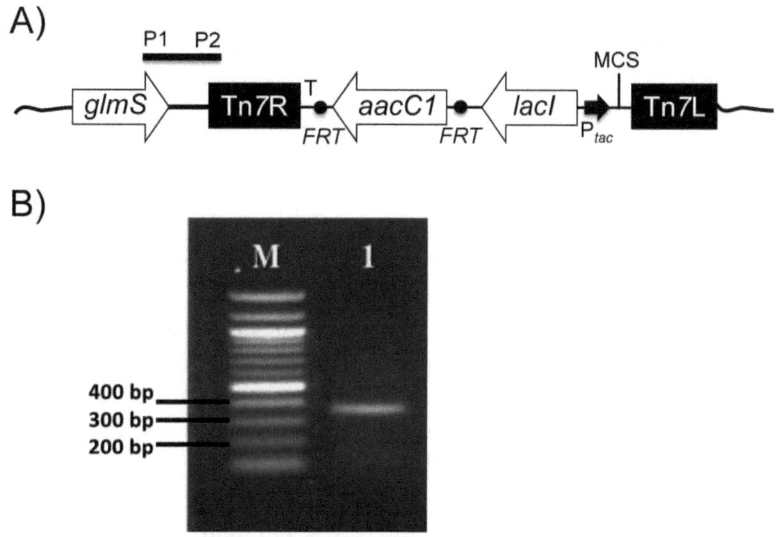

Fig. 3 Insertion of mini-Tn*7*T-LAC-Gm in *A. baumannii*. (**a**) Map of mini-Tn*7*T-LAC-Gm inserted at the *A. baumannii att*Tn*7* site. The horizontal bar indicates the PCR fragment amplified with *A. baumannii glmS*-specific primer ABglmS_F (P1) and Tn*7*R-specific primer (P2). Other genes and other features are explained in the Fig. 2 legend. Genes and other features are not drawn to scale. (**b**) Agarose gel electrophoresis of the PCR product from the screening of the mini-Tn*7*T insertion in *A. baumannii*. The expected band for mini-Tn*7*T insertion is 368 bp shown in lane 1. The lane marked with "M" is a 100 bp molecular weight marker

3.4 Removal of the Gm^R Marker

3.4.1 Day 1: Inoculation (Time Required: 15 min)

1. Inoculate the two strains listed below in 3 mL LB broth in a 13-mL culture tube (with appropriate antibiotic, if applicable, at concentrations indicated in parentheses), and incubate overnight at 37 °C with 250 rpm shaking.
   (a) ATCC 17978::mini-Tn*7*T-LAC-Gm (50 μg/mL gentamicin, optional).
   (b) *E. coli* SM10/pFLP2^{ori1266} (100 μg/mL ampicillin).

3.4.2 Day 2: Setting Up the Conjugation (Time Required: ~40 min)

1. Set up a conjugation as described in Subheading 3.2 using the recipient (ATCC 17978::mini-Tn*7*T-LAC-Gm) and the donor (*E. coli* SM10/pFLP2^{ori1266}) strains.

3.4.3 Day 3: Plating the Conjugation Mixture (Time Required ~30 min)

1. After the overnight incubation, remove the LB agar plates from the 30 °C incubator, and proceed to plate the conjugation mixture as described in Subheading 3.2 on Simmons citrate Cb200 plates.
2. Incubate the Simmons citrate agar plates for 40–48 h at 37 °C or until colonies appear.

3.4.4 Days 5 and 6: Selecting Isolated Colonies Transformed with pFLP2 and Loss of Gm^R Marker (Time Required: ~15 min Each Day)

1. Pick single colonies from the Simmons citrate agar plate, and crosspatch them onto LB Cb200 and LB Gm50 plates and incubate for 16–18 h in a 37 °C incubator.
2. After the incubation period, remove the plates from the incubator, and screen for patches that grow on the LB Cb200 plate but not on the LB Gm 50, i.e., Cb-resistant but Gm-susceptible patches, as that would indicate the loss of Gm^R marker (*see* **Note 14**).
3. Streak out the Cb-resistant but Gm-susceptible patches onto a LB agar plate supplemented with 10% (w/v) sucrose, and incubate for 16–18 h in the 30 °C incubator. (Addition of sucrose to the growth media provides counterselection based on the levansucrase toxicity conferred by the *sacB* gene in pFLP2.)

3.4.5 Day 7: Screening Colonies for the Loss of pFLP2^{ori1266} Plasmid (Time Required ~15 min)

1. Remove the LB agar +10% sucrose plates from the incubator, and observe for isolated colonies.
2. Pick the isolated colonies using a sterile toothpick, and cross-patch them onto LB Cb200 and LB agar plates.
3. Incubate the plates at 37 °C for 16–18 h, and select the colonies that are carbenicillin sensitive. These colonies represent the cells with unmarked mini-Tn*7*T insertions.

3.5 Applications

Mini-Tn*7*-based systems can be used for diverse genetic manipulations, including complementation, fluorescent protein tagging, as well as for reporter assays in diverse Gram-negative bacterial species, including *A. baumannii* [4, 13]. Genetic complementation achieved using mini-Tn*7* plasmids is likely to provide biologically relevant data because of the single-copy insertion of the gene. Expression of the gene can be tightly regulated using the appropriate promoter. We illustrate this using an example shown in Fig. 4 and Table 1. We deleted *adeN* that encodes a transcriptional repressor of the AdeIJK pump in *A. baumannii* ATCC17978 [14]. Upon the deletion of *adeN*, the expression of AdeIJK pump is derepressed (Fig. 4b). We then inserted *adeN* using the mini-Tn*7* plasmid in *A. baumannii* AB141. The expression of this gene being under the control of P_{tac} promoter can be induced using 1 mM IPTG as seen in Fig. 4a. By inducing the expression of *adeN*, we were able to suppress the expression of AdeIJK pump to the wild-type levels (Fig. 4b). There is no suppression of *AdeIJK* observed in the absence of *adeN* expression. Utility of this system is also illustrated in Table 1, where we observe that the susceptibility of AB141 to the antibiotics that are a substrate of the AdeIJK pump (moxifloxacin and ciprofloxacin) is restored to the wild-type levels when *adeN* expression is induced, whereas the susceptibility to gentamicin remains unchanged as it is not a substrate of the AdeIJK. Therefore,

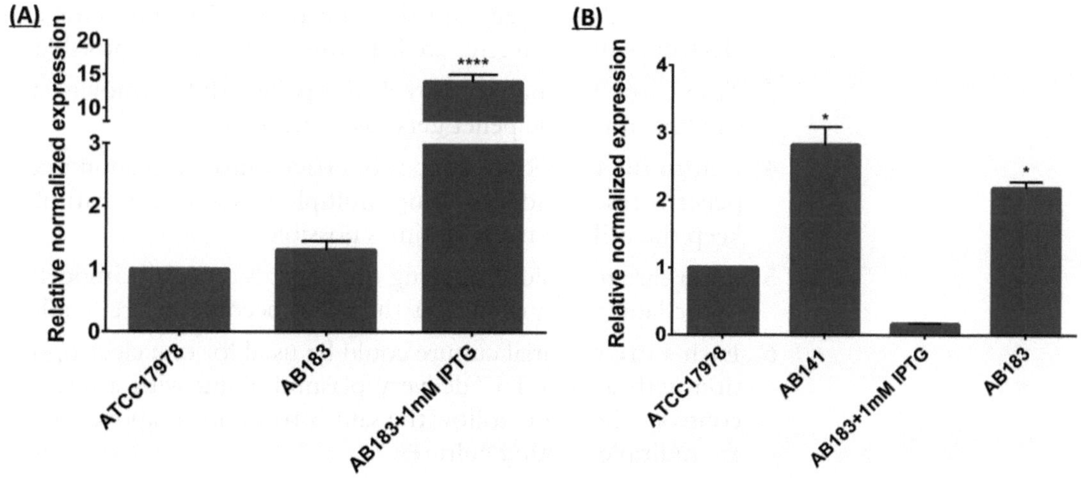

Fig. 4 Relative normalized expression of *adeN* complemented in *A. baumannii* Δ*adeN* strain (AB141) using a mini-Tn*7*T-LAC vector expressing the *adeN* gene from the IPTG-inducible P_{tac} yielding strain AB183 (**a**). *adeN* expression in AB183 was induced using 1 mM IPTG. Relative normalized expression of *adeJ*, the RND component of AdeIJK RND efflux pump showing the repression caused by *adeN* complementation (**b**). ****$P < 0.0001$, *$P < 0.05$ (statistical analyses were carried out with GraphPad Prism software v6.07 [La Jolla, CA, USA] using two-way ANOVA to determine the statistical significance)

Table 1
Antibiotic susceptibility of *A. baumannii* strains

Strain	Characteristic	Minimum inhibitory concentration (μg/mL)		
		Moxifloxacin	Ciprofloxacin	Gentamicin
ATCC17978	Type strain	0.032	0.094	0.5
AB141	ATCC17978:Δ*adeN*	0.094	0.19	1.0
AB183	AB141::mini-Tn*7*T-P_{tac}-*adeN*	0.094	0.19	1.0
AB183 (+1 mM IPTG)		0.016	0.032	1.0

this system can be effectively used to characterize proteins, such as RND efflux pumps, whose expression from a multi-copy vector is often not well tolerated by bacteria. A mini-Tn*5/7-lux* element has been shown to be a useful tool for promoter identification and capture in *A. baumannii* [13].

4 Notes

1. 4 mL of culture is sufficient for electroporating one mini-Tn*7*-based plasmid along with a negative control. Therefore, determine the appropriate number of culture tubes to inoculate based on the number of plasmids.

2. It is critical to make sure that the bacterial suspensions and electroporation cuvettes are kept on ice at all times possible.
3. Take extra care not to disturb the pellet while removing the supernatant as the pellet gets easily dislodged.
4. Return the microcentrifuge tubes back onto ice as soon as cell pellet is resuspended if using multiple tubes as it is critical to keep the cells on ice at all times possible.
5. Extra care to avoid disturbing the pellet is especially important in the later washing steps as the pellet becomes more fragile.
6. Each 4 mL of initial culture could be used for one electroporation with a mini-Tn*7* delivery plasmid along with a negative control. Therefore, follow the same procedure as above if there are multiple starting cultures.
7. Do not use more than a combined volume of 5 μL of the plasmids, and take extra care to make sure the plasmids are properly mixed with the cell suspension avoiding air bubbles.
8. It is critical to add the pre-warmed LB broth to the electroporation cuvette as it facilitates the recovery of the electrocompetent cells from the electric shock.
9. Make sure to check the time constant displayed on the electroporator for each round of electroporation (a time constant of 5–6 is optimal).
10. Electroporation efficiencies can vary with different strains. When using a strain for the first time, it is advisable to try multiple dilutions to determine the optimal dilution necessary to obtain isolated colonies.
11. Pipette the cell suspension gently and carefully onto the nitrocellulose paper to avoid any splashing.
12. As an alternative to four-parental mating, one can also employ a triparental mating strategy in which the pRK2013 helper plasmid is not needed and the citrate metabolic counterselection against *E. coli* strains is replaced with metabolic counterselection on rich medium. This method employs the SM10(λ*pir*)-based strain RHO3 that was engineered to contain an *asd* deletion that makes the strain auxotrophic for diaminopimelic acid (DAP) [15]. DAP is absent from virtually all bacterial growth media, including rich media like LB. Triparental conjugation mixtures containing the recipient strain and the RHO3 strains containing the mini-Tn*7* delivery and pTNS2 helper plasmids are incubated on a filter on an LB plate with DAP. For transformant selection, the cell mixture is then washed free of DAP with LB medium and plated on LB agar with the appropriate antibiotic [13].

13. Distilled water in the PCR tube should become visually turbid as the cells are resuspended, and it should be an indication of proper mixing of the cells.
14. The typical frequency of the loss of Gm^R marker when using pFLP2 was generally 100%; as we observe, almost all the colonies are Gm-sensitive.

Acknowledgments

This work was supported by grants from the Natural Sciences and Engineering Council of Canada (NSERC, 2015-05550) and the Canadian Institutes of Health Research (CIHR, 201610PJT-376486) to AK. Original work in the HPS laboratory was funded in part by the National Institute of Allergy and Infectious Diseases of the National Institutes of Health under Award Number U54 AI065357. The content is solely the responsibility of the authors and does not necessarily represent the official views of the National Institutes of Health.

References

1. Clark NM, Zhanel GG, Lynch JP 3rd (2016) Emergence of antimicrobial resistance among *Acinetobacter* species: a global threat. Curr Opin Crit Care 22:491–499
2. World Health Organization (2017) Global priority list of antibiotic-resistant bacteria to guide research, discovery, and development of new antibiotics. In: WHO. http://www.who.int/medicines/publications/global-priority-list-antibiotic-resistant-bacteria/en/#.WpBPsTSZ1qo.mendeley. Accessed 23 Feb 2018
3. Biswas I (2015) Genetic tools for manipulating *Acinetobacter baumannii* genome: an overview. J Med Microbiol 64:657–669
4. Kumar A, Dalton C, Cortez-Cordova J, Schweizer HP (2010) Mini-Tn7 vectors as genetic tools for single copy gene cloning in *Acinetobacter baumannii*. J Microbiol Methods 82:296–300. https://doi.org/10.1016/j.mimet.2010.07.002
5. Choi K-H, Gaynor JB, White KG et al (2005) A Tn7-based broad-range bacterial cloning and expression system. Nat Methods 2:443
6. Choi K-H, Schweizer HP (2006) mini-Tn7 insertion in bacteria with single attTn7 sites: example *Pseudomonas aeruginosa*. Nat Protoc 1:153
7. Peters JE, Craig NL (2001) Tn7: smarter than we thought. Nat Rev Mol Cell Biol 2:806
8. Choi K-H, DeShazer D, Schweizer HP (2006) mini-Tn7 insertion in bacteria with multiple glmS-linked attTn7 sites: example *Burkholderia mallei* ATCC 23344. Nat Protoc 1:162
9. Choi K-H, Mima T, Casart Y et al (2008) Genetic tools for select-agent-compliant manipulation of *Burkholderia pseudomallei*. Appl Environ Microbiol 74:1064–1075. https://doi.org/10.1128/AEM.02430-07
10. Choi K-H, Schweizer HP (2006) mini-Tn7 insertion in bacteria with secondary, non-*glmS*-linked *att*Tn*7* sites: example *Proteus mirabilis* HI4320. Nat Protoc 1:170
11. Waddell CS, Craig NL (1989) Tn*7* transposition: recognition of the *att*Tn*7* target sequence. Proc Natl Acad Sci U S A 86:3958–3962
12. Hunger M, Schmucker R, Kishan V, Hillen W (1990) Analysis and nucleotide sequence of an origin of an origin of DNA replication in *Acinetobacter calcoaceticus* and its use for *Escherichia coli* shuttle plasmids. Gene 87:45–51. https://doi.org/10.1016/0378-1119(90)90494-C
13. Bruckbauer ST, Kvitko BH, Karkhoff-Schweizer RR, Schweizer HP (2015) Tn5/7-

lux: a versatile tool for the identification and capture of promoters in Gram-negative bacteria. BMC Microbiol 15:17. https://doi.org/10.1186/s12866-015-0354-3

14. Rosenfeld N, Bouchier C, Courvalin P, Périchon B (2012) Expression of the resistance-nodulation-cell division pump AdeIJK in *Acinetobacter baumannii* is regulated by AdeN, a TetR-type regulator. Antimicrob Agents Chemother 56:2504–2510. https://doi.org/10.1128/AAC.06422-11

15. López CM, Rholl DA, Trunck LA, Schweizer HP (2009) Versatile dual-technology system for markerless allele replacement in *Burkholderia pseudomallei*. Appl Environ Microbiol 75:6496–6503. https://doi.org/10.1128/AEM.01669-09

Chapter 14

Distinguishing Colony Opacity Variants and Measuring Opacity Variation in *Acinetobacter baumannii*

Sarah E. Anderson and Philip N. Rather

Abstract

Many strains of *Acinetobacter baumannii*, including the highly virulent strain AB5075, undergo a high-frequency switch that results in two cell types that are distinguished by their opaque or translucent colony opacities when viewed by oblique lighting. Opaque (VIR-O) and translucent (AV-T) colonies exhibit multiple phenotypic differences, including virulence. Here we describe how to distinguish between VIR-O and AV-T colony variants and how to generate highly pure stocks of each variant. We also describe methods for measuring opacity switching frequencies of cells grown on agar plates and in liquid cultures.

Key words *Acinetobacter baumannii*, Opacity, Phenotypic switch, Opaque, Translucent

1 Introduction

Acinetobacter baumannii strain AB5075, as well as many clinical isolates, undergoes a high-frequency switch that generates two cell types distinguished by their colony opacity. Opaque (VIR-O) and translucent (AV-T) colonies of *A. baumannii* appear identical under room lighting, but are easily distinguished using a dissecting microscope with oblique illumination [1]. VIR-O and AV-T variants of AB5075 exhibit a number of distinct phenotypic properties, including differing levels of motility, biofilm formation, and virulence [1, 2]. While VIR-O variants are virulent in animal models, AV-T variants are unable to cause disease and are selected against in vivo [1, 2] Opacity switching in AB5075 occurs at a high rate in both liquid cultures and on agar plates [1, 2]. Although the mechanisms underlying opacity switching remain incompletely understood, it is thought that distinct mechanisms control the VIR-O to AV-T and AV-T to VIR-O switches, as mutations affecting switching frequencies in one direction often have no effect on frequencies in the other direction [3, 4]. Because the VIR-O and AV-T cells exhibit differential phenotypes, it is important to control

Indranil Biswas and Philip N. Rather (eds.), *Acinetobacter baumannii: Methods and Protocols*, Methods in Molecular Biology, vol. 1946, https://doi.org/10.1007/978-1-4939-9118-1_14, © Springer Science+Business Media, LLC, part of Springer Nature 2019

for opacity when comparing different strains of *A. baumannii* to ensure that differential phenotypes observed are actually due to the mutations being tested and not to differences in opacity between strains [5, 6]. Herein we describe how to distinguish opacity variants in *A. baumannii* (Subheading 3.1), how to generate highly pure stocks of each variant (Subheading 3.2) and two easy methods for quantifying switching frequencies from colonies and liquid cultures (Subheadings 3.3 and 3.4, respectively).

2 Materials

1. 0.5× LB agar: 2.5 g NaCl, 5 g tryptone, 2.5 g yeast extract, 8 g agar in 1 L distilled water.
2. Dissecting microscope with oblique illumination (Fig. 1).
3. LB broth: 5 g NaCl, 10 g tryptone, 5 g yeast extract in 1 L distilled water.
4. LB agar: 5 g NaCl, 10 g tryptone, 5 g yeast extract, 15 g agar in 1 L distilled water.

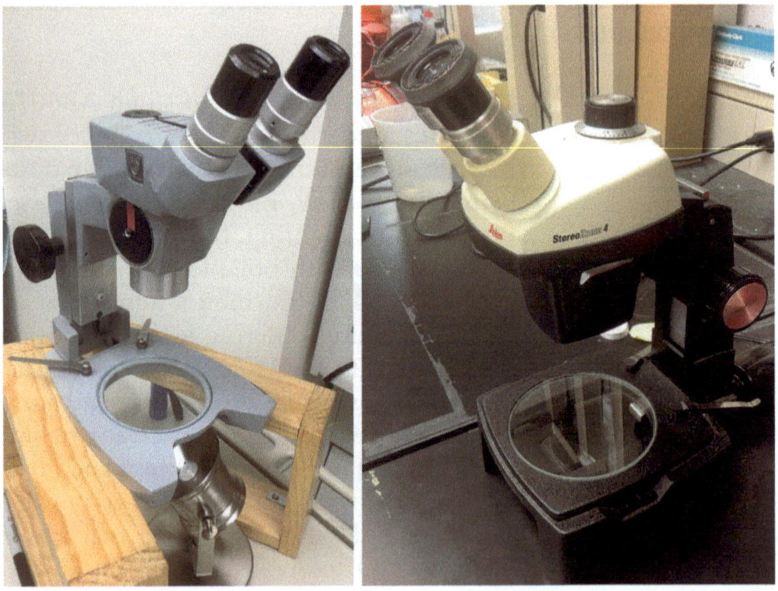

Fig. 1 Examples of stereo microscopes with oblique illumination that can be used to visualize colony opacity. Two separate microscopes that are used in our lab to visualize differences in colony opacity are shown. It is essential to have a light source with an angle that can be adjusted to visualize the colony opacity phenotypes

3 Methods

3.1 Distinguishing Opaque and Translucent Colony Types

1. Plate tenfold serial dilutions of a glycerol stock, resuspended colony, or culture of *A. baumannii* onto 0.5× LB so that individual colonies can be distinguished (*see* **Note 1**).
2. Incubate plates overnight at 37 °C.
3. Visualize plates on a dissecting microscope with oblique lighting (*see* **Note 2**). Colonies are easier to distinguish at lower magnification. The differences between opaque and translucent colony phenotypes are shown in Fig. 2.

3.2 Generating Pure Stocks of Opaque and Translucent Cells

1. Streak a stock of the strain of interest onto 0.5× LB.
2. Incubate plate overnight at 37 °C.
3. Select a well-isolated colony of the desired opacity phenotype (visualized using a dissecting microscope) and passage onto a 1× LB plate (*see* **Note 3**).
4. Incubate the plate at 37 °C for 6–8 h.
5. Using a dissecting microscope to visualize small colonies, pick 6–12 colonies and inoculate separately into LB broth (*see* **Note 4**).
6. Incubate cultures overnight at room temperature.
7. Cultures should be slightly turbid ($OD_{600} < 0.1$); if not shake for a short period of time (*see* **Note 5**).

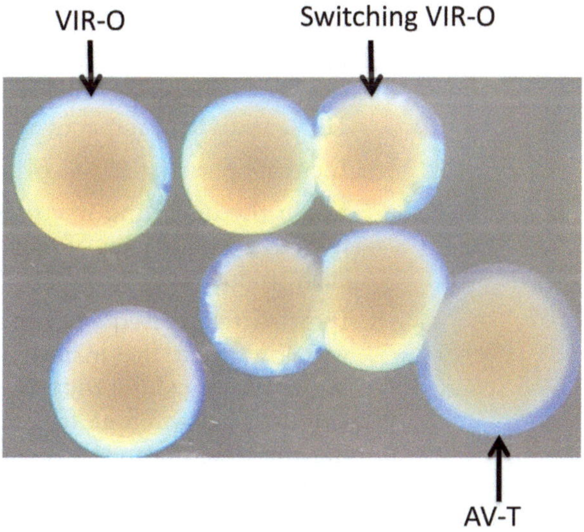

Fig. 2 Examples of translucent (AV-T), opaque (VIR-O), and switching VIR-O colonies. A 24-h 0.5× LB agar plate with a mixture of VIR-O, AV-T, and VIR-O colonies that are switching to AV-T variants is shown. AV-T colonies typically do not sector

8. Using a sterile loop, streak cells from each tube onto a 0.5× LB plate to assess purity. Store broth cultures from **step 5** above at 4 °C overnight, or make glycerol stocks of all cultures and freeze at −80 °C.
9. Incubate plates overnight at 37 °C.
10. Using a dissecting microscope, visualize streaks of cultures to check for purity of the opacity phenotype. Only cultures that have an estimated purity of greater than 99% should be selected for further use. If broth cultures were stored overnight at 4 °C, use sufficiently pure cultures to make a glycerol stock and freeze at −80 °C. If all cultures were used to make glycerol stocks in **step 8**, dispose of stocks derived from cultures that are insufficiently pure.

3.3 Measuring Switching Frequencies of Individual Colonies

1. Plate tenfold serial dilutions of pure VIR-O or AV-T glycerol stocks of the strains of interest onto 0.5× LB so that individual colonies can be isolated (*see* **Note 6**).
2. Incubate plates for the desired length of time at 37 °C (*see* **Note 7**).
3. Isolate individual colonies by using a sterile toothpick to cut the agar around the colony so that a small section of agar containing the complete colony can be lifted from the plate. Transfer individual colonies to tubes containing 1 mL LB broth and vortex to resuspend cells. Typically six individual colonies should be tested for each strain (*see* **Note 8**).
4. Perform tenfold serial dilutions with the isolated colonies down to a dilution of 10^{-6}.
5. Plate 100 μL of the 10^{-3}, 10^{-4}, and 10^{-6} serial dilutions for each colony onto 0.5× LB to obtain final dilutions of 10^{-4}, 10^{-5}, and 10^{-7}, respectively.
6. Incubate plates at 37° C overnight.
7. Using the dissecting microscope with oblique lighting, determine the frequencies of opaque and translucent colonies from each strain.
8. If the switching frequency of the original colony taken in **step 3** was low, then the majority of the colonies from the serial dilutions will exhibit the same opacity phenotype as the original colony. The switching frequency can be determined by counting the total number of colonies of the opposite opacity phenotype on the 10^{-4} or 10^{-5} dilution plates. The total number of colonies of both phenotypes on the 10^{-7} plate should also be counted. The switching frequency would then be given by the following equation:

$$\text{Frequency} = \frac{(\text{total colonies opposite phenotype})(\text{plate dilution factor})}{(\text{total colonies both phenotypes})(10^7)} \times 100$$

9. If the switching frequency of the original colony taken in **step 3** was high (>10%), the switching frequency can more easily be determined by counting the number of opaque and translucent colonies on sectors of the 10^{-5} plate. Sectors of the plate should be chosen that exhibit clearly defined colonies that can be easily counted. To ensure that colonies are not counted multiple times, crush colonies with a toothpick as they are being counted. At least 100 total colonies should be enumerated to ensure accuracy. Using this counting method, the switching frequency would be.

$$\text{Frequency} = \frac{\text{Total colonies oppposite phenotype}}{\text{Total colonies both phenotypes}} \times 100\%$$

3.4 Measuring Switching Frequencies from Liquid Cultures

1. Inoculate cultures directly from highly pure frozen glycerol stocks into LB broth (*see* **Note 9**).
2. Incubate cultures overnight at room temperature (*see* **Note 10**).
3. Grow cultures at 37 °C in a shaking incubator until they reach the desired OD (*see* **Note 11**).
4. Perform tenfold serial dilutions of cultures. Typically a lowest dilution of 10^{-6} is sufficient, depending upon the density of the culture.
5. Plate 100 μL of serial dilutions on 0.5× LB plates. Depending on the density of the original culture, final dilutions of 10^{-4} through 10^{-7} are typically sufficient.
6. Incubate 0.5× LB plates overnight at 37 °C.
7. Enumerate colonies and determine switching frequencies as in **steps 7–9** of Subheading 3.3.

4 Notes

1. Colony density has a profound effect on the ability to distinguish VIR-O and AV-T colonies. For unknown reasons, VIR-O and AV-T colonies are almost impossible to distinguish at low colony density. The easiest way to ensure that colonies can be easily distinguished is to plate serial dilutions yielding a range of colony densities. Similarly, it is also very hard to distinguish VIR-O and AV-T colonies grown on 1× LB plates, so it is important that 0.5× LB is used.
2. Even with oblique illumination, it can initially be difficult for the untrained eye to distinguish opaque and translucent

colonies. However, adjusting the angle of the light source can often dramatically improve visualization of the two colony types. It is also important to have VIR-O and AV-T stocks as a reference.

3. For unknown reasons, colonies often switch less readily on 1× LB than on 0.5× LB plates. Therefore, when trying to generate pure VIR-O or AV-T stocks, it is better to use 1× LB so that colonies will not switch while growing.
4. It is important to use very young and small colonies to inoculate when generating stocks, as switching occurs in a density-dependent manner [1]. Picking multiple colonies increases the chances that at least one culture will be mostly pure and of the desired opacity phenotype (6–8 h colonies are too small to distinguish colony opacity).
5. Because switching occurs at higher cell densities, it is important not to grow cultures past early-log phase, i.e., barely turbid to the naked eye.
6. For unknown reasons, plate thickness and moisture content can affect switching frequencies. Therefore, it is important to use plates that are roughly the same thickness and dryness across experiments to ensure reproducibility of results. We typically perform switching assays with plates made using 20 mL of 0.5× agar. Best results are also obtained if all plates used in a single experiment were generated from the same batch of media.
7. We typically perform switching assays after 24 or 48 h, but other incubation times may be used. It is important to keep the length of incubation consistent between experiments, as cell density and the age of the colony can affect switching.
8. For unknown reasons, colony density can affect the degree of switching, particularly for VIR-O colonies. Colonies at high density often switch less and are more difficult to cut out of plates. Therefore, it is typically preferable to pick colonies off of very low-density plates for switching assays. Plating serial dilutions in **step 1** allows a range of colony densities to be obtained such that colonies from similar densities can be used across all strains.
9. It is important to culture cells directly from pure stocks rather than streaking stocks onto plates and inoculating from isolated colonies. Switching will occur once colonies are grown up on plates, so the inoculum from a plate will no longer be pure. Since no stock is 100% pure, it is generally advisable to qualitatively compare that the purity of stocks to be compared are similar, so that observed differences in switching are not actually due to differences in the purity of the inoculum.

10. Minimal switching should occur during incubation at room temperature overnight. To ensure that cultures are similarly pure before growing at 37 °C, a loopful of each culture can be struck onto a 0.5× LB plate to qualitatively compare purity. If there are concerns about cultures switching differently during the room temperature incubation, switching frequencies can also be quantified from overnight cultures. Switching following incubation at 37 °C could then be compared to the level of overnight switching for each sample using a fold change.
11. The OD_{600} of the overnight cultures at room temperature should be roughly similar so that the cultures spend comparable amounts of time shaking before the desired OD is reached. Typically cultures should be grown at 37 °C shaking to an OD_{600} between 1.1 and 1.8 to ensure that appreciable switching will have occurred. It is important that all cultures should be taken at a similar OD and that cultures should not be grown past the desired OD and then back-diluted, as this will affect the switching frequency.

Acknowledgment

This work is supported by the following awards to P.N.R., VA Merit Award I01 BX001725 and Research Career Scientist Award IK6BX004470, both from the Department of Veterans Affairs, and R01AI072219 from the National Institutes of Health. S.E.A. was supported by a T32 training grant AI106699 from the National Institutes of Health.

References

1. Tipton KA, Dimitrova D, Rather PN (2015) Phase-variable control of multiple phenotypes in Acinetobacter baumannii strain AB5075. J Bacteriol 197(15):2593–2599. https://doi.org/10.1128/jb.00188-15
2. Chin CY, Tipton KA, Farokhyfar M, Burd EM, Weiss DS, Rather PN (2018) A high-frequency phenotypic switch links bacterial virulence and environmental survival in Acinetobacter baumannii. Nat Microbiol 3(5):563–569. https://doi.org/10.1038/s41564-018-0151-5
3. Tipton KA, Farokhyfar M, Rather PN (2017) Multiple roles for a novel RND-type efflux system in Acinetobacter baumannii AB5075. Microbiologyopen 6(2):e00418. https://doi.org/10.1002/mbo3.418
4. Tipton KA, Rather PN (2017) An ompR/envZ two-component system ortholog regulates phase variation, osmotic tolerance, motility, and virulence in Acinetobacter baumannii strain AB5075. J Bacteriol 199(3). https://doi.org/10.1128/jb.00705-16
5. Harding CM, Tracy EN, Carruthers MD, Rather PN, Actis LA, Munson RS Jr (2013) Acinetobacter baumannii strain M2 produces type IV pili which play a role in natural transformation and twitching motility but not surface-associated motility. MBio 4(4). https://doi.org/10.1128/mBio.00360-13
6. Clemmer KM, Bonomo RA, Rather PN (2011) Genetic analysis of surface motility in Acinetobacter baumannii. Microbiology 157 (Pt 9):2534–2544. https://doi.org/10.1099/mic.0.049791-0

Chapter 15

A Simple Static Biofilm Assay for *Acinetobacter baumannii*

Indranil Biswas and Joshua Mettlach

Abstract

Acinetobacter baumannii and other *Acinetobacter* spp. display biofilm formation on abiotic surfaces. The degree of biofilm formation varies considerably depending on the isolates. While culturing, bacteria tend to attach to abiotic surfaces, and some surfaces perform better than others. Numerous methods are available to study in vitro biofilm formation by bacteria under static growth, and they basically rely upon culturing the bacteria in an ampule followed by staining the bacterial growth attached to the abiotic surface with a suitable dye such as crystal violet to visualize the biofilm. Subsequently, the dye is eluted with a suitable solvent, and optical density is measured to quantify the attached bacterial mass. In this chapter, we provide a basic and simple yet a powerful method to study biofilm formation by *A. baumannii* grown under static conditions.

Key words *Acinetobacter baumannii*, Biofilm formation, Static biofilm, Air-liquid interface

1 Introduction

Like many other bacteria, *Acinetobacter baumannii* display a robust biofilm formation on many abiotic surfaces. The ability to form biofilm on fomites and other medical devices is an important virulence trait of this pathogen [1]. Biofilm formation appears to facilitate *A. baumannii* to survive under desiccation and nutritional limiting conditions. This trait also explains why some clinical isolates are more successful in hospitals than others. Biofilm formation is a multifactorial process involving proteins, polysaccharides, and extracellular nucleic acids [2]. In the case of *A. baumannii*, appendages such as fimbriae encoded by the cus operon are the major player for initial attachment during biofilm development [3]. Two other proteins, BAP (biofilm-associated protein) and the major outer membrane protein OmpA, are also very important [4, 5]. Other extracellular appendages are also required for maintenance of biofilm structures such as those formed at the air-liquid interface. In addition, extracellular matrix polysaccharide (PBNA)

Indranil Biswas and Philip N. Rather (eds.), *Acinetobacter baumannii: Methods and Protocols*, Methods in Molecular Biology, vol. 1946, https://doi.org/10.1007/978-1-4939-9118-1_15,

helps to mature the biofilm by providing adhesion between bacterial cells to generate a multilayered structure [6].

In vitro, *A. baumannii* is able to attach to a variety of abiotic surfaces including polystyrene, polypropylene, glass, and metal surfaces such as titanium. Biofilm formation can be studied under static or continuous-flow conditions [7]. There are several formats for static biofilm cultivations such as stand-alone individual tubes, microtiter plates (various formats), or colony biofilm grown on filter disks. Continuous-flow biofilms can be grown in a chemostat (such as a CDC reactor) or under drip-flow conditions. However, static biofilm growth offers certain advantages over continuous-flow growth in respect to analyzing early attachment events and can be easily performed using common laboratory supplies. Static biofilm can be adapted to high-throughput studies especially when grown in microtiter plates [8].

In this chapter, we provide a simple stand-alone static biofilm assay that can be performed in any standard microbiological laboratory [9]. This stand-alone biofilm cultivation method offers some advantages over microtiter plate methods. Since relatively larger culture volume is used (>2.0 mL), the biomass obtained by this "macro" growth method provides sufficient materials for downstream proteomic and transcriptomic studies. Cultures grown in larger volumes also increase reproducibility. Furthermore, some *A. baumannii* isolates form pellicles due to air-liquid biofilm formation [6], and the stand-alone method offers easy harvest of these pellicles from the culture tubes for downstream studies.

2 Materials

Use ultrapure deionized water while preparing the solutions and media. After preparation of solutions, autoclave at 120 °C for 30 min prior to use to maintain sterility. Store the media and the solutions at room temperature. Consider all the waste as biohazardous and dispose appropriately.

1. Nutrient Broth (NB): 0.75 g beef extract, 1.25 g peptone, 250 mL water. Add 1.5 gm agar to 100 mL NB to prepare NBA plate (*see* **Note 1**).
2. Phosphate-buffered saline (PBS): dissolve 8 g NaCl, 0.2 g KCl, 1.44 g Na_2HPO_4, and 0.24 g KH_2PO_4 in 800 mL water, adjust pH to 7.4 with HCl, and bring the final volume to 1 L with water.
3. Crystal violet stain, 0.1% (CV): dissolve 0.1 g in 20 mL of 95% ethanol and then add 80 mL water (*see* **Note 2**).
4. 95% ethanol.
5. 30% acetic acid.

6. Polystyrene tube: 75 mm × 12 mm, sterile (*see* **Note 3**).
7. Absorbent paper (any kind).
8. Autoclavable bottle: 500 mL.
9. Sterile disposable serological pipettes: 10 mL.
10. Sterile disposable pipette tips: 200 μL, 1 mL.
11. Pipette boy (pipette controller).
12. Pipettes: P200, P1000.
13. Bunsen burner.
14. Mini Vortex.
15. Temperature-controlled incubator (preferably water jacketed).
16. Shaker incubator.
17. Digital camera and white light box for documentation (*see* **Note 4**).

3 Methods

1. Streak *A. baumannii* strains on NBA for isolated colonies. Incubate the plates at 37 °C overnight.
2. Take a single isolated colony and inoculate in fresh 3 mL NB in polystyrene tube. Incubate at 37 °C overnight with shaking at 220 rpm.
3. Take 25 μL of fresh overnight culture and add to 2.5 mL NB in polystyrene tube (1:100 dilution). Incubate the tubes 37 °C for 24–48 h without shaking (static) (*see* **Note 5**).
4. Discard the liquid medium by gently inverting the tubes (*see* **Note 6**).
5. To the tubes, using a 10 mL pipette, gently add 4 mL PBS. Care should be taken so that the adherent cells are not disturbed (*see* Fig. 1).
6. Discard PBS by gently inverting the tubes. Tap dry on a piece of absorbent paper until all the liquids are gone (*see* **Note 7**).
7. Add 5 mL 0.02% of crystal violet to each tube and allow it to stain for 10–15 min.
8. Carefully decant the stain from the tubes by gently inverting the tubes. Sometimes the biomass will be loosely attached to the tubes. In this case, aspirate the liquid slowly with a pipette.
9. Wash excess stains by adding 3 mL of PBS to the tubes. Decant the liquid by inverting the tubes. Repeat this step one more time.
10. Air dry the tubes for 30 min until all the liquids are evaporated (*see* **Note 7**).

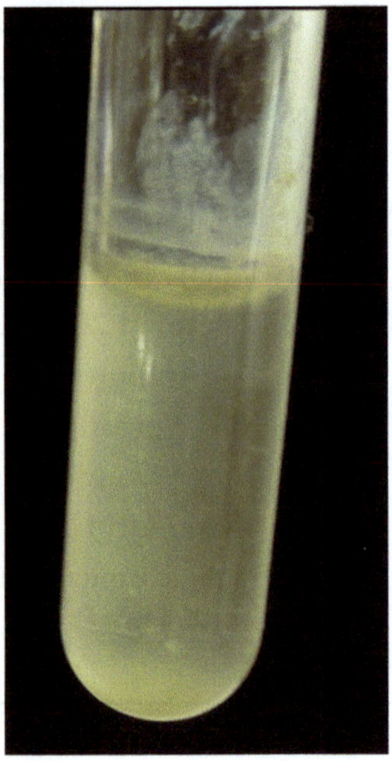

Fig. 1 Example of biofilm formation along with the pellicle after 24 h of growth. Photograph was taken before discarding the culture liquid for crystal violet staining

11. After staining and washing, you will be able to see the stained biomass rings on the inner surface of the tubes. Depending on the quality of the polystyrene tubes, the background should be clear or lightly purplish (*see* Fig. 2) (*see* **Note 8**).
12. At this point, either proceed for imaging or for quantitative measurement.
13. For image taking, either place the stained tubes on the light box or take suitable image with a digital camera. Be careful not to alter the reference frame (field or depth of vision) when photographing multiple tubes. Photographs with different frames may create problems during cropping and assembly for creating presentation.
14. For quantitation, add 3 mL 30% acetic acid to each tube. This will extract the dye from the biomass. Incubate the tubes for 30 min. Gently vortex the tubes to ensure the biomass is well solubilized (*see* **Note 9**).
15. Allow it to settle for 5 min, and then take out a 1 mL sample from the top avoiding the particulate materials.
16. Measure optical density of all the samples at 590 nm. Gather all the data and create a graph (*see* **Note 10**).

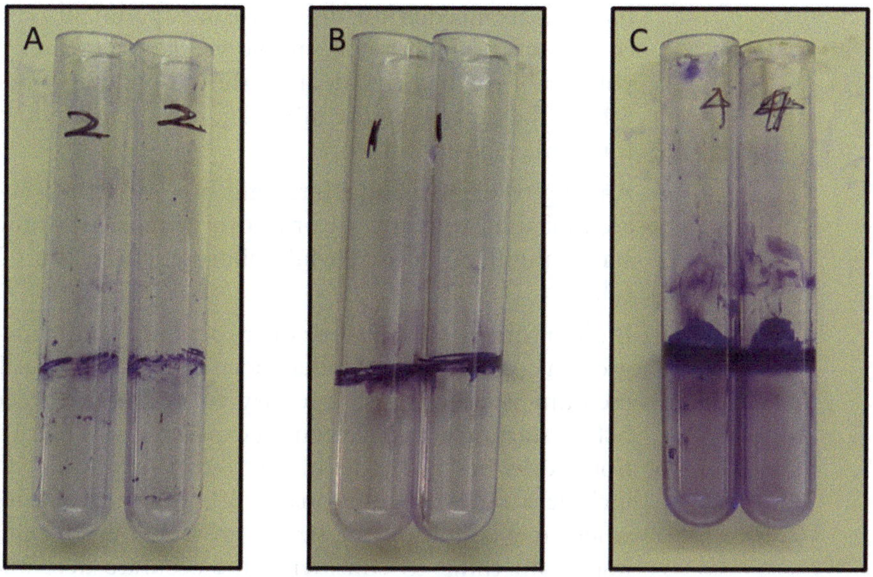

Fig. 2 Examples of biofilm formation by various *A. baumannii* clinical isolates. Biofilms were grown for 48 h and stained with crystal violet. Shown here are polystyrene tubes with varying degrees of biofilms. Weak biofilm (**a**), strong biofilm (**b**), and strong biofilm with pellicle formation (**c**). For each strain, two independent tubes were cultured

4 Notes

1. Some of the other media commonly used for static biofilm assays are LB, Mueller-Hinton broth (MHB), and M9 minimal medium [10]. Nutrient-rich media such as LB and MHB are preferred since they provide enhanced growth of *A. baumannii.* Since NB is used for motility and other phenotypic studies, it is often used. MHB is used for antibiotic susceptibility testing; therefore, it is generally preferred while analyzing clinical samples. However, if the aim of the experiment is to determine the effect of a particular metabolic pathway on biofilm formation, chemically defined media such as M9 with appropriate supplements should be used.
2. Crystal violet is considered a toxic substance. An alternative to crystal violet is to use safranin, which is relatively much safer and works as well as crystal violet [11].
3. Some polystyrene tubes are better than others. Before starting a lengthy experiment, initially try several different types of tubes from different vendors, and then select the one that gives the best result. Tubes can be of any dimension and are suitable as long as they can hold 3–10 mL of liquid. We found that narrower tubes work better for biofilm than the fatter tubes. The fatter tubes are a better choice when air-liquid

biofilms are needed for downstream experiments (proteomics or transcriptomics) since sample yields are greater. The final volume of the growth medium should preferably be less than 50% of the capacity of the tube (i.e., use 2.5 mL culture volume for 5 mL capacity tubes).

4. Any digital camera and white light box will work. These boxes are available from any art or photo store. A test tube holding stand might be needed while photographing the CV-stained tubes.
5. Biofilm formation varies greatly depending on the media, time, growth conditions, and initial inoculum. If a chemically defined medium is used, it is necessary to optimize the conditions before starting a large screening experiment. While 1:100 dilution works best for most of the routine studies, it is useful to include a range of dilutions (1:20, 1:50, 1:100, and 1:500) when new conditions or strains are being evaluated. Another important thing to consider is the accurate determination of the initial inoculum. The density of the initial inoculum should be adjusted to $OD_{600} = 0.5$. This initial adjustment of the culture density is particularly necessary when multiple strains are compared for biofilm-forming capacity. Also, some isolates form robust biofilm within 24 h, while other isolates require a longer time of incubation. Thus, the optimum length of incubation should be empirically determined.
6. Some strains, in addition to forming robust biofilm on an abiotic surface, produce a dense pellicle with bacterial cells and other extracellular matrices at the air-liquid interface. Several factors including nutrient composition of the culture media can enhance pellicle formation [6, 12]. If the overall goal is to study the composition of the pellicles or to analyze for transcriptomic or proteomic expression, pellicles can be harvested with the help of a sterile loop. It is also important to note that, if a strain forms pellicle, extreme care should be taken during decanting the culture; otherwise, there is a risk of losing the biomass. Generally, pellicles get trapped on the side of the tubes (Fig. 1).
7. Always dry the tubes at room temperature. Do not overdry the tubes. If the tubes are overdried, the staining with crystal violet will be uneven.
8. Crystal violet often stains low-quality plastic surfaces and produces a purplish background. Sometimes composition of the media or extracellular factors secreted by bacteria also interfere with the crystal violet staining. If too much background is observed, consider washing additional rounds with PBS until the background fades.

9. While 30% acetic acid is generally used for elution of the crystal violet stain from biomass, other solvents such as DMSO, ethanol, or ethanol/acetone mixture can also be used [13].
10. The λ_{max} for crystal violet is 592 nm. However, optical density of the eluted samples can be measured in a range from 550 to 600 nm [8].

Acknowledgment

Work reported in this publication was supported in part by the National Institute of General Medical Sciences of the National Institutes of Health under Award Number P20 GM113117.

References

1. McConnell MJ, Actis L, Pachon J (2013) Acinetobacter baumannii: human infections, factors contributing to pathogenesis and animal models. FEMS Microbiol Rev 37(2):130–155
2. Flemming HC et al (2016) Biofilms: an emergent form of bacterial life. Nat Rev Microbiol 14(9):563–575
3. Tomaras AP et al (2003) Attachment to and biofilm formation on abiotic surfaces by Acinetobacter baumannii: involvement of a novel chaperone-usher pili assembly system. Microbiology 149(Pt 12):3473–3484
4. Brossard KA, Campagnari AA (2012) The Acinetobacter baumannii biofilm-associated protein plays a role in adherence to human epithelial cells. Infect Immun 80(1):228–233
5. Gaddy JA, Tomaras AP, Actis LA (2009) The Acinetobacter baumannii 19606 OmpA protein plays a role in biofilm formation on abiotic surfaces and in the interaction of this pathogen with eukaryotic cells. Infect Immun 77 (8):3150–3160
6. Nait Chabane Y et al (2014) Characterisation of pellicles formed by Acinetobacter baumannii at the air-liquid interface. PLoS One 9(10): e111660
7. Franklin MJ et al (2015) New technologies for studying biofilms. Microbiol Spectr 3(4). https://doi.org/10.1128/microbiolspec.MB-0016-2014
8. Coffey BM, Anderson GG (2014) Biofilm formation in the 96-well microtiter plate. Methods Mol Biol 1149:631–641
9. Vijayakumar S et al (2016) Biofilm formation and motility depend on the nature of the Acinetobacter baumannii Clinical Isolates. Front Public Health 4:105
10. O'Toole GA, Kolter R (1998) Initiation of biofilm formation in Pseudomonas fluorescens WCS365 proceeds via multiple, convergent signalling pathways: a genetic analysis. Mol Microbiol 28(3):449–461
11. Ommen P, Zobek N, Meyer RL (2017) Quantification of biofilm biomass by staining: Non-toxic safranin can replace the popular crystal violet. J Microbiol Methods 141:87–89
12. Paytubi S et al (2017) Nutrient composition promotes switching between pellicle and bottom biofilm in Salmonella. Front Microbiol 8:2160
13. Merritt JH, Kadouri DE, O'Toole GA (2005) Growing and analyzing static biofilms Curr Protoc Microbiol Chapter 1:Unit 1B.1

Chapter 16

BioFlux™ 200 Microfluidic System to Study *A. baumannii* Biofilm Formation in a Dynamic Mode of Growth

Bertrand Naudin, Anja Heins, Stéphane Pinhal, Emmanuelle Dé, and Marion Nicol

Abstract

The ability of *A. baumannii* to develop biofilms on a wide range of surfaces can be associated to its persistence in hospital settings and the emergence of recalcitrant and chronic infections. Few compounds are available to eradicate *A. baumannii* biofilms, and most of them have been tested for their antibiofilm properties in static conditions. Microfluidics systems as BioFlux™ system are now available for studying *A. baumannii* biofilm formation in dynamic conditions. Here, we described the use of this system for studying the biofilm development of the reference strain *A. baumannii* ATCC 17978 in a dynamic mode. We showed how to test the activity of an antibiotic (colistin at the MIC concentration, 0.5 μg/mL) in these conditions of growth.

Key words *A. baumannii*, Biofilm, Microfluidic system, BioFlux™, Colistin, Antibiofilm

1 Introduction

To date, *Acinetobacter baumannii* is one of the most problematic pathogens in the world. This bacterium is responsible for worldwide epidemic outbreaks associated with increasing mortality rates and possesses high capacities to evade the immune host system and to resist to numerous available antibacterial agents [1]. *A. baumannii* is also able to persist in hospital environment due to high adhesion abilities which induce community development as biofilms [2, 3]. This process is associated to enhance survival rates especially because very few molecules are developed to eradicate *A. baumannii* in sessile lifestyles. Most studies on antibiofilm compounds are performed in static growth conditions, but dynamic environments should be considered as they can surround biofilms on medical devices and have an impact on bacterial behavior and/or on the success of therapeutic treatment [4–6]. Microfluidic systems, as the BioFlux™ system, are now available to provide this particular environment and access to new

Indranil Biswas and Philip N. Rather (eds.), *Acinetobacter baumannii: Methods and Protocols*, Methods in Molecular Biology, vol. 1946, https://doi.org/10.1007/978-1-4939-9118-1_16, © Springer Science+Business Media, LLC, part of Springer Nature 2019

data on bacterial biofilms (flow of fluids in biofilms, biofilm complex topographies and its behavioral responses to temporal variations of physicochemical conditions) [4]. In this method, biofilm development in dynamic mode of growth was performed on the reference strain *A. baumannii* ATCC 17978 with the microfluidic BioFlux™ 200 from Fluxion Biosciences Incorporation. This system is a unique microfluidic system that offers the biological relevance of a laminar flow cell with the throughput and convenience of standard microplates. It allows to (1) simultaneously grow up to 96 biofilms under controlled shear conditions, temperature, and gas content and (2) drastically reduce media/reagents consumption with no messy pumps or tubing. We will only highlight here some of the functions of this microfluidic system. It could serve as well for screening of antibiofilm compounds as previously showed with virstatin [7]. We will describe here the steps necessary to form *A. baumannii* biofilms via this technology. We will follow and quantify the 24 h biofilm formation of ATCC 17978 by a time-lapse method. This function of the BioFlux™ system is usually used to compare the dynamics of biofilm formation between different bacterial strains. Finally, we will evaluate the activity of an antibacterial agent. We used colistin, a cationic peptide considered as last-line option for the treatment of *A. baumannii*-resistant strains [8]. Its effect will be tested at 0.5 μg/mL (corresponding to minimal inhibitory concentration) to highlight the impact of flow on biofilm eradication process. Here we found that colistin has lost its antibacterial effect after 2 h of growth under continuous 0.5 dyn/cm^2 flow.

2 Materials

2.1 Bacterial Cultures and Reagents

1. Overnight bacterial culture of *A. baumannii* ATCC 17978.
2. Bacterial culture of *A. baumannii* ATCC 17978 at 10^7 CFU/mL.
3. Mueller-Hinton Broth (MHB) and Mueller-Hinton Agar (MHA).
4. Colistin sulfate salt.
5. 0.01 M phosphate-buffered saline (PBS) solution.
6. SYTO® 9 nucleic acid stain.
7. Dimethyl sulfoxide.

2.2 Equipment (Fig. 1)

1. BioFlux™ 200 system (Fluxion Biosciences).
2. BioFlux™ software version 2.5.0.8.
3. BioFlux™ 200 48-well low shear plate (0–20 dyn/cm^2) (910-0004, Fluxion Biosciences, Inc.).

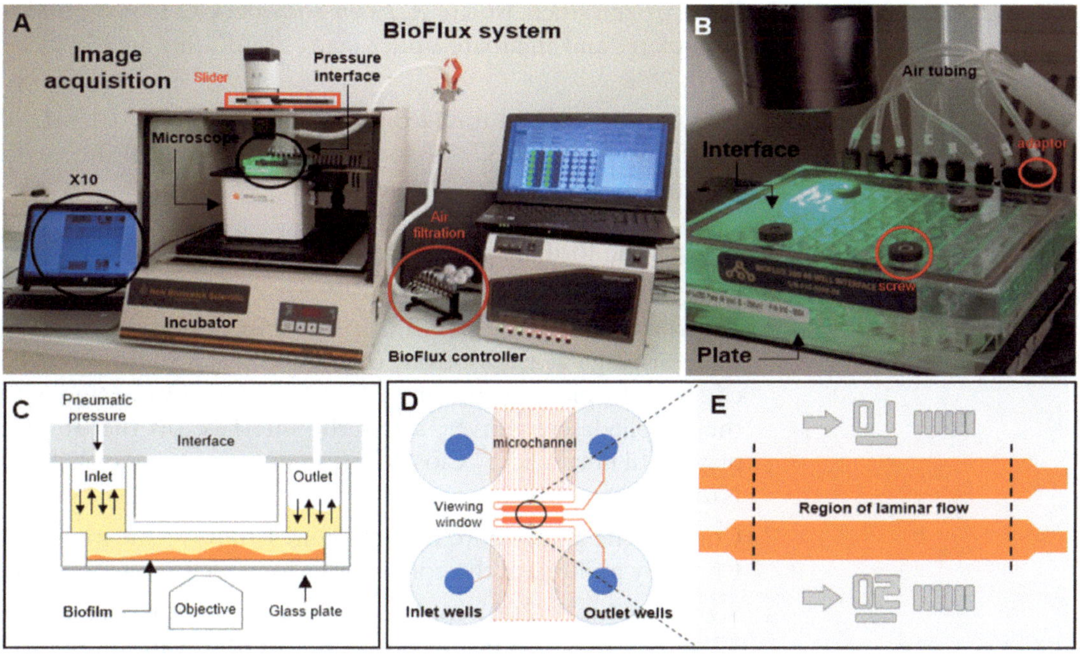

Fig. 1 Equipment. (**a**) General organization. The BioFlux™ 200 system is composed of a BioFlux controller that controls shear flow, temperature, and flow direction. It is connected to the pressure interface for airflow induction into the BioFlux plate. (**b**) Interface and plate assembly. The interface is screwed on the BioFlux plate and connected to air pressure pipes via black adaptors. (**c**) Schematic view of the microchannel. Pneumatic pressure pushes the medium from inlet to outlet wells through microchannel. Biofilm is formed in the microchannels and can be observed with the Lumaview image acquisition system. (**d**) Scheme of the 48-well low shear BioFlux plate. Inlet and outlet wells are connected by microchannels. The biofilm will be formed in the widest section in the viewing window. The black circle represents window observed with a 10× objective. (**e**) Region of laminar flow. This region is corresponding to a uniform laminar flow within the viewing window

4. LS560 microscope (Lumascope, Etaluma, Inc.) with a 10× objective.
5. Microscopy software Lumaview-Series version 17.11.04.0.
6. Two computers with Windows 7 or other compatible versions.
7. Incubation chamber at 37 °C without shaking.

3 Methods

Prepare all media, culture, and reagents described in deionized water. Each step requiring the BioFlux plate opening is done in *sterile conditions*. The method described here is made of three main steps, "priming," "inoculation," and "biofilm formation." For each step, flow settings could be automated and programmed for more comfort using the "Autorun" mode in BioFlux software (*see* **Note 1**).

3.1 Bacterial Culture

1. From frozen stock, isolate *A. baumannii* ATCC 17978 on MH agar plate, and incubate under static conditions at 37 °C for 24 h.
2. Prepare bacterial culture from a single colony in 10 mL of MHB. Incubate at 37 °C for 15 h (overnight culture) with shaking at 140 rpm.

3.2 Priming Channels

1. Unpack the BioFlux plate from its sterile pocket, and place the plate on its foam support (*see* **Note 2**).
2. Put 200 μL of sterile MHB in *inlet* wells beginning with the smallest rings at the bottom of the well. Tap slightly on the edges of the plate to raise any air bubbles. One by one and with the supplied screwdriver, screw the interface on the 48-well plate until first "click." Close the system with a second click (*see* **Note 3**).
3. Connect the air pressure tubing (from 1 to 8) to the interface. Check by pulling on the connections for a secured installation.
4. Start BioFlux™ 200, computers, and incubator chamber at 37 °C without orbital shaking. Indicate the plate number as requested on the screen of the computer controlling the BioFlux (*see* **Note 4**).
5. In the BioFlux software, enter settings of the "priming" step (Table 1) as *forward shear* (direction from the inlet to the outlet) at 5 dyn/cm² (corresponding to 629 μL/h) for 4 min in "LB at 37 °C" as default fluid (*see* **Note 5**), and activate the shear manually or automatically (*see* **Note 1**).
6. In the BioFlux software, active wells are colored in green depending on the direction of the shear. Check parameters. After 4 min, stop the shear by clicking on inlet wells if you do not use automated program. Wells become uncolored.
7. With image acquisition system, focus on channels with bright field light illumination and diaphragm at AS position (middle). Use an objective at 10× to visualize two channels in the same viewing window (*see* **Note 6**). Check with image acquisition system that *no bubbles* are present in the channels (*see* **Note 7**).
8. Take off pressure tubing by pressing adaptors to handle the following inoculation step under sterile conditions (Fig. 1b).

3.3 Inoculation

1. Screw off the interface to load 100 μL of bacterial culture in MHB medium at 10^7 UFC/mL (OD_{600} = 0.01) in *outlet* wells. As previously, close the system and connect pressure pipes.
2. Enter settings of the "inoculation" step (Table 1) as *reverse shear* at 5 dyn/cm² (corresponding to 629 μL/h) for 4 s in "LB at 37 °C" (default fluid), and activate the shear manually

Table 1
Protocol interval settings

Step	Duration	Direction	Shear (dyn/cm^2)
Priming	4 min	Forward	5
Inoculation	4 s	Reverse	5
	1-h incubation without shear		
Biofilm formation	24 h	Forward	0.5

or automatically (*see* **Note 1**). With image acquisition system, focus on viewing channels. Check bacteria flowing and the absence of bubbles in the channels. Stop the shear by clicking on outlet wells if you do not use automated program.

3. Incubate for 1 h at 37 °C without shear (*see* **Note 8**).
4. Take off pressure tubing by pressing adaptors (Fig. 1b).

3.4 Biofilm Formation

3.4.1 Kinetic of Biofilm Formation

1. Screw off the interface and load 1 mL of MHB in *inlet* wells. Close the system and connect pressure pipes.
2. Incubate the plate at 37 °C and activate manually or automatically (*see* **Note 1**), a *forward shear* at 0.5 dyn/cm^2 (corresponding to 63 μL/h) for 24 h (biofilm formation protocol, Table 1) (*see* **Note 9**).
3. With image acquisition system, focus on channels and take a picture each hour for 24 h for a time-lapse method of biofilm formation (*see* **Note 10**, Fig. 2a).
4. After 24 h of growth, the autorun is stopped.
5. To quantify biofilm formation, use the "Analyze" mode of the BioFlux software, and download a set of time-lapse images (*see* **Note 11**).

Calibrate your assay (pixels/μm) by drawing a μm line corresponding to a given number of pixels, and indicate the length of this μm line (*see* **Note 12**).

Draw a "representative region of interest" (ROI is a rectangle) on the first image (*see* **Note 13**).

Analyze images using "area coverage" module in "grayscale" mode. Exclude gray level corresponding to background noise on control channel image using "threshold pixel" cursor (*see* **Note 14**; Fig. 2a).

3.4.2 Activity of Antibacterial Agent Against *A. baumannii* Biofilm

1. In inlet wells, put 1 mL of MHB, with or without colistin at 0.5 or 32 μg/mL (Fig. 3a). Close the system and connect pressure pipes.

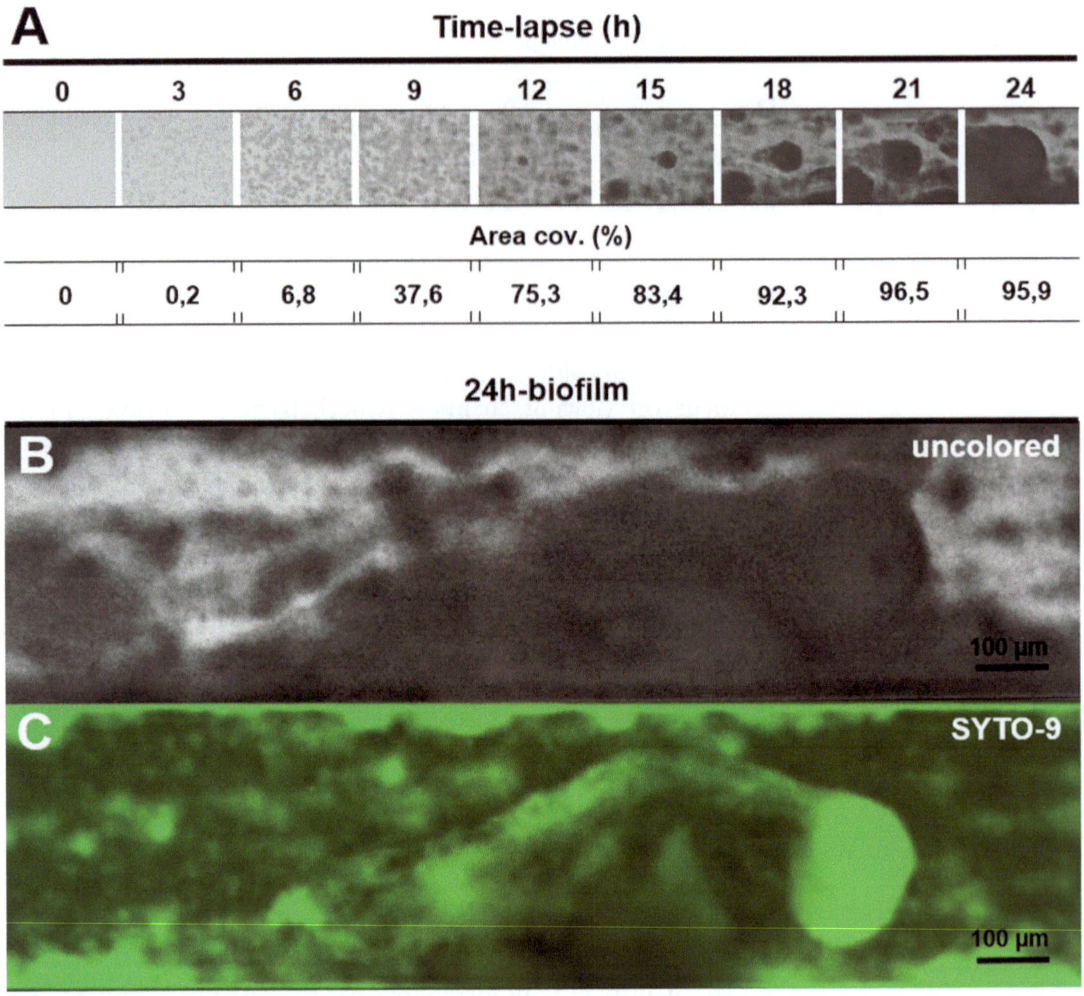

Fig. 2 *A. baumannii* ATCC 17978 biofilm development in microfluidic channel of BioFlux 200. (**a**) Time-lapse of 24 h-biofilm formation. Percentages of area coverage were calculated using the "Analyze" mode of BioFlux software in grayscale mode. These percentages were estimated with a representative window and by an average value of coverage (μm^2) depending on time (from 0 to 24 h). (**b**, **c**) 24-h biofilm of *A. baumannii* ATCC 17978. Images **b** and **c** represented the same biofilm colored (**c**) or not (**b**) with a SYTO-9 staining (*see* **Note 15**)

2. Incubate the plate at 37 °C, and activate manually or automatically (*see* **Note 1**) a *forward shear* at 0.5 dyn/cm² for 24 h (biofilm formation protocol, Table 1).
3. Addition of colistin can be done after 1, 2, 3, or 4 h of this shear (Fig. 3b–e) on untreated neoformed biofilms. Incubate up to 24 h.
4. With image acquisition system, focus on microchannels, take picture of each channel (containing 24-h biofilm growth in the

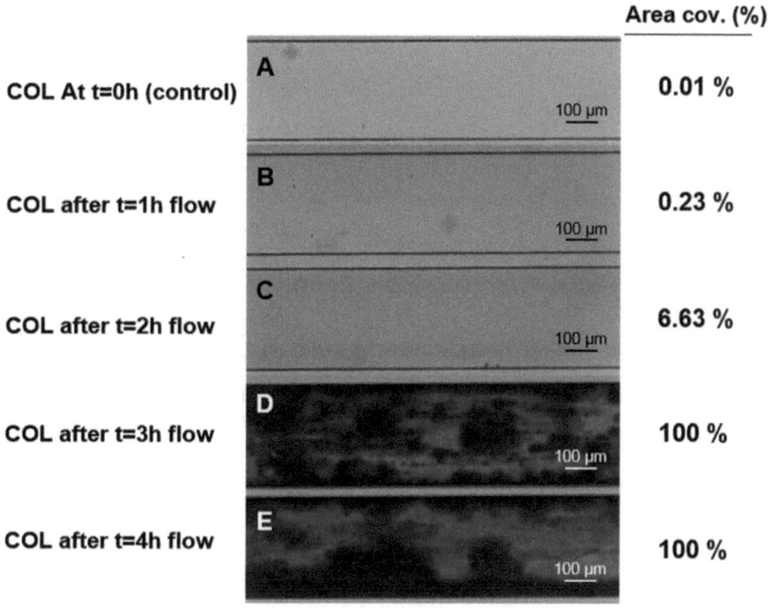

Fig. 3 Colistin effect on *A. baumannii* ATCC 17978 dynamic biofilm formation. Colistin used as an antimicrobial agent was introduced directly after inoculation (**a**) as control or after 1 (**b**), 2 (**c**), 3 (**d**), or 4 h (**e**) of flow. Percentages of area coverage (Area cov.) were estimated on 24-h biofilm as described in Subheading 3

presence of colistin), and quantify biofilms formed as described in Subheading 3.4.1, **step 5** (Fig. 3).

4 Notes

1. Different protocols corresponding to the three main steps of this method are created: (1) priming, (2) inoculation, and (3) biofilm formation. Protocol interval settings are detailed in Table 1.

   To create a protocol:

   (a) In "Edit AutoRun mode" and "Protocol setup page" of the BioFlux software, confirm the well plate type and "Add" a protocol line.

   (b) Enter interval settings (direction, duration, and shear, Table 1) and "Save as" a new protocol.

   (c) In "Sequence setup page," "Add" sequence steps. Four lines appear and represent four groups of channels (channels [1–6], [7–12], [13–18], and [19–24]).

   (d) Enter "LB at 37 °C" as default fluid and choose "newly-created protocol." Select channels used in your study to

"Enable" and "Apply" your parameters. Confirm the sequence in the four lines.

(e) "Save as" a new sequence and name it.

(f) In "AutoRun mode," select your newly created sequence and press "Start" to activate your protocol. Active wells are colored in green in function of the direction of the shear. Check parameters.

In manual mode, you should select each time, the fluid, the shear, and its duration. The activation of the shear and its direction can be triggered by clicking directly on the inlet or outlet well. The shear direction is color coded "from light green to dark green" (example given in Fig. 1a on the computer screen). To stop the shear, repeat a click on the same well.

2. The glass plate that covers the bottom of the microchannels (Fig. 1c) is particularly fragile/breakable. It is recommended to keep the BioFlux plate on the foam support to prevent damages.

3. Volumes chosen throughout the study are adapted to parameters used for the fluid circulation. With 200 μL of MHB at 5 dyn/cm^2 for 4 min, channels and microcups (inlet and outlet) are filled. With these parameters, we advise to not use less than 100 μL.

4. A barcode on the BioFlux plate informs the software of the characteristics of the plate (number of wells, maximum shear tolerance). At this step of the study, you must notify the framed code (4 digits type AA99) placed at the end of the barcode in the software.

5. In "Sequence" setup, we select "LB at 37 °C" as the default fluid because it has similar specifications as MHB at 37 °C.

6. To change objectives on the Lumascope 560 microscope, it requires to remove the plate, particularly fragile. Throughout our study, we conserved thus the same objective. We advise to choose a small magnification objective (4× or 10×) to visualize a major part of the viewing window (Fig. 1d) and to check the state of the microchannels throughout the study. It helps also to orient yourself on the plate if you do not have a motorized platform. For better contrast of the pictures, a phase objective including the Olympus phase slider could be used.

7. At each step of adding or removing medium in wells, it is possible to generate bubbles. Removing air bubbles is imperatively required because otherwise it will modify the flow and thus the biofilm formation. To prevent bubble formation, avoid emptying the wells completely and leave a minimum volume in the inner part of the well (microcups) to limit air

entry. If an air bubble enters the channel, it is recommended to add medium in inlet well and induce a step-by-step increasing shear (1–10 dyn/cm^2) to push out the bubble.

8. The inoculation step allows bacteria to move into the viewing channel. With the distance between the outlet well and the viewing channel being very short, the inoculation must be brief. If bacterial colonization beyond the viewing window is observed, the time and the shear must be reduced at this step.
9. In the BioFlux software, notifying the "well volumes" at the beginning of the test allows to know if the added volume will be enough for the duration of the sequence. In this method, wells are filled with 1 mL of medium. It is just enough to apply a shear at 0.5 dyn/cm^2 for 24 h. With different duration, shear, or fluid, it is possible to fill wells a second time.
10. Parameters used for the image acquisition (illumination, gain, exposure) are dependent on the diaphragm and slider position (Fig. 1a). To configure a time-lapse protocol with the acquisition system used here (Lumaview), click on P icon (protocol). In "Acquisition" page, select "Time-lapse" mode and enter settings: Interval (01:MM:SS), Duration (24:MM:SS). In "Image" page, select "Brightfield" light, "Illumination" at 2.7%, "Gain" at 1.000, and "Exposure" at <2 ms. "Save as" a new protocol image acquisition and "Run this protocol." These parameters allow both to visualize colonies formation in first steps of biofilm formation and mature biofilms without image saturation.
11. It is more convenient to create a folder containing all the images to be analyzed, renaming them in either alphabetical or numeric order, and download the entire folder and "sort [the images] by title name." If images are not renamed and "sort by time stamp," the software will not analyze them according to the incubation time but according to the o'clock time. If they are not kept in the same file and you analyze them one by one, you will lose some information as % change (mean) between image and reference.
12. Calibration can be performed by drawing a line (with the icon available in BioFlux™ software) similar to the μm scale bar present on the image. The software indicates that this drawn line corresponds to certain number of pixels. This gives the conversion rate which can be notified in the software. Here, calibration gives 100 μm = 76 pixels.
13. It is necessary to select a region representing as much as possible the visible channel (in width and in height) and to check that there is no shift between all images.
14. The grayscale corresponding to background noise is selected with the image corresponding to a "control" obtained before

biofilm development, at incubation $T = 0$ h. To fix this threshold more easily, use "highlight thresholded pixels." Here, pixels between 141 (low) and 255 (high) were excluded.

15. It is also possible to stain biofilms with different fluorophores and thus determine cell viability in the community [9]. For example, in Fig. 2c, we stained the 24-h biofilm using a solution containing 500 μL of SYTO-9 in DMSO at 1/1000 and 500 μL of sterile MHB for 20 min with a shear at 0.5 dyn/cm². Then, we rinsed the biofilm for 1 h with sterile PBS at 0.5 dyn/cm [5].

References

1. Lee C-R, Lee JH, Park M, Park KS, Bae IK, Kim YB, Cha C-J, Jeong BC, Lee SH (2017) Biology of Acinetobacter baumannii: pathogenesis, antibiotic resistance mechanisms, and prospective treatment options. Front Cell Infect Microbiol 7:55. https://doi.org/10.3389/fcimb.2017.00055
2. Espinal P, Martí S, Vila J (2012) Effect of biofilm formation on the survival of Acinetobacter baumannii on dry surfaces. J Hosp Infect 80:56–60. https://doi.org/10.1016/j.jhin.2011.08.013
3. Longo F, Vuotto C, Donelli G (2014) Biofilm formation in Acinetobacter baumannii. New Microbiol 37:119–127
4. Yawata Y, Nguyen J, Stocker R, Rusconi R (2016) Microfluidic studies of biofilm formation in dynamic environments. J Bacteriol 198:2589–2595. https://doi.org/10.1128/JB.00118-16
5. Dafopoulou K, Xavier BB, Hotterbeekx A, Janssens L, Lammens C, Dé E, Goossens H, Tsakris A, Malhotra-Kumar S, Pournaras S (2015) Colistin-resistant Acinetobacter baumannii clinical strains with deficient biofilm formation. Antimicrob Agents Chemother 60:1892–1895. https://doi.org/10.1128/AAC.02518-15
6. Feng SH, Stojadinovic A, Izadjoo M (2013) Distinctive stages and strain variations of A. baumannii biofilm development under shear flow. J Wound Care 22:173–174, 176–178, 180–181. https://doi.org/10.12968/jowc.2013.22.4.173
7. Nait Chabane Y, Mlouka MB, Alexandre S, Nicol M, Marti S, Pestel-Caron M, Vila J, Jouenne T, Dé E (2014) Virstatin inhibits biofilm formation and motility of Acinetobacter baumannii. BMC Microbiol 14:62. https://doi.org/10.1186/1471-2180-14-62
8. Da Silva GJ, Domingues S (2017) Interplay between colistin resistance, virulence and fitness in Acinetobacter baumannii. Antibiot Basel Switz 6:28. https://doi.org/10.3390/antibiotics6040028
9. Tremblay YDN, Vogeleer P, Jacques M, Harel J (2015) High-throughput microfluidic method to study biofilm formation and host-pathogen interactions in pathogenic Escherichia coli. Appl Environ Microbiol 81:2827–2840. https://doi.org/10.1128/AEM.04208-14

Chapter 17

In Vitro Motility Assays for *Acinetobacter* Species

Indranil Biswas, Alexandra Machen, and Joshua Mettlach

Abstract

Bacteria belonging to genus *Acinetobacter* do not encode flagella and are characterized as a non-motile organism. However, several recent studies have demonstrated that *Acinetobacter* spp. display two types of motility: twitching and swarming. Twitching motility requires the presence of functional type IV pili and occurs on solid surfaces under humid conditions. The extension and the retraction of the pili allows the bacterium to move and spread. In contrast the exact molecular mechanism for swarming motility in *Acinetobacter* spp. is currently unknown; but it requires semisolid surfaces and humid conditions. Motility has been associated with the bacterial virulence; however, motility appears to be highly dependent on the experimental conditions and the isolate understudy. In this chapter we attempt to provide detailed methods for twitching and swarming motilities that are specific for *Acinetobacter* spp.

Key words Twitching motility, Swarming motility, Type IV pili, *Acinetobacter* spp.

1 Introduction

Although *Acinetobacter* spp. are defined as non-motile, many isolates are shown to exhibit both twitching and swarming motility in vitro. The exact purpose for bacterial motility is debatable; however, for pathogens, motility is often associated with virulence. Twitching motility is mediated by retractile type IV pili and involves the translocation of bacteria on interstitial space between solid and semisolid surfaces such as the space between nutrient agar medium and the rigid plastic of a petri dish (subsurface). The overall process of twitching involves the extension of pili forming an anchor point on a surface and retraction of the pili, which will translocate a bacterium to the anchor point, similar to a grappling hook [1]. When inoculated on the plastic surface of the agar plate, bacteria migrate in the interstitial space between agar and the petri plates by twitching motility. This migration can be visualized by a characteristic blooming or halo appearance on the base of a petri dish after staining with a suitable stain (Fig. 1, [2, 3]). Swarming motility is defined as the ability of a bacterium to move in a

Indranil Biswas and Philip N. Rather (eds.), *Acinetobacter baumannii: Methods and Protocols*, Methods in Molecular Biology, vol. 1946, https://doi.org/10.1007/978-1-4939-9118-1_17,

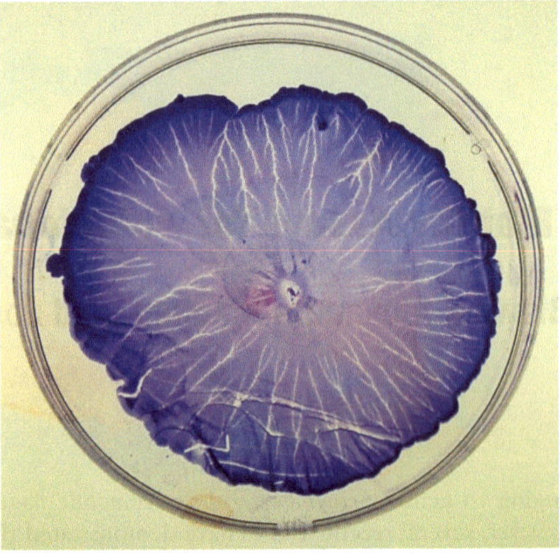

Fig. 1 Macroscopic visualization of subsurface twitching motility. *A. baumannii* B8342 strain was inoculated on LBNS agar plate and incubated at 37 °C [11]. The plate was stained with crystal violet

multicellular coordinated fashion across a semisolid surface through the use of a flagella complex. Since *Acinetobacter* spp. do not encode flagella, the exact molecular mechanisms behind swarming motility are currently unknown [4]. However, unlike twitching motility, swarming motility in *Acinetobacter* spp. is depended on several factors including temperature, darkness, and the presence of surfactants [5–7]. Swarming motility is also highly variable and often produces inconsistent results. Media composition and freshness are two important factors that affect swarming motility. Even for a particular strain, the degree of swarming motility varies from time to time due to other unknown compounding factors. Characteristically there is no universal pattern that occurs with a given strain making quantitative comparison between two strains difficult [8].

In this chapter, we provide two methods to study twitching motility in *Acinetobacter* spp. The first method is the classical subsurface twitching motility assay, and the second method is "on-surface" twitching motility that we adapted from a recently described method [9]. The swarming motility method is modified from the available published methods to improve the reproducibility and to satisfy specific needs for *Acinetobacter* spp. The methods described here have been tested with three *Acinetobacter* spp.: *A. baumannii* (at least ten different strains), *A. baylyi*, and *A. radioresistens*.

2 Materials

Prepare all solutions and media using ultrapure deionized water (conductivity 18 mΩ/cm at 25 °C). Always use analytical grade reagents purchased from a reputed source for reproducibility. After preparation of solutions, autoclave at 120 °C for 30 min prior to use to maintain sterility. Ultrapure water is also autoclaved. Consider all the waste as biohazard and dispose appropriately.

1. Lysogeny broth (LB): 2.5 g NaCl, 1.25 g yeast extract, 2.5 g tryptone, 1 tablet NaOH, 250 mL water.
2. Bacterial strains streaked and grown overnight at 37 °C on LB agar plate (LBA): 1.5% agar (*see* **Note 1**).
3. LB without NaCl containing 1% agar (LBNSA) (*see* **Note 1**).
4. Nutrient broth (NB): 0.75 g beef extract, 1.25 g peptone, 250 mL water.
5. NB containing various concentrations of Eiken agar (*see* **Note 2**).
6. Phosphate-buffered saline (PBS): dissolve 8 g NaCl, 0.2 g KCl, 1.44 g Na_2HPO_4, and 0.24 g KH_2PO_4 in 800 mL water, adjust pH to 7.4 with HCl, and bring the final volume to 1 L with water.
7. 95% ethanol.
8. Triphenyl tetrazolium chloride, 10% (TTC): dissolve 0.1 g in 1 mL water, and store at 4 °C (*see* **Note 3**).
9. Resuzurin, 10% (RZN): dissolve 0.1 g in 1 mL water (*see* **Note 3**).
10. Crystal violet stain, 0.1% (CV): dissolve 0.1 g in 20 mL of 95% ethanol, and then add 80 mL water.
11. Petri plates: 100 mm × 15 mm sterile (*see* **Note 4**).
12. Cellophane paper, cut into circular discs (2-in. diameter) and autoclaved (*see* **Note 5**).
13. Parafilm.
14. Forceps, metal.
15. Absorbent paper (any kind).
16. Autoclavable bottle: 500 mL.
17. Sterile disposable cotton swab.
18. Sterile disposable serological pipettes: 10 mL.
19. Sterile disposable pipette tips: 10 μL, 200 μL, 1 mL.
20. Pipette boy (pipette controller).
21. Pipettes: P10, P200, P1000.
22. Sterile disposable 50 mL polypropylene tubes with graduations.

23. Bunsen burner.
24. Mini-Vortex.
25. Circular paper puncher (2-in. diameter) or scissor (for cutting cellophane paper).
26. Temperature-controlled incubator (water jacketed; *see* **Note 6**).
27. Laminar hood.
28. Digital camera for documentation.
29. White light box for documentation (*see* **Note 7**).
30. Image J software for analysis (freely available from NIH; *see* **Note 8**).

3 Methods

3.1 Subsurface Twitching Motility Assay

1. Autoclave LBNS agar medium at 121 °C for 30 min.
2. Allow medium to cool until able to handle with ease.
3. Underneath laminar hood, using 50 mL graduated tubes, pour exactly 25 mL of LBNS agar in 100 mm petri plates. Allow it to solidify and dry under ambient conditions for few hours with the lids off. Store the plates in stack of five plates in an airtight bag. Use within 24 h (*see* **Note 9**).
4. Prepare samples by simply picking four to six colonies from a freshly streaked plate, and resuspend in 100 μL of LB.
5. Vortex gently for 5 s to ensure even mixture of cells.
6. Take a LBNS agar plate, and using a sterile 1 mL pipette tip, gently stab the middle of the agar of a petri dish until you reach the base, swirl the tip slightly, and then remove the agar simply by pulling up, forming a well all the way to the interface between the basal surface of the agar and the petri plate.
7. Dispense 5 μL of the cell suspension into well of agar, ensuring that the cell suspension reaches the base of the petri plate (*see* **Note 10**).
8. Cover the plate with lid, label and seal the plate with parafilm, and place the petri plate in an upright position in the incubator, and allow the bacteria to grow 16–24 h (*see* **Note 11**).
9. After desired time of incubation, remove the parafilm from petri plate.
10. Remove agar by swirling a sterile 1 mL pipette tip around the edge of the petri plate in an attempt to dislodge the agar medium from the plate.
11. After agar removal, add 5 mL of crystal violet stain; allow to incubate at room temperature for 1 h (*see* **Note 12**).

12. Gently remove the stain out of petri plate with a 10 mL pipette. Wash at least twice with 10 mL of PBS to remove the excess dye. Air-dry the plate on a piece of absorbent paper by inverting the plates.
13. Place the plates in an inverted position on the light box without the lids, and take a photograph using a digital camera.
14. Transfer the image to a computer, and analyze the image with Image J software for area computation and comparison (*see* **Note 13**).

3.2 On-Surface Twitching Motility Assay

1. Autoclave LBNS agar medium at 121 °C for 30 min.
2. Allow medium to cool until able to handle with ease.
3. Add 50 μL of 10% TTC or 10 μL of RZN to 50 mL of LBNS agar in 50 mL graduated tube. Mix gently by inverting the tubes.
4. Pour exactly 25 mL of LBNS agar (containing TTC or RZN) in 100 mm petri dish. Avoid trapped air bubbles while pouring. Allow it to solidify and dry under ambient condition for few hours with the lids off. Store the plates in stack of five plates in an airtight bag for future use. Use within 24 h (*see* **Note 9**).
5. Once the plates are dried, with the help of a forceps, gently place a cellophane disc on the agar surface. Spread the disc on the surface with the tip of cotton swab. Remove any trapped air bubbles (*see* **Note 14**).
6. Prepare samples by picking four to six colonies from a freshly streaked plate, and resuspend in 100 μL of LBNS.
7. Spot 2 μL of cell suspension at the center of the cellophane disc. Let the droplet air-dry for 10 min.
8. Stack the plates in set of two to four plates, and place them in upright position in 37 °C incubator, and grow overnight (or as desired).
9. Remove the plate from the incubator. The bacteria that display twitching motility can be visualized by development of red color on TTC-containing plate or purple color on RZN-containing plates (*see* **Note 3**).
10. Place the plates in an inverted position on the light box without the lids, and take photograph using a digital camera.
11. Transfer the image to a computer, and analyze the image with Image J software for area computation and comparison (*see* **Note 15**).

3.3 Swarming Motility Assay

1. Prepare 0.5 L NB medium and add 2.5 g Eiken agar (final concentration 0.5%). Autoclave NB agar medium at 121 °C for 30 min (*see* **Note 16**).

2. Allow NB agar to cool until able to handle with ease.
3. Add 500 μL of 10% TTC to NB agar.
4. Underneath laminar hood, using 50 mL tube, pour exactly 25 mL of NB agar in each 100 mm petri dish, and allow to air-dry (20 plates total). Do not overdry the plates. Use the plates within couple of hours (*see* **Note 17**).
5. Using a sterile 10 μL pipette tip, gently touch the middle of the agar to form a small well on the top surface of the agar.
6. Dispense 5 μL of cell suspension into the well of agar, ensuring that the cell suspension does not pool out of the well (*see* **Note 18**).
7. Label, seal with parafilm, and place the petri dishes upright in a 37 °C incubator, and grow overnight (or as desired).
8. Monitor the growth for up to 48 h. Avoid unnecessary exposure to light (*see* **Note 19**).
9. After 48 h, remove the plates from the incubator, detach the parafilm seal, and record the motility.
10. Remove the lids from the plates, place the plates inverted on the light box, and take pictures with a digital camera.
11. Transfer the image to a computer and analyze with Image J software (*see* **Note 20**).

4 Notes

1. For routine bacterial propagation, use either 1.0% or 1.5% agar in LB medium (LBA). Streak bacterial cultures from stored glycerol sources directly on to LBA. For routine bacterial growth, avoid NBA or LBNSA. NBA or LBNSA is used only when motility assays are carried out. The compositions of commercially available NB medium vary drastically. Make sure the composition is similar to our NB.
2. Eiken agar is a specialized agar that tends to produce consistent results during swarming motility. This agar is more "wettable" than commonly available agar. Eiken agar is only available from Eiken Chemical Co. Ltd. (Japan). Commonly available agar works and some sources are better than the other. We recommend to run a preliminary assay with agar from different sources using ATCC 17978 as positive control. Some brands of agar may inhibit motility. Alternatively, one can add Triton X 100 or similar nonionic surfactants to enhance the motility. The concentration of the surfactants should be empirically determined (generally <0.2%).

3. TTC or RZN is added for visualization purposes. Metabolically active bacterial cells are able to reduce the colorless TTC to a red formazan precipitate, which is visible in the naked eye. Similarly, active bacterial cells reduce the blue resazurin (non-fluorescent) to the pink resorufin, which is fluorescent. Filter sterilize the solutions by passing through 0.22 μm filter, and store at 4 °C for a maximum of 15 days after preparation.
4. We strongly recommend the size of the petri dishes should be at least 100 × 15 mm. Some *A. baumannii* isolates display vigorous motility phenotypes (both types) covering almost the entire plate (*see* Figs. 1, 2, and 3). For high-throughput screening experiments, we recommend larger-size plates (as well as reduced inoculum size) to provide ample space between each inoculum for differentiating the mutants or isolates. The quality of the petri dishes is also important, specifically for the subsurface twitching motility. While we generally use standard petri plates available from various sources, some perform better than the other. We found that tissue culture-treated plates drastically enhance the subsurface motility. For on-surface twitching or swarming motility, plate quality is not important (only the size is important).
5. Cellophane paper can be purchased from any craft store. Regular quality works the best. Avoid glossy or moisture proof paper as they will not work. We found that regular gel-drying cellophane paper works the best.
6. We recommend using a water-jacketed incubator since humidity loss is lower compared to an air-jacketed incubator. However, placing a 500 mL beaker containing water in an air-jacketed incubator works well. Use a double door incubator.
7. Any white light box will work. These boxes are available from any art or photo store.
8. Download Image J software from the NIH (https://imagej.nih.gov/ij/download.html). Read the instruction carefully while installing and the manual to familiarize with the analysis.
9. Dry the plates under ambient air to ensure uniform drying until no water is visible on the surface. If the plates are too dry, twitching will be inhibited. While it is best to use the plates within the same day, we have used the plates within 24 h without any noticeable decline in motility.
10. When developing the well for the twitching assay, be sure that you displaced the agar after puncture with the 1 mL sterile pipette tip. It is important to make sure inoculum reaches to the bottom of the petri plate. While inoculating directly dispense the culture once the tip gently touches to the bottom surface. Dispensing culture prematurely will result in

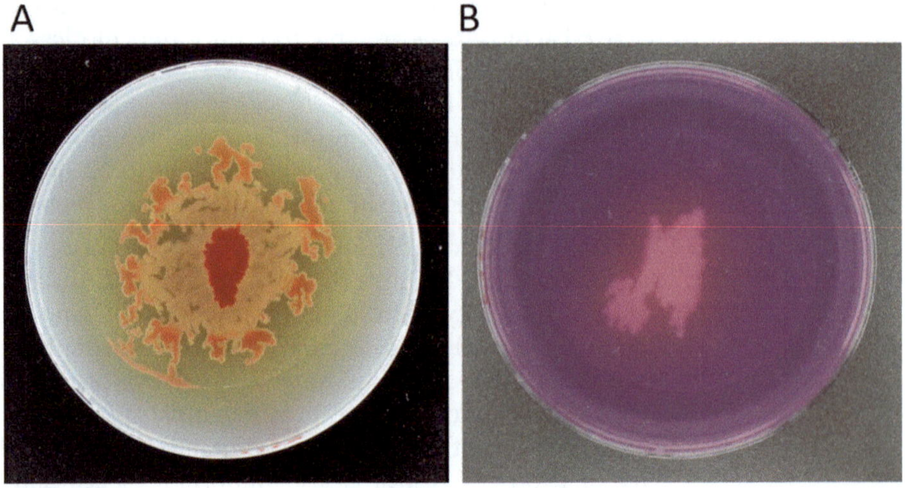

Fig. 2 Visualization of on-surface motility using cellophane coupons. (**a**) *A. baumannii* ATCC 17978 strain was spotted on LBNS agar (0.5%) containing TTC. (**b**) *A. baumannii* B8342 strain was spotted on LBNS agar (0.5%) containing RZN. Plates were incubated at room temperature for 24 h before taking the images

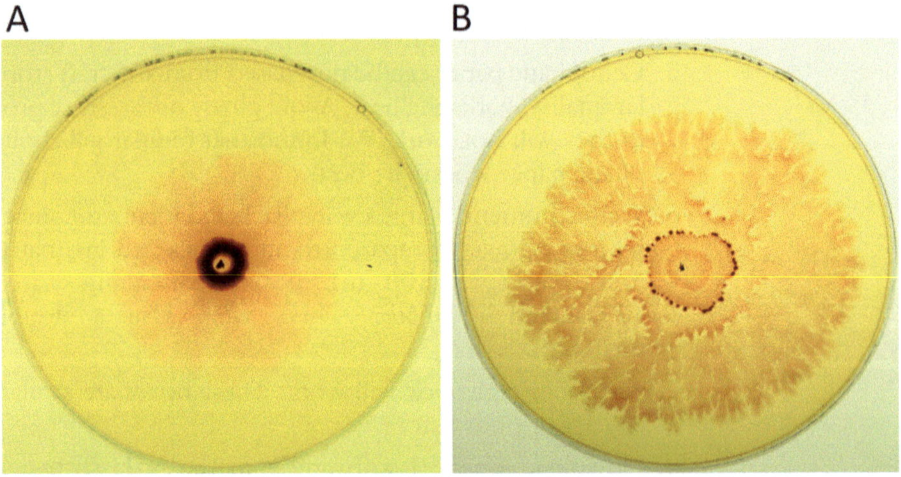

Fig. 3 Swarming-like motility. *A. baumannii* strains B8342 (**a**) and ATCC 17978 (**b**) were inoculated on LBNS containing 0.5% Eiken agar. Plates were incubated at 37 °C for 24 h before taking the images

inconsistency and large bacterial growth on the agar surface. Care must be taken so that unnecessary culture droplets are not scattered on the plate during inoculation. Although we prepare fresh bacterial suspension from colonies grown on LBA plate, one can use fresh overnight grown liquid culture in NB medium. We have found no significant difference between the cell suspension and freshly grown liquid culture.

11. Although we recommend plates should be kept in upright position during incubation, we found no significant difference when the plates are kept in inverted position. If there is excess

moisture on the agar surface or if visible moisture condensation occurs on the lids, it is better to incubate the plates in inverted position to avoid droplets from falling on the agar surface from the lids.

12. We recommend the use of crystal violet for staining the bacteria because it gives very good contrast. Alternatively, safranin or Coomassie can be used as well. The concentration of safranin is generally between 0.2% and 0.5%. The composition of the Coomassie stain solution is the same to that which is used for protein gel staining.
13. We recommend to use control strains while comparing unknown clinical isolates. Some strains are weakly motile and display reduced zone sizes. A negative control is necessary to distinguish between motile and non-motile isolates. ATCC 19606 is a good negative control for subsurface twitching motility.
14. We recommend to individually wrap cellophane disc with aluminum foil and autoclave. If multiple discs are autoclaved together, it is very difficult to separate them during spreading. The cellophane discs tend to wrinkle during spreading on the agar surface. The degree of wrinkle depends on the moisture content of the agar. If the agar is dry and the discs are wrinkled too much, dip a cotton swab in LBNS, and then roll over the discs to spread them flat on the agar surface. Avoid too much liquid.
15. We notice that some isolates that displayed robust subsurface motility exhibited moderate level of on-surface motility. On the other hand, some other isolates that showed no or low level of subsurface motility with our method exhibited enhanced on-surface motility. The exact reason for this discrepancy is currently unknown.
16. We recommend using 0.5% agar concentration for swarming assay for initial experiments. Concentration ranging from 0.3% to 0.8% can be used for swarming studies. If Eiken agar is not used, we recommend setting the initial agar concentration to 0.3% and increasing gradually until the desire results are obtained.
17. Swarming motility is highly sensitive to unevenness of the surface, especially when non-Eiken agar is used. Agar media that is kept at 60 °C and gently mixed will avoid the presence of unnecessary air bubbles. Quickly running an open flame over a freshly poured plate will remove any small air bubbles by popping them. If the agar is too hot, moisture condensation can occur. The plate thickness is an important parameter, and we use 25 mL of medium for each 100 mm petri plate. Since thicker plates will produce better swarming motility perhaps

due to higher level of moisture retention, we recommend to standardize the thickness if a different size plate or different media is used. Always avoid drying under air flow since it produces uneven dryness. We recommend plates to be dried under room temperature and ambient condition with lids off.

18. Extreme care must be taken during inoculation step. The well formed in the swarming assay is meant to be very small; do not puncture too far into the agar or varying unexpected growth patterns may occur. The tips containing bacterial suspension should be close enough to the well. Avoid direct contact with the well during deposition of the inoculum that could result in squirting the inoculum over the agar surface producing unwanted growth. On the other hand, if the tips are far away from the well during inoculum deposition, splashing may occur producing multiple inoculation centers.
19. Swarming motility by *Acinetobacter* spp. is highly sensitive to light, mainly when the incubation is carried out below 25 °C [10]. Exposure to light is not a serious issue when the plates are incubated at 37 °C. Nevertheless, we recommend to place the plates in a small cardboard box during incubation to avoid accidental exposure to light.
20. Addition of TTC to the growth media greatly improves contrast during bacterial swarming. However, we noticed that often the edges of the swarming growth do not produce color growth. Care should be taken while photographing for optimum contrast.

Acknowledgment

Work in this publication was supported in part by the National Institute of General Medical Sciences of the National Institutes of Health under Award Number P20 GM113117.

References

1. Mattick JS (2002) Type IV pili and twitching motility. Annu Rev Microbiol 56:289–314
2. Burrows LL (2012) Pseudomonas aeruginosa twitching motility: type IV pili in action. Annu Rev Microbiol 66:493–520
3. Vijayakumar S et al (2016) Biofilm formation and motility depend on the nature of the Acinetobacter baumannii clinical isolates. Front Public Health 4:105
4. Murray TS, Kazmierczak BI (2008) Pseudomonas aeruginosa exhibits sliding motility in the absence of type IV pili and flagella. J Bacteriol 190(8):2700–2708
5. Clemmer KM, Bonomo RA, Rather PN (2011) Genetic analysis of surface motility in Acinetobacter baumannii. Microbiology 157 (Pt 9):2534–2544
6. McQueary CN, Actis LA (2011) Acinetobacter baumannii biofilms: variations among strains and correlations with other cell properties. J Microbiol 49(2):243–250
7. McQueary CN et al (2012) Extracellular stress and lipopolysaccharide modulate Acinetobacter baumannii surface-associated motility. J Microbiol 50(3):434–443

8. Kearns DB (2010) A field guide to bacterial swarming motility. Nat Rev Microbiol 8 (9):634–644
9. Xu B, Wozniak DJ (2015) Development of a novel method for analyzing Pseudomonas aeruginosa twitching motility and its application to define the AmrZ regulon. PLoS One 10(8): e0136426
10. Mussi MA et al (2010) The opportunistic human pathogen Acinetobacter baumannii senses and responds to light. J Bacteriol 192 (24):6336–6345
11. Vijaykumar S, Balaji V, Biswas I (2015) Complete genome sequence of Acinetobacter baumannii strain B8342, a motility-positive clinical isolate. Genome Announc 3(4):e00925-15

Chapter 18

Desiccation Tolerance Assays for *Acinetobacter baumannii*

Xun Wang, M. Stephen Trent, and Bryan W. Davies

Abstract

Acinetobacter baumannii is a hospital-associated pathogen of growing importance and is a paradigm for endemic hospital contamination. Desiccation tolerance has been implicated as an important characteristic that potentiates the spread of *A. baumannii* in clinical settings through contaminated healthcare equipment and personnel. Desiccation is a multifaceted stress, and many physiological and environmental factors can influence its impact on bacterial survival. This chapter provides a protocol for assessing desiccation survival that facilitates comparisons among *A. baumannii* strains under various environmental conditions.

Key words *Acinetobacter baumannii*, Desiccation, Environmental stress, Percent survival

1 Introduction

Acinetobacter baumannii is a hospital-associated pathogen of growing importance [1–3]. The success of *A. baumannii* in clinical settings relies largely on its ability to rapidly acquire antibiotic resistance and to spread to new patients through contaminated healthcare equipment and personnel [4–9]. The latter is attributed in part to the ability of *A. baumannii* to survive desiccation for extended periods of time. Despite its potential importance, there is limited understanding of the molecular mechanisms supporting desiccation tolerance of *A. baumannii* and other bacterial species. Molecular factors that have been implicated in bacterial desiccation tolerance include lipopolysaccharide modifications, biofilm formation, and hydrophilic stress proteins [10–17]. Non-reducing sugars, especially trehalose, have also been implicated in promoting desiccation tolerance of several organisms [10–16, 18].

Assaying the desiccation tolerance of microorganisms requires control of both environmental and microbial factors that can influence desiccation survival [19–22]. These factors include assay temperature, humidity, desiccation medium, desiccation surface, desiccation time, initial growth phase, and cell density. Desiccation tolerance is commonly reported as the time required for a bacterial

Indranil Biswas and Philip N. Rather (eds.), *Acinetobacter baumannii: Methods and Protocols*, Methods in Molecular Biology, vol. 1946, https://doi.org/10.1007/978-1-4939-9118-1_18,

population to lose total viability or as the percent survival of a bacterial population after a specific period of time [5, 19, 20, 22]. This protocol aims to assay desiccation percent survival of *A. baumannii* under conditions that are potentially important in clinical settings. The input colony-forming units (CFU) and the output CFU are assayed before and after desiccation, respectively. The survival percentage is calculated by the ratio of the output CFU density to the input CFU density. With this method, the percent survival for an *A. baumannii* stain after 48 h of desiccation is usually obtained within a 5-day experimental period.

2 Materials

1. 96-well polystyrene plate.
2. Aerating hood and (de)humidifier. Alternatively, humidity chamber.
3. Humidity meter and thermometer.
4. Lysogeny broth (LB): 10 g tryptone, 5 g yeast extract, 5 g NaCl dissolved in 1 L H_2O and autoclaved.
5. LB agar plates: 10 g tryptone, 5 g yeast extract, 5 g NaCl, 15 g agar dissolved in 1 L H_2O and autoclaved.
6. *A. baumannii* or other bacterial strains of interest.
7. Desiccation medium. As bacteria are generally suspended in liquid before desiccation, we refer to the initial liquid as the desiccation medium. Desiccation medium could be water, LB, or any liquid of interest.
8. Spectrophotometer or plate reader.
9. 37 °C incubator.

3 Methods

3.1 Sample Preparation

1. On day 0, streak out *A. baumannii* on an LB plate, or grow the bacteria in liquid LB medium at 37 °C.

3.2 Suspend Bacteria in Desiccation Medium and Adjust Population Density (See Note 1)

1. On day 1, scrape *A. baumannii* from the LB plate into 50 mL of desiccation medium, and resuspend the bacteria. For liquid cultures, spin down the bacteria at 1800 rcf for 10 min, discard supernatant, and add 50 mL of desiccation medium to resuspend the bacteria.
2. Wash the bacteria by centrifuging at 1800 rcf for 10 min, discarding the supernatant, and resuspending the pellet in 50 mL of the desiccation medium. Repeat the washing step twice (*see* **Note 2**).

3. After the bacteria are suspended into the desiccation medium, measure the population density by assaying the optical density at 600 nm (OD_{600}) of the suspension. Adjust OD_{600} to 1 in the desiccation medium. An OD_{600} of 1 is approximately 10^8–10^9 CFU/mL for most *A. baumannii* strains we have studied (*see* **Note 3**).

3.3 Desiccation

1. Spot 10 μL of *A. baumannii* suspension into a well of a 96-well polystyrene plate in at least triplicate. Put the plate in a chamber where temperature is kept at 25 °C and relative humidity is kept at 40% (*see* **Note 4**). Alternatively, put the plate into an aerating hood, and control the temperature as well as the relative humidity as needed (*see* **Note 5**). We have found that desiccation medium evaporates in 3–4 h under these conditions (*see* **Note 6**).
2. Serially dilute the remaining bacterial suspension in LB medium in tenfold increments seven times. After seven dilutions, the population density generally becomes <1 CFU/10 μL. Spot 5 μL of each dilution onto an LB agar plate, and incubate overnight to numerate the input CFU density (Fig. 1) (*see* **Note 7**).
3. Input CFU density = ($X_{in} \times 10^\wedge N_{in}$) CFU/5 μL. X_{in} represents the CFU of the first numerable spot, and N_{in} represents its fold of dilution from the original bacterial suspension.

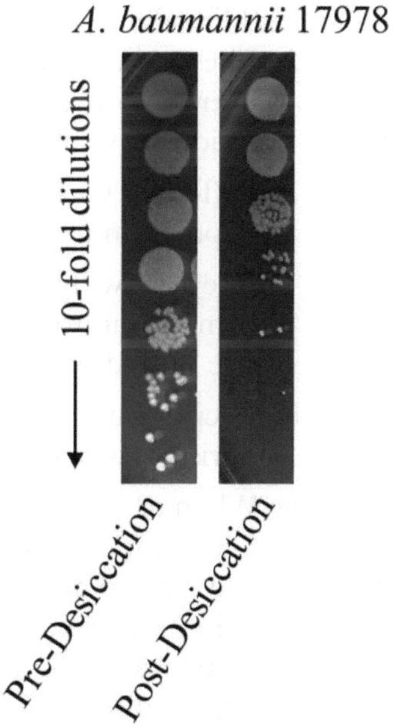

Fig. 1 *A. baumannii* desiccation survival. Shown is *A. baumannii* strain 17978 spotted for CFU density pre- and post desiccation. Adapted from [17]

3.4 Rehydration

1. On day 3, after 48 h of desiccation, take out the plate from the chamber, and add 300 μL of water into each well to rehydrate bacteria. Mix by gently rocking the plate for 5 min.
2. Serially dilute the rehydrated bacteria suspension with LB medium in tenfold increments as was done for determining input CFU density. Spot 5 μL of each dilution onto an LB agar plate, and incubate overnight to numerate the output CFU density.
3. Output CFU density = (X_{out} × 10^N_{out}) CFU/5 μL. X_{out} represents the colony-forming unit of the first numerable spot, and N_{out} represents its fold of dilution from the rehydrated bacteria suspension (Fig. 1) (*see* **Note** 7).

3.5 Statistical Analysis

1. The percent survivals are calculated by (output CFU density)/(input CFU density).
2. Mean percent survival with error bars is calculated from the three replicates. More replicates increase the statistical power of the analysis. ANOVA could be used for examining the influence of a factor on percent survival (*see* **Note 8**).

4 Notes

1. The following parameters can influence desiccation percent survival and need to be controlled: desiccation medium, growth condition, population density, and desiccation duration.
2. The washing step is crucial because the nutrient and salt carried over from the growth medium could strongly influence desiccation survival.
3. Different CFU densities can be used to assay the impact of population density on desiccation survival.
4. The following parameters are usually kept constant to mimic common indoor environments: temperature (25 °C) and relative humidity (40%).
5. Keep a humidity meter and a thermometer beside the plates during desiccation, even working with a humidity chamber.
6. When using 10 μL as the desiccation volume, the drops usually dry out in 3–4 h at the given condition. Increasing the volume will lead to an increased drying time. When desiccating larger volumes, using a larger petri dish and spotting the bacteria suspension in smaller droplets will help to shorten the drying time.
7. The LB agar plates for CFU spotting shall be dried in a 37 °C incubator for ~2 h before spotting to avoid the merging of

liquid drops while spotting. Wait until the liquid drops dry out before moving the plates. A volume of 5 μL instead of 10 μL is used for spotting as larger drops tend to merge and hard to numerate.

8. If the percent survivals vary from day to day for unknown reasons, it would be useful to keep a wild-type strain from a certain growth condition as a control group. Relative percent survivals normalized to the control group could be compared among samples processed on different dates. More replicates will give more statistical power.

Acknowledgment

This work was supported by ARO grant W911NF-16-1-0146 to B. W.D., NIH grant AI125337 to B.W.D, and NIH grants AI064184, AI76322, AI119879, and AI129940 to M.S.T.

References

1. Fournier PE, Richet H (2006) The epidemiology and control of Acinetobacter baumannii in health care facilities. Clin Infect Dis 42 (5):692–699
2. Rodríguez-Baño J, Cisneros JM, Fernández-Cuenca F, Ribera A, Vila J, Pascual A et al (2004) Clinical features and epidemiology of Acinetobacter baumannii colonization and infection in Spanish hospitals. Infect Control Amp Hosp Epidemiol 25(10):819–824
3. Weisenberg SA, Schuetz AN, Alexander EA, Eiss B, Behta M, Saiman L et al (2011) Endemic Acinetobacter baumannii in a New York hospital. PLoS One 6(12):e28566
4. Giannouli M, Antunes LC, Marchetti V, Triassi M, Visca P, Zarrilli R (2013) Virulence-related traits of epidemic Acinetobacter baumannii strains belonging to the international clonal lineages I-III and to the emerging genotypes ST25 and ST78. BMC Infect Dis 13:282
5. Jawad A, Seifert H, Snelling AM, Heritage J, Hawkey PM (1998) Survival of Acinetobacter baumannii on dry surfaces: comparison of outbreak and sporadic isolates. J Clin Microbiol 36 (7):1938–1941
6. Kramer A, Schwebke I, Kampf G (2006) How long do nosocomial pathogens persist on inanimate surfaces? A systematic review. BMC Infect Dis 6:130
7. Musa EK, Desai N, Casewell MW (1990) The survival of Acinetobacter calcoaceticus inoculated on fingertips and on formica. J Hosp Infect 15(3):219–227
8. Rao GG (1998) Risk factors for the spread of antibiotic-resistant bacteria. Drugs 55 (3):323–330
9. Roca I, Espinal P, Vila-Farrés X, Vila J (2012) The Acinetobacter baumannii oxymoron: commensal hospital dweller turned pan-drug-resistant menace. Front Microbiol [Internet] 3:148. https://www.ncbi.nlm.nih.gov/pmc/articles/PMC3333477/
10. Espinal P, Martí S, Vila J (2012) Effect of biofilm formation on the survival of Acinetobacter baumannii on dry surfaces. J Hosp Infect 80(1):56–60
11. García AH (2011) Anhydrobiosis in bacteria: from physiology to applications. J Biosci 36 (5):939–950
12. Garmiri P, Coles KE, Humphrey TJ, Cogan TA (2008) Role of outer membrane lipopolysaccharides in the protection of Salmonella enterica serovar Typhimurium from desiccation damage. FEMS Microbiol Lett 281(2):155–159
13. Ophir T, Gutnick DL (1994) A role for exopolysaccharides in the protection of microorganisms from desiccation. Appl Environ Microbiol 60(2):740–745
14. Vanderlinde EM, Muszyński A, Harrison JJ, Koval SF, Foreman DL, Ceri H et al (2009) Rhizobium leguminosarum biovar viciae 3841, deficient in 27-hydroxyoctacosanoate-modified lipopolysaccharide, is impaired in

desiccation tolerance, biofilm formation and motility. Microbiology 155(9):3055–3069
15. Vanderlinde EM, Harrison JJ, Muszyński A, Carlson RW, Turner RJ, Yost CK (2010) Identification of a novel ABC transporter required for desiccation tolerance, and biofilm formation in Rhizobium leguminosarum bv. viciae 3841. FEMS Microbiol Ecol 71(3):327–340
16. Welsh DT, Herbert RA (1999) Osmotically induced intracellular trehalose, but not glycine betaine accumulation promotes desiccation tolerance in Escherichia coli. FEMS Microbiol Lett 174(1):57–63
17. Boll JM, Tucker AT, Klein DR, Beltran AM, Brodbelt JS, Davies BW et al (2015) Reinforcing lipid A acylation on the cell surface of Acinetobacter baumannii promotes cationic antimicrobial peptide resistance and desiccation survival. mBio 6(3):e00478-15
18. Slade D, Radman M (2011) Oxidative stress resistance in Deinococcus radiodurans. Microbiol Mol Biol Rev 75(1):133–191
19. Chaibenjawong P, Foster SJ (2011) Desiccation tolerance in Staphylococcus aureus. Arch Microbiol 193(2):125–135
20. Jawad A, Heritage J, Snelling AM, Gascoyne-Binzi DM, Hawkey PM (1996) Influence of relative humidity and suspending menstrua on survival of Acinetobacter spp. on dry surfaces. J Clin Microbiol 34(12):2881–2887
21. Neely AN (2000) A survey of Gram-negative bacteria survival on hospital fabrics and plastics. J Burn Care Rehabil 21(6):523–527
22. Wendt C, Dietze B, Dietz E, Rüden H (1997) Survival of Acinetobacter baumannii on dry surfaces. J Clin Microbiol 35(6):1394–1397

Chapter 19

Measuring Intracellular Metal Concentration via ICP-MS Following Copper Exposure

Caitlin L. Williams, Heather M. Neu, Sarah L. J. Michel, and D. Scott Merrell

Abstract

This chapter describes protocols for measuring fluctuation in intracellular metal concentration of *A. baumannii* isolates in response to copper exposure using inductively coupled plasma mass spectrometry (ICP-MS).

Key words *Acinetobacter*, *A. baumannii*, Copper, Heavy metal, Intracellular, ICP-MS, Mass spectrometry

1 Introduction

Heavy metals, such as copper, are essential for bacterial life but are also inherently toxic to the cell when present in excess quantities [1]. Thus, bacteria possess complex homeostasis systems to regulate the concentration of these metals within the cell. Because these systems are essential, there is potential for novel metal-based drug targets and therapeutics. Indeed, bacterial strains with mutations in crucial copper homeostasis mechanisms have been shown to be attenuated in in vivo infection models, highlighting the importance of these systems for survival and virulence [2]. Protocols presented describe methods to measure the changes in intracellular metal concentration following metal exposure using inductively coupled plasma mass spectrometry (ICP-MS); copper is used as the primary example. Comparing strains with and without intact homeostasis systems may give insight into the functionality of these systems.

Indranil Biswas and Philip N. Rather (eds.), *Acinetobacter baumannii: Methods and Protocols*, Methods in Molecular Biology, vol. 1946, https://doi.org/10.1007/978-1-4939-9118-1_19, © Springer Science+Business Media, LLC, part of Springer Nature 2019

2 Materials

1. 5× M9 salts: dissolve 64 g of $Na_2HPO_4 \cdot 7H_2O$, 15 g of KH_2PO_4, 2.5 g of NaCl, and 5 g of NH_4Cl in 1 L of deionized distilled H_2O (di-dH_2O), and autoclave [3]. Store at room temperature.
2. 20% glucose: dissolve 10 g of D-glucose in 50 mL of di-dH_2O, and filter sterilize with a 0.2 μm filter. Store at room temperature.
3. 1 M $MgSO_4$: dissolve 12.32 g of $MgSO_4 \cdot 7H_2O$ in 50 mL of di-dH_2O, and filter sterilize. Store at room temperature.
4. 1 M $CaCl_2$: dissolve 1.47 g of $CaCl_2 \cdot 2H_2O$ (Millipore Sigma) in 10 mL of di-dH_2O, and filter sterilize. Store at room temperature.
5. Bacto™ Casamino Acids (BD Biosciences).
6. 1 M $CuSO_4$: dissolve 1.60 g of anhydrous copper sulfate in 10 mL of di-dH_2O. Store at 4 °C (*see* **Note 1**).
7. LB agar: add 40 g/L to di-dH_2O. Autoclave, and cool to approximately 55 °C. Pour approximately 16–20 mL of agar into each petri dish. Let cool and solidify at room temperature. Store at 4 °C in closed bags. Before use, dry the plates by incubating at 37 °C for approximately 10 min with the lids open.
8. Petri dishes.
9. 1× PBS: add 10 mL of 10× PBS liquid concentrate (Millipore Sigma) to 90 mL of di-dH_2O, and autoclave or filter sterilize. Store at room temperature.
10. 0.5 M EDTA: add 9.30 g of EDTA dihydrate to 50 mL of di-dH_2O and autoclave. Store at room temperature.
11. Metal-free centrifuge tubes.
12. 6% HNO_3: add 90 mL of nitric acid (67–70% (w/w)) (Trace-Metal™ Grade) to 910 mL of MilliQ H_2O.
13. Copper reference standard solution: 1000 mg/L ± 1% (Fisher Scientific).
14. 500 μg/L ICP-MS internal standard solution: add 25 μL of scandium standard for ICP (1000 mg/L) (Millipore Sigma) and 25 μL of germanium standard for ICP (1000 mg/L) (Millipore Sigma) to 50 mL of 6% HNO_3.

3 Methods

3.1 Collection of Sample Cells

1. Prepare 250 mL of growth medium: In a large beaker on a stir plate, combine 194 mL of di-dH_2O, 50 mL of 5× M9 salts, 0.5 mL of 1 M $MgSO_4$, 5 mL of 20% glucose, 25 μL of 1 M $CaCl_2$, and 0.25 g of Casamino Acids [3]. Sterilize with a 0.2 μm filter. Use immediately or store at 4 °C for up to 7 days (*see* **Notes 2** and **3**).
2. Prepare the wash solution, 1× PBS containing 0.5 mM EDTA: Add 500 μL of sterile 0.5 M EDTA to 500 mL of sterile 1× PBS. Swirl to mix. Store at 4 °C. Keep on ice when on the bench.
3. Inoculate 5 mL of growth medium in a large glass test tube with a single bacterial colony (*see* **Note 4**) (*see* Fig. 1).
4. Incubate the culture at 37 °C, shaking at 190 rpm overnight (*see* **Note 5**).
5. Pipette 100 mL of growth medium into a 250 mL glass flask to be used for the copper exposure. Pipette 50 mL of medium into a 125 mL glass flask to be used later on for the recovery. Warm both to 37 °C (*see* **Note 6**) (*see* Fig. 1).
6. Measure the OD_{600} of the overnight cultures, and determine the volume required to start the 100 mL subcultures at an OD_{600} of 0.05. Pipette the required volume into 1.5 mL tubes, and pellet the bacteria by spinning at max speed (21,130 × *g*) for 1 min in a tabletop centrifuge (*see* **Notes 7** and **8**).
7. Discard the supernatant, resuspend each bacterial pellet in about 100 μL of growth medium taken directly from the 100 mL subculture, and then add the resuspended bacteria back to that same subculture. Swirl to evenly mix the bacteria. Start a timer to record the length of exposure.
8. Incubate the flask at 37 °C, 190 rpm.
9. Grow until the desired growth phase is reached (*see* **Note 9**).
10. Take a time zero (T0) sample of the culture as a measurement of basal copper levels (*see* **Note 10**).

3.2 Cell Sampling Procedure (See Fig. 2)

1. Take 0.5 mL of the culture and dilute it into 0.5 mL of growth medium in a cuvette.
2. Pipette 6 mL of culture into three 2 mL tubes.
3. Pellet the cells by spinning for 30 s at max speed in a 4 °C tabletop centrifuge.
4. While waiting for the spin to finish, return the flask to the incubator (*see* **Note 11**).

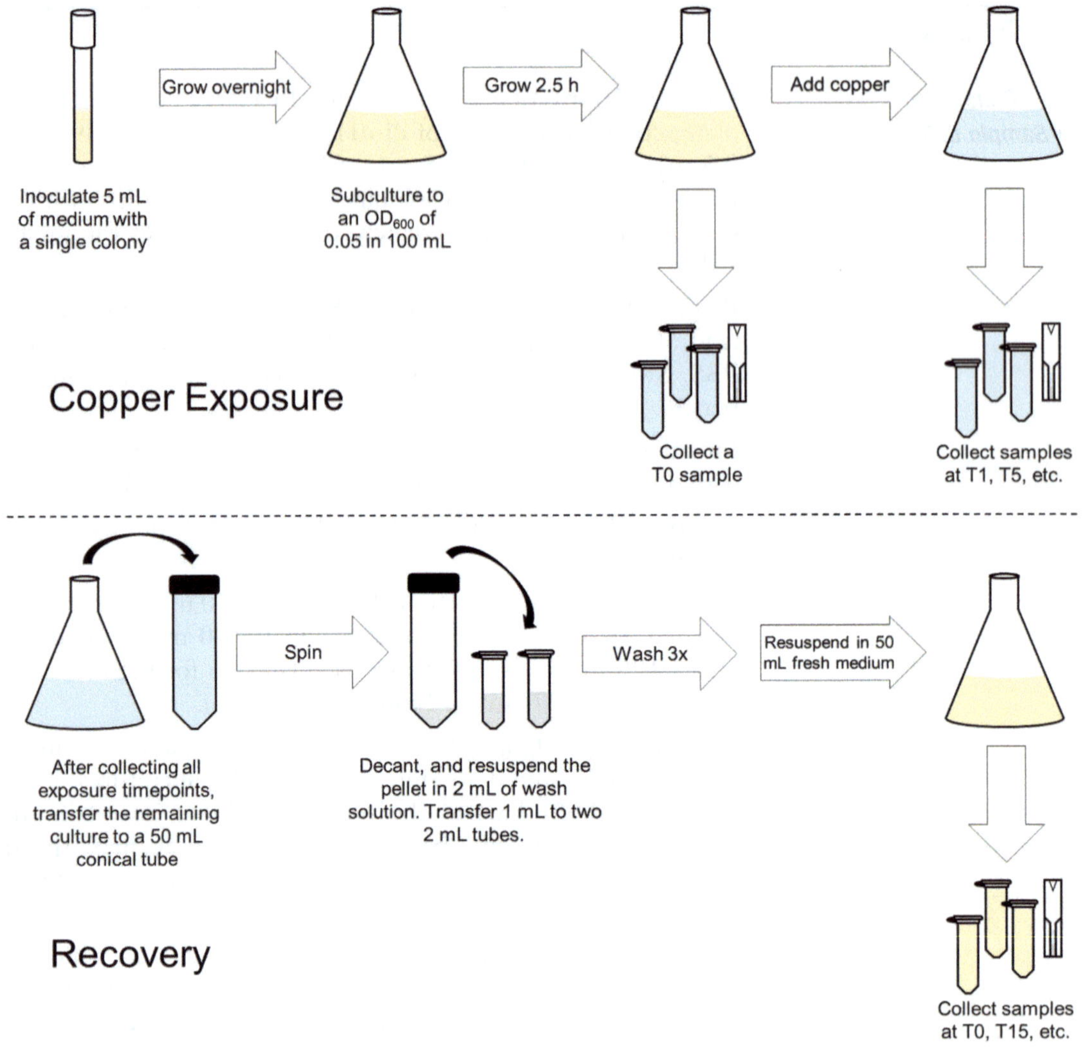

Fig. 1 Schematic overview of copper exposure and recovery techniques. A large subculture is started at an OD_{600} of 0.05 from an overnight culture. The subculture is grown 2.5 h to mid log phase. A time zero (T0) sample is collected to determine baseline metal concentrations. Metal exposure is begun by adding copper sulfate, and additional samples are collected over a time course. Once all of the exposure samples have been collected, the entire culture is pelleted, washed, and resuspended in fresh growth medium to begin the recovery period. Additional samples are collected over the recovery phase

5. Carefully aspirate the supernatants (*see* **Note 12**).
6. Combine the three pellets by resuspending them in 1 mL total of cold wash solution in a single tube (wash 1 of 3) (*see* **Note 13**).
7. Pellet the cells again.
8. While waiting for the spin to finish, record the OD_{600} of the diluted sample collected in **step 1**.

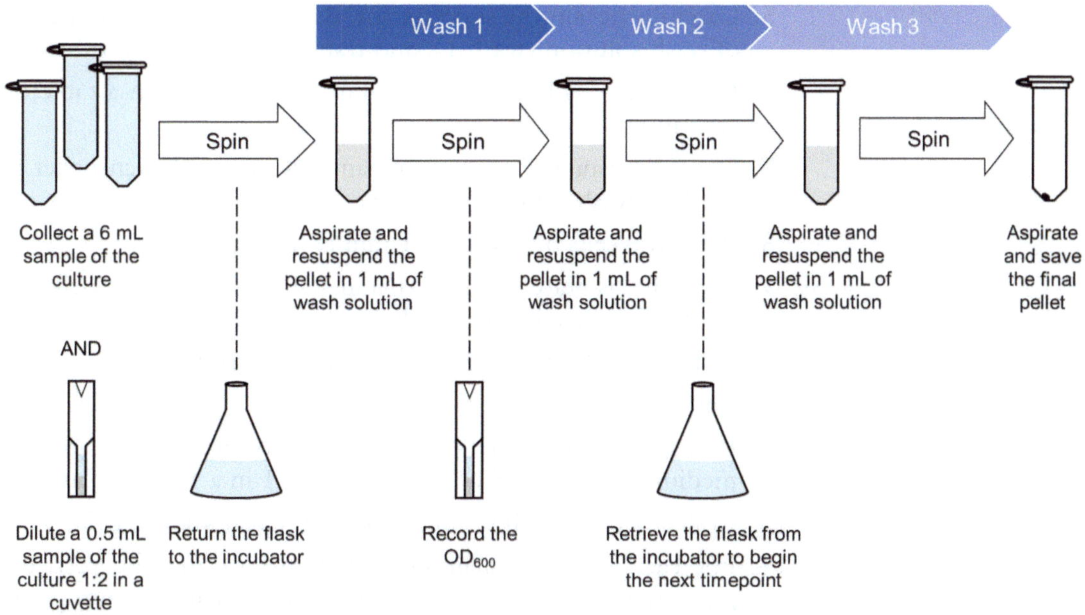

Fig. 2 Schematic overview of the sample collection protocol. At each timepoint, a 6 mL sample of the culture is collected for subsequent metal concentration analysis. At the same time, a 0.5 mL sample is collected for OD_{600} measurement. Because each wash step includes a 30 s spin, additional tasks (shown below) are performed during the incubation, indicated by the dashed lines

9. Carefully aspirate the supernatant.
10. Resuspend the cells in 1 mL of wash solution (wash 2 of 3).
11. Pellet the cells again.
12. Carefully aspirate the supernatant.
13. Resuspend the cells in 1 mL of wash solution (wash 3 of 3) (*see* **Note 14**).
14. Pellet the cells again (*see* **Note 15**).
15. Carefully aspirate the supernatant.
16. Set the tube aside with the lid open to allow liquid to begin evaporating.

3.3 Copper Exposure and Recovery (See Fig. 1)

1. Add 23.38 μL of 1 M $CuSO_4$ to the remaining 93.5 mL of culture, and swirl for approximately 30 s to mix (*see* **Note 16**).
2. Take a 1-min (T1) sample of the culture, and repeat steps described in Subheading 3.2 (*see* **Notes 17** and **18**).
3. For each of the remaining timepoints during the shock (T5, T10, T15, T20, T25, and T30), collect samples at the appropriate time, and repeat steps described in Subheading 3.2.
4. After collecting the T30 timepoint, pipette all of the remaining culture into a 50 mL conical tube. Spin for 15 min at maximum speed (3,700 × *g*) to pellet the cells (*see* **Notes 19** and **20**).

5. Carefully decant the supernatant. Resuspend the pellet in 2 mL of wash solution (wash 1), and transfer to two 2 mL tubes.
6. Pellet the cells by spinning 1 min at max speed in a tabletop centrifuge.
7. Carefully aspirate the supernatants. Resuspend each pellet in 1 mL of wash solution (wash 2).
8. Pellet the cells by spinning 1 min at max speed.
9. Carefully aspirate the supernatants. Resuspend each pellet in 1 mL of wash solution (wash 3).
10. Pellet the cells by spinning 1 min at max speed.
11. Carefully aspirate the supernatants.
12. Resuspend both pellets in the 50 mL of pre-warmed growth medium from **step 5** of Subheading 3.1 in a 125 mL flask.
13. Collect a post-wash sample (p0) and repeat **steps 11–26**.
14. For each of the remaining timepoints during the recovery phase (p15, p30, p45, p60), collect samples at the appropriate time, and repeat steps described in Subheading 3.2 (*see* **Note 18**).
15. Dry all of the cell pellets in a SpeedVac for 1 h at 80 °C.
16. Store the dried cell pellets at −20 °C (*see* **Note 21**).

3.4 ICP-MS Sample Preparation

1. To prepare cell pellets for ICP-MS analysis, add 200 μL of concentrated HNO_3 to each tube containing a cell pellet. Vigorously vortex to coat the cell pellet in acid. This will begin to break down the cell pellet and solubilize the contents.
2. Transfer the capped tubes to an oven set to 80 °C and incubate overnight (*see* **Note 22**).
3. The next morning, examine the tubes to ensure that all of the pellets have dissolved completely. Allow the tubes to cool to room temperature.
4. Add 2 mL of MilliQ water to the cooled samples so that the final HNO_3 concentration is 6%.
5. Transfer each acidified sample to a metal-free 15 mL conical tube.

3.5 ICP-MS Copper Analysis

1. Prepare a series of copper samples to be used to generate a calibration curve. Typically, samples that range between 10 and 2500 μg/L should be prepared, using serial dilution of a copper atomic absorption standard with 6% HNO_3 (*see* **Note 23**).
2. Turn on the ICP-MS, allow the instrument to warm up, and complete the manufacturer's recommended tuning protocol.
3. Samples are run in "batch" meaning that all calibration samples, internal standard samples, and unknown samples are run

sequentially. Suggested batch analysis parameters are octopole reaction system cell in helium mode, peristaltic pump speed of 0.1 rps, RF power of 1550 W, argon carrier gas flow of 1.01 L/min, argon makeup gas flow of 0.1 L/min, helium gas flow of 4.3 mL/min, octopole RF of 200 V, and OctP bias of −18 V.

4. The recommended order of analysis is:
   (a) Blank sample containing 6% HNO_3.
   (b) Calibration samples containing a range of copper atomic absorption standard.
   (c) Blank sample containing 6% HNO_3.
   (d) Unknown samples of interest.
   (e) Blank sample containing 6% HNO_3 (to ensure no carry-over of unknown samples into system).
5. Set up the ICP-MS sample introduction system with a mixing T, so that, when the sample is drawn up with the peristaltic pump, it is mixed with the internal standard solution prior to infusion into the ICP-MS by the MicroMist nebulizer. This allows for a consistent concentration of internal standard solution to be present in each sample. If a mixing T is not available, the internal standard solution can be spiked independently into each calibration and unknown sample. Care should be taken to ensure accuracy if this latter method is used.
6. Load the calibration sample and the unknown samples, housed in the 15 mL metal-free tubes, into the autosampler, and start the ICP-MS measurement. A minimum volume of 1 mL of sample is required for ICP-MS analysis.
7. Employ Agilent's Mass Hunter software to analyze data.
8. Construct a calibration curve by plotting the concentration of copper in the calibration samples versus the ratio of counts of copper (63) to the counts of germanium (72).
9. To determine the copper concentration in each unknown sample, compare the ratio of counts of copper to counts of germanium to the calibration curve.

3.6 Calculations

1. The raw data acquired is the quantity of copper atoms present in parts per billion (ppb, μg/L) in the liquid sample. Divide the ppb by the molecular weight of copper (63.546 g/mol) to determine the μM concentration of copper in the diluted sample. For example, 5 ppb Cu ÷ 63.546 g/mol = 0.07868 μM Cu.
2. Multiply by the number of mL of liquid in the sample to quantify the nanomolar amount of Cu in the sample. For example, 0.07868 μM × 2.2 mL = 0.1731 nmol of Cu in the sample.

3. Divide by the number of milliliters of sample collected to determine the number of nmol/mL of cells of the culture. For example, 0.1731 nmol Cu ÷ 6 mL = 0.0289 nmol/mL.
4. Divide by the OD_{600} to determine nmol/ODU of cells (1 mL of cells at an OD_{600} of 1.0). For example, 0.0289 nmol/mL ÷ 0.575 = 0.0502 nmol/ODU.
5. Graph the data as nmol/ODU vs. time. Mark or break the x-axis where appropriate to identify the exposure and recovery phases.

4 Notes

1. We have noticed differences between fresh stock solutions of $CuSO_4$ and solutions stored at room temperature for many months. We recommend making small batches of fresh solution every 6–12 months and storing at 4 °C.
2. We have noticed reduced bacterial growth in M9 growth medium that is greater than 7 days old; therefore, we make fresh batches as often as possible and store the media at 4 °C.
3. We chose M9 medium with Casamino Acids for these experiments because it is a defined, minimal growth medium in which *A. baumannii* can still grow. It is important to use a minimal medium to limit sequestration of the added metal ions by components that are typically found in rich media. Metal binding to these components reduces the effective concentration of metal.
4. Some strains of *A. baumannii* undergo phase variation [4]. Our lab carefully checks each plate of colonies struck from a freezer stock and chooses a single, opaque colony for each experiment.
5. Our lab has defined an overnight culture as 16–19 h of growth. This length of time may be changed, but it is important to keep consistent timing between replicate experiments in order to maximize reproducibility.
6. It is important that the flasks have been previously acid washed to remove all heavy metals. We submerge flasks in a 10% nitric acid bath overnight, rinse 10× with dH_2O, cover with lids or foil, and autoclave.
7. Our overnight cultures typically have an OD_{600} of 1.0–1.7. Therefore, the volume of culture pelleted for a 100 mL subculture is typically 3–5 mL. For example, [(0.05 × 100 mL)/ 1.5 = 3.33 mL].

8. To reliably measure high OD_{600}, we dilute 1:2–1:5, so that the measured optical density is no higher than 0.7 and within the linear range of the spectrophotometer.
9. For our experiments, we grow the bacteria for 2.5 h to reach log phase before beginning the copper exposure.
10. The amount of copper and other metals present in the growth medium will depend on the water supply. We have found that our M9 medium routinely contains 1–2 μM copper [5]. If the growth medium has high concentrations of copper, the baseline amount of intracellular copper may be elevated. We recommend measuring the metal concentration of growth medium prepared with various water samples collected over time to ensure that the water source has consistently low levels of heavy metals.
11. Because the T1 and T5 timepoint are so close together, we do not return the flask to the incubator after collecting the T1 sample, but wait until after the T5 sample is collected.
12. The final copper measurements are adjusted for the number of cells collected using the OD_{600} measurement and a consistent collection volume (6 mL). Therefore, it is very important to ensure accurate collection volumes and to minimize loss of cells during the wash steps.
13. It is important to make sure that the tubes used are acid resistant and metal-free before collecting all the samples. For acid digestion and ICP-MS analysis, VWR metal-free tubes are utilized.
14. If many samples are being collected, it may be beneficial to collect the cell pellets in 15 mL conical tubes so that they do not have to be transferred when preparing the samples for subsequent ICP-MS analysis. To make this change, at **step 13** (Subheading 3.2) of the sampling procedure, when resuspending the cells in wash solution for the final time, transfer the cells to a metal-free, acid-resistant conical tube. Pellet the cells by spinning for 3 min at max speed in a Sorvall Legend RT centrifuge (4150 rpm/3,700 × *g*). Be aware that we observed much greater variability in the results when using this modified procedure. Also note that because of the increased spin time, it is more difficult to collect samples quickly and efficiently.
15. We've found that this process runs rather smoothly despite the fact that the procedure overlaps for two consecutive timepoints. It is likely that the flask will need to be retrieved to begin the next timepoint during the pelleting step of wash two (Subheading 3.2, **step 11**). The initial pelleting of the next timepoint can be combined with the final spin of wash 3 of the previous sample (Subheading 3.2, **steps 3** and **14**). We have found it helpful to mark each tube lid with small lines to track

the number of washes completed. Also, we have found that we can process three strains simultaneously on a given day; processing more samples was not possible due to the required steps and precise timing.

16. We chose a concentration of copper for the shock that did not kill any of the strains in the growth curve experiments: 250 μM. We did not want to kill any of the bacteria during this experiment; however we did see signs of stress, e.g., halting of growth. The appropriate concentration of copper or any other chosen metal will have to be determined. Note that concentrations of $CuSO_4$ above 1.5 mM do not reliably stay in solution in this minimal medium and therefore are not recommended.
17. It is important to account for the OD_{600} of the added $CuSO_4$ and make the proper adjustments. For example, when measuring the OD_{600} of a sample containing 250 μM $CuSO_4$, you must either blank the spectrophotometer with medium containing 250 μM $CuSO_4$ *or* subtract the OD_{600} of medium containing 250 μM $CuSO_4$ after the data is collected. When measuring the OD_{600} of a sample containing 250 μM $CuSO_4$ that was diluted 1:2, you must make adjustments for 125 μM $CuSO_4$. Note that we have found the OD_{600} of 125 μM $CuSO_4$ to be negligible.
18. The appropriate time course length and sampling interval required to observe changes in intracellular metal concentration will have to be empirically determined for your chosen strain and metal.
19. The purpose of returning the cells to media without supplemented metal is to test the ability of each strain to recover intracellular metal concentrations to baseline.
20. We determined that a 15 min spin at 3,700 × *g* was sufficient to pellet the cells as quickly as possible while still leaving a pellet that did not fall apart when the supernatant was removed via decanting. The spin time and speed can be modified depending on the centrifuge used; minimizing the time without losing cells is most important.
21. If dried cells need to be shipped, send on dry ice.
22. We determined that acid alone was not sufficient to fully digest the cell pellet. If the sample was only acidified, a solid cell pellet remained in the tube. Heating the acidified samples overnight at 80 °C was required to accomplish complete digestion of the cells. We recommend always inspecting each of the samples after overnight digestion to verify that each cell pellet has been completely digested.

23. The series of copper samples used to generate a calibration curve contained 10–2500 μg/L. It is preferable that the analyte of interest lies at the midpoint of the standard copper atomic absorption samples; however, our samples contained 1–250 ppb copper, which is on the low end of the calibration curve. Collecting larger samples of cells increases the final concentration of copper that is measured and is the reason for the 6 mL sample volume. Further increases in sample size would likely be too technically cumbersome to collect given the rapid timepoints; additionally, our results showed consistency across experiments, and therefore we did not feel that the measurements were inaccurate due to low measured values.

Acknowledgment

Disclaimer and Funding: The opinions or assertions contained herein are the private ones of the authors and are not to be construed as official or reflecting the views of the Department of Defense, the Uniformed Services University of the Health Sciences, the University of Maryland, or any other agency of the US Government.

Work in the Merrell lab is supported by funds from the NIH and DoD. S.L.J.M. would like to thank the NSF (CHE-10020356) for support.

References

1. Ladomersky E, Petris MJ (2015) Copper tolerance and virulence in bacteria. Metallomics 7 (6):957–964. https://doi.org/10.1039/c4mt00327f
2. Djoko KY, Ong CL, Walker MJ, McEwan AG (2015) The role of copper and zinc toxicity in innate immune defense against bacterial pathogens. J Biol Chem 290(31):18954–18961. https://doi.org/10.1074/jbc.R115.647099
3. Sambrook J, Russell DW (2001) Molecular cloning: a laboratory manual, 3rd edn. Cold Spring Harbor Laboratory Press, Cold Spring Harbor, NY
4. Tipton KA, Dimitrova D, Rather PN (2015) Phase-variable control of multiple phenotypes in *Acinetobacter baumannii* strain AB5075. J Bacteriol 197(15):2593–2599. https://doi.org/10.1128/JB.00188-15
5. Williams CL, Neu HM, Gilbreath JJ, Michel SL, Zurawski DV, Merrell DS (2016) Characterization of copper resistance in *Acinetobacter baumannii*. Appl Environ Microbiol 82 (20):6174–6188. https://doi.org/10.1128/AEM.01813-16

Chapter 20

PBP Isolation and DD-Carboxypeptidase Assay

Shilpa Pal and Anindya S. Ghosh

Abstract

Penicillin-binding proteins (PBPs) share the namesake because of their ability to bind penicillin or any beta-lactam antibiotic. In other words, PBPs are the targets of β-lactam antibiotics that hold nearly 60% of the global antibiotic market. These enzymes catalyze the final stages of peptidoglycan (PG) biosynthesis by acting as transglycosylases and transpeptidases. PBPs are also involved in PG remodeling by catalyzing DD-carboxypeptidase (DD-CPase) and endopeptidase reactions. Though the cross-linking abilities of PBPs are well known, the process of remodeling is still unclear, thereby drawing attention toward the DD-CPase enzymes. Here, we describe the step-by-step procedures for isolation of the bacterial cell membrane and detection of PBPs in it, followed by the purification of PBPs (DD-CPases) by both ampicillin-affinity and nickel-nitrilotriacetic acid (Ni-NTA) chromatography. The protocols to determine the enzymatic efficiency are also elucidated. The assays are aimed to determine the kinetic parameters for the interaction of the PBP with BOCILLIN, to evaluate its acylation and deacylation rates, and with its peptide substrates, to assess its DD-CPase activity.

Key words Penicillin-binding protein (PBP), Ampicillin-affinity chromatography, Ni-NTA chromatography, BOCILLIN, Enzyme kinetics, Acylation, Deacylation, DD-Carboxypeptidase assay

1 Introduction

Penicillin-binding proteins (PBPs) are periplasmic proteins (ectoproteins) that are tethered to the cytoplasmic membrane. They comprise of four different types of enzymes having the unique ability to bind penicillin. PBPs that catalyze the final stages of peptidoglycan (PG) synthesis by linking *N*-acetyl glucosamine and *N*-acetyl muramic acid are called transglycosylases, whereas the PBPs that join two adjacent glycan strands of the opposite nascent PG, through peptide linkages, are called transpeptidases. Also, PBPs are involved in PG remodeling, which prevents unwanted cross-link formation (DD-carboxypeptidases) as well as the removal of aberrantly formed cross-links (endopeptidases) [1]. Not much research has been done on PG remodeling as compared to PG cross-linking; therefore, DD-CPases are relatively unexplored. DD-CPases act on the pentapeptide side chains of the

Indranil Biswas and Philip N. Rather (eds.), *Acinetobacter baumannii: Methods and Protocols*, Methods in Molecular Biology, vol. 1946, https://doi.org/10.1007/978-1-4939-9118-1_20, © Springer Science+Business Media, LLC, part of Springer Nature 2019

muramic acids in the PG layer and remove the terminal D-alanine to generate tetrapeptides. Unlike pentapeptides, these tetrapeptides are unable to participate in cross-link formation [2]. The D-alanyl-D-alanine moiety, which structurally resembles the beta-lactam ring, is present at the end of the pentapeptide stem and is the substrate of these enzymes.

So far, most of the work on PBPs have been carried out in *Escherichia coli*, possessing 12 known PBPs, and their functions are partially identified. However, little is known about the role of individual PBPs of *Acinetobacter baumannii*, which is reported to have six PBPs [3]. There are four DD-CPases in *E. coli*, PBP4, PBP5, PBP6, and DacD [4], whereas only two have been annotated as DD-CPases in *A. baumannii*, namely, PBP5/6 (*dacC*) and PBP6b (*dacD*) [3].

PBPs are ectoproteins tethered to the outer facet of the cytoplasmic membrane with a C-terminal amphipathic membrane anchor that maintains the flexibility of the enzymatic domain while working on the PG layer placed just above. As PBPs are membrane proteins, we have described the process of membrane isolation and detection of PBPs using the fluorescent penicillin BOCILLIN [5]. However, being a membrane protein, it is difficult to purify these proteins in their active form. The conventional method of solubilization for such proteins is to use different detergents, but in most of the cases, PBPs undergo denaturation upon detergent treatment. Renaturation of the protein is not guaranteed and might result in loss of activity. As PBPs can bind to ampicillin, ampicillin-affinity chromatography is a suitable and convenient procedure for their purification [6]. PBPs can also be purified using immobilized metal-affinity (Ni-NTA) chromatography. Both the methods have been described in this chapter. The binding efficiency of PBPs with penicillin is determined by calculating their acylation and deacylation rates upon interaction with BOCILLIN. The related protocols are discussed here, and we have also explained the methodology of the DD-carboxypeptidase assay that is helpful to determine the kinetic parameters of the interaction of the PBP with its peptide substrates.

2 Materials

Solvents and reagents used should be of analytical grade or equivalent. Prepare all solutions using ultrapure Milli-Q water. Allow the mixtures to dissolve in a beaker using a magnetic stirrer. Prepare and store all reagents at room temperature (unless indicated otherwise).

2.1 Isolation of Membrane-Anchored PBPs from the Cell Membrane

1. 10 mM potassium phosphate buffer, pH 7.4:
   1 M K_2HPO_4: Dissolve 1.74 g dipotassium phosphate (K_2HPO_4) in 10 mL water.

1 M KH_2PO_4: Dissolve 1.36 g potassium diphosphate (KH_2PO_4) in 10 mL water.

To prepare 100 mM phosphate buffer, pH 7.4, mix 8.02 mL of 1 M K_2HPO_4 and 1.98 mL of 1 M KH_2PO_4, and adjust the volume to 100 mL with water. Finally, dilute 100 mM phosphate buffer with water to prepare 10 mM phosphate buffer.

2.2 Cloning of Soluble Form of PBPs

1. pT7-7K and pET28a(+) plasmids.
2. *E. coli* BL21(DE3) strain.

2.3 Expression of Soluble PBPs

1. Inducer: 100 mM IPTG. Dissolve 238 mg IPTG (isopropyl-1-thio-β-D-galactopyranoside) in 8 mL of water. Make up the volume to 10 mL. Filter the solution through a 0.22 μm syringe-driven filter and store at −20 °C. (Protect from light.)
2. Lysis buffer: 10 mM Tris–HCl, 300 mM NaCl, 0.2 mM PMSF, pH 8.0:

   1 M Tris–HCl, pH 8.0, stock solution. To prepare a 1 M stock solution of Tris–HCl, dissolve 121.14 g Tris base in 800 mL of water. Adjust the pH to 8.0 by adding concentrated HCl, and make up the volume to 1000 mL.

   1 M NaCl stock solution: Dissolve 58.44 g sodium chloride in 1000 mL water.

   100 mM PMSF stock solution: Dissolve 1.74 mg PMSF (phenylmethylsulfonyl fluoride) in 10 mL isopropanol. Store at −20 °C.

To prepare 100 mL of lysis buffer, mix 1 mL of 1 M Tris–HCl, pH 8, and 30 mL of 1 M NaCl. Finally, make up the volume to 100 mL with water. Add PMSF to the buffer just before sonication such that the final concentration is 0.2 mM (*see* **Note 1**).

2.4 Purification of Soluble PBPs

Affinity chromatography is popularly used for protein purification. It exploits the biological function of the biomolecules to separate them. The process involves a specific and reversible interaction of a protein or peptide (mobile phase) with a ligand which has been immobilized on a solid support (stationary phase).

2.4.1 Coupling of Ampicillin into Activated CH-Sepharose 4B

Ampicillin-coupled CH-Sepharose 4B is used for purification of PBPs. Activated CH-Sepharose 4B is synthesized in several steps. Firstly, aminocaproic acid is immobilized to cyanogen bromide-activated Sepharose 4B to form CH-Sepharose 4B, which is further activated by esterification of its carboxyl group with *N*-hydroxysuccinimide (Fig. 1). As ampicillin [6-(D(−)-alpha-aminophenylacetamido)penicillanic acid] contains a primary amino group, it can easily be coupled with the activated CH-Sepharose 4B.

1. Affinity resin: Use activated CH-Sepharose 4B.

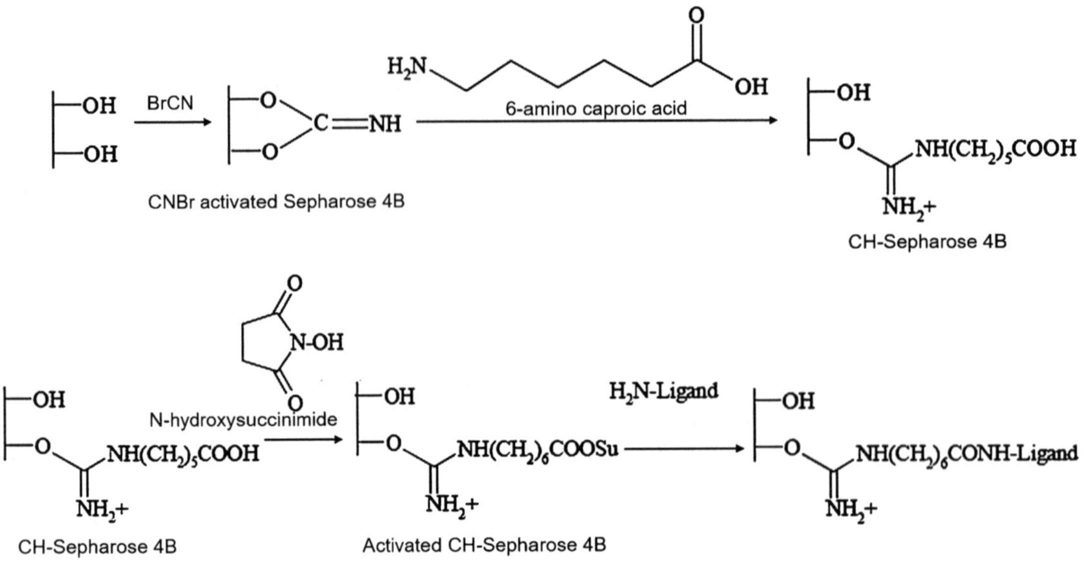

Fig. 1 Structure of activated CH-Sepharose 4B and its attachment to the ligand

2. HCl (0.001 N): Dilute 1 N HCl 1000 times in water. Be careful while handling concentrated HCl.
3. Coupling buffer: 100 mM $NaHCO_3$, 500 mM NaCl, pH 8.0. Dissolve 4.2 g sodium bicarbonate ($NaHCO_3$) in 200 mL of water. Add 250 mL of 1 M NaCl to it. Adjust the pH to 8.0. Make up the volume to 500 mL with water.
4. Blocking buffer: 100 mM Tris–HCl, pH 8.0. Add 25 mL of 1 M Tris–HCl, pH 8.0, to water, and make up the volume to 250 mL.
5. Wash buffer A: 100 mM acetate buffer, 500 mM NaCl, pH 4.0. Dissolve 3.4 g sodium acetate ($C_2H_3NaO_2$) in 100 mL of water. Add 125 mL of 1 M NaCl to it. Adjust the pH to 4 using glacial acetic acid. Finally, make up the volume to 250 mL with water.
6. Wash buffer B: 100 mM Tris–HCl, 500 mM NaCl, pH 8.0. Mix 25 mL of 1 M Tris–HCl, pH 8.0, and 125 mL of 1 M NaCl to water, and make up the volume to 250 mL.

2.4.2 Purification Through Ampicillin-Affinity Chromatography

1. Wash buffer C: 50 mM Tris–HCl, 1 M NaCl, pH 7.5. Dissolve 29.22 g NaCl in 300 mL water. To this, add 25 mL of 1 M Tris–HCl, pH 8.0. Adjust the pH to 7.5 with concentrated HCl, and make up the volume to 500 mL with water.
2. Wash buffer D: 50 mM Tris–HCl, pH 7.5. Add 25 mL of 1 M Tris–HCl, pH 8.0–400 mL water. Adjust the pH to 7.5 with

concentrated HCl. Finally, make up the volume to 500 mL with water.

3. Elution buffer: 1 M NH_2OH, 500 mM Tris–HCl, pH 7.0. Dissolve 20.8 g hydroxylamine (NH_2OH) in 100 mL water. To this, add 150 mL of 1 M Tris–HCl, pH 8.0. Adjust the pH to 7.0 with concentrated HCl. Finally, make up the volume to 300 mL with water.
4. Dialysis buffer: 20 mM Tris–HCl, 150 mM NaCl, pH 7.5. Mix 20 mL of 1 M Tris–HCl, pH 8.0, and 150 mL of 1 M NaCl with 700 mL water. Adjust the pH to 7.5 with concentrated HCl. Finally, make up the volume to 1 L with water.

2.4.3 Ni-NTA Affinity Chromatography

Immobilized metal-affinity chromatography is widely used for the purification of histidine-tagged proteins. The $(histidine)_6$ tag is one of the most common tags used to facilitate the purification of recombinant proteins. It binds to Ni-NTA and is eluted with high concentrations of imidazole. Nitrilotriacetic acid (NTA) is a tetradentate chelating adsorbent that engages four of the six ligand binding sites in the coordination sphere of the nickel ion, and the remaining two sites interact with the His tag. NTA binds metal ions in a stable condition and inhibits the loss under strict washing conditions [7]. The optimal placement of the tag can be either at the C- or the N-terminus and is protein specific. At pH 8.0 the tagging is small and uncharged and does not affect the folding of the protein of interest. If necessary, the tag can be removed.

1. Binding buffer: 10 mM Tris–HCl, 300 mM NaCl, 25 mM imidazole, pH 8.0,

   1 M imidazole. Dissolve 6.808 g imidazole in 100 mL water. Mix 2.5 mL of 1 M Tris–HCl, pH 8.0, 75 mL of 1 M NaCl, and 6.25 mL of 1 M imidazole. Make up the volume to 250 mL with water.
2. Elution buffer: 10 mM Tris–HCl, 300 mM NaCl, 500 mM imidazole, pH 8.0. Mix 2.5 mL of 1 M Tris–HCl, pH 8.0, 75 mL of 1 M NaCl, and 125 mL of 1 M imidazole. Make up the volume to 250 mL with water.

2.5 BOCILLIN Binding Assay

BOCILLIN FL penicillin is a fluorescent penicillin and is used to determine penicillin-binding protein profiles of a number of bacteria [5]. It is synthesized from penicillin V and BODIPY FL dye (which is spectrally similar to fluorescein). The excitation and emission wavelengths are 488 and 526 nm, respectively.

1. 10 mM potassium phosphate buffer, pH 7.4 (*see* Subheading 2.1, **item 1**).
2. 5 mM BOCILLIN stock solution: To prepare a solution of 5 mg/mL (7.5 mM), add 1 mg of BOCILLIN FL penicillin

and sodium salt (mol. wt. 661.46) to 200 μL of water. From this, dilute 133.3–200 μL to get 5 mM BOCILLIN. (Protect from light.)

2.6 Kinetic Analysis of the Interaction of PBPs with BOCILLIN

2.6.1 Determination of the Acylation Rate Constant

1. 10 mM potassium phosphate buffer, pH 7.4 (*see* Subheading 2.5, **item 1**).
2. 5 mM BOCILLIN (*see* Subheading 2.5, **item 2**).
3. Denaturing buffer: 250 mM Tris–HCl, pH 6.8, 10% SDS, 0.02% bromophenol blue. The SDS-PAGE gel loading dye is used as the denaturing buffer. For 10 mL buffer, mix 2.5 mL of 1 M Tris–HCl with 6.5 mL water. Adjust the pH to 6.8. To this add 1 g SDS (sodium dodecyl sulfate) and 2 mg bromophenol blue, and let it dissolve. Finally, make up the solution to 10 mL with water.

2.6.2 Determination of the Deacylation Rate Constant

1. 10 mM potassium phosphate buffer, pH 7.4 (*see* Subheading 2.5, **item 1**)
2. 5 mM BOCILLIN (*see* Subheading 2.5, **item 2**).
3. Penicillin G (sodium salt).
4. Denaturing buffer (*see* Subheading 2.6.1, **item 3**).

2.7 Kinetic Analysis of DD-Carboxypeptidase Activity of PBP

1. 50 mM Tris–HCl, pH 8.5: Add 5 mL of 1 M Tris–HCl to 70 mL water. Adjust the pH to 8.5 with concentrated HCl. Finally, make up the volume to 100 mL with water.
2. 1 mg/mL flavin adenine dinucleotide (FAD) stock solution: Weigh 1 mg FAD and dissolve in 1 mL water. Store the stock solution at −20 °C. (Protect from light.)
3. 0.3 mg/mL FAD working solution: Dilute 300 μL of 1 mg/mL FAD with water to prepare 1 mL of 0.3 mg/mL FAD solution. Prepare freshly before use. (Protect from light.)
4. 5 mg/mL D-amino acid oxidase: Weigh 5 mg of D-amino acid oxidase and dissolve in 1 mL water. Store the solution at −20 °C. (Protect from light.)
5. 10 μg/mL horseradish peroxidase: Dissolve 5000 units of horseradish peroxidase in 3.2 M ammonium sulfate suspension containing potassium phosphate buffer of pH 6.0. The protein concentration in this suspension is 13 mg/mL. Dilute 0.77 μL of the horseradish peroxidase solution to 1 mL with water to form peroxidase solution of the desired concentration.
6. Enzyme-coenzyme mixture: Prepare a mixture containing horseradish peroxidase, D-amino acid oxidase, and FAD in the following ratio: 50 mM Tris–HCl, pH 8.5; 0.3 mg/mL FAD; 10 μg/mL horseradish peroxidase; 5 mg/mL D-amino acid oxidase = 20:10:5:1. (Prepare freshly before use.)

7. 10 mg/mL *o*-dianisidine dihydrochloride: Weigh 10 mg *o*-dianisidine dihydrochloride and dissolve in 1 mL water. (Prepare freshly before use.)
8. Artificial tripeptide substrate: N_α,N_ε-diacetyl-Lys-D-Ala-D-Ala. Prepare a stock solution of 50 mM of the peptide, and dilute it to the desired concentration required. (Prepare freshly before use.)
9. Peptidoglycan mimetic pentapeptide substrate: L-Ala-γ-D-Glu-L-Lys-D-Ala-D-Ala (custom synthesized). Prepare a stock solution of 50 mM of the peptide, and dilute it to the desired concentration required. (Prepare freshly before use.)

3 Methods

Unless otherwise specified, carry out all the procedures at room temperature.

3.1 Detection of Membrane-Anchored PBPs in the Cell Membrane

1. Inoculate 500 mL of LB with 1% of an overnight bacterial culture, and incubate at 37 °C with shaking until it reaches an OD_{600} ~ 0.4.
2. Harvest the cells by centrifugation (5000 × *g* for 10 min) (*see* **Note 2**).
3. Wash the pellet twice, and finally resuspend it in 100 mM phosphate buffer, pH 7.0.
4. Lyse the homogenous suspension of cells by sonication (*see* **Note 3**).
5. Centrifuge the lysate at 5000 × *g* for 10 min to remove the cellular debris, and transfer the supernatant to a fresh tube.
6. Centrifuge the supernatant at 100,000 × *g* for 60 min at 4 °C to precipitate the membranes by using an ultracentrifuge.
7. Resuspend the pelleted membranes in the same buffer, and estimate the total protein by Bradford protein estimation kit. Plot a standard curve of the amount of BSA (μg) versus absorbance at 595 nm, and determine the concentration of the protein from this standard curve.

3.2 Cloning of Soluble Form of PBPs

3.2.1 pT7-7K and pET28a(+) Expression Systems

When ampicillin-affinity chromatography is used for purifying PBPs, the plasmid pT7-7K is used as the expression vector, whereas the pET28a(+) expression system is used for purification using Ni-NTA chromatography. Both the vectors have the T7 promoter. T7 RNA polymerase is very active and is used for high-level transcription of the gene of interest. The expression of the PBP gene is induced by providing a source of T7 RNA polymerase which is provided by using a host that contains a chromosomal copy of the T7 RNA polymerase gene. There are many such host strains.

However, the use of *E. coli* BL21 (DE3) as the host is specified here. It contains the λDE3 lysogen that carries the gene for T7 RNA polymerase under control of the lacUV5 promoter, which can be induced by IPTG [8].

3.2.2 Detection of Sequences for N-Terminal Signal Peptide and C-Terminal Amphipathic Helix

PBPs are membrane-anchored proteins which make their purification difficult. Conventionally membrane proteins are solubilized using detergents, but PBPs get denatured upon detergent treatment. However, for the ease of purification and subsequent characterization, the PBPs can be genetically engineered to express them in their soluble forms within the cytoplasm. For this, the nucleotide sequences corresponding to the amino acids at the N-terminal signal peptide and C-terminal amphipathic regions have to be removed. The presence and location of the signal peptide cleavage site in the amino acid sequence of the gene are detected by using the "SignalP 4.1" server [9]. Signal peptides range from 21 to 29 amino acids. A representative image of a typical prediction result from the SignalP server is shown in Fig. 2. Here, the first 21 amino acids from the N-terminal are detected as the signal peptide sequence. The C-terminal amphipathic regions can be detected using different servers, e.g., "HELIQUEST" [10]. These regions can also be detected by three-dimensional modeling of the protein.

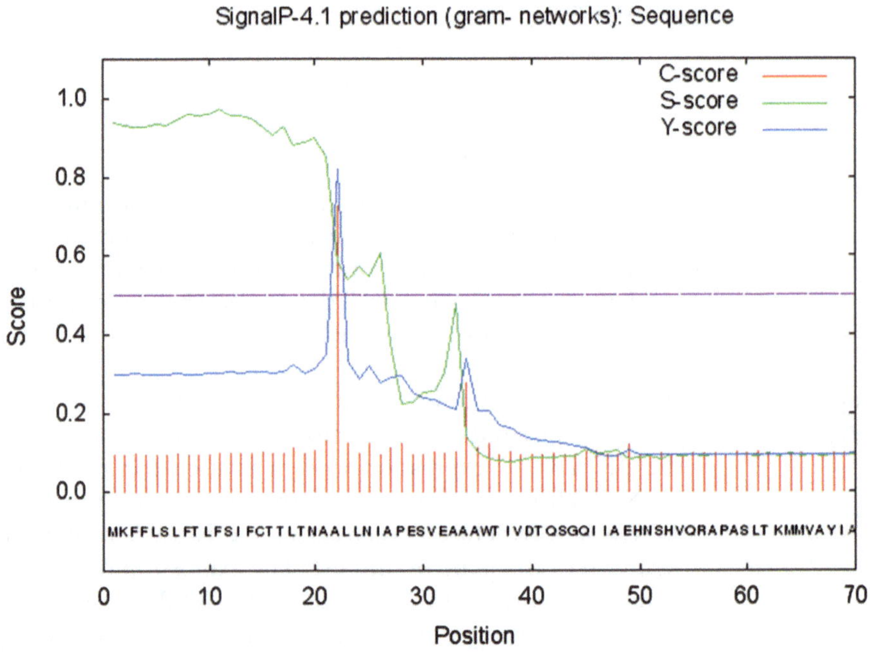

Fig. 2 SignalP 4.0 server prediction result for N-terminal signal peptide of a PBP

3.2.3 Construction of the Clones and Transformation

1. Design the primers such that the resultant gene would express in its soluble form, devoid of the aforementioned terminal sequences.
2. Amplify the gene encoding the soluble form of the PBP, and clone it in the vector. The choice of vector depends on the method of purification to be used (*see* Subheading 3.2.1). Confirm the sequence of the gene by sequencing.
3. Prepare chemical competent cells of *E. coli* BL21 (DE3) by following the standard protocol (*see* **Note 4**).
4. Transform BL21 (DE3) chemical competent cells with the cloned constructs, through the heat-shock method. Select the recombinant clones against 50 μg/mL kanamycin. (The expression vectors pT7-7K and pET28a (+) have the kanamycin resistance gene.)
5. Make sure the transformants are correct by digesting with the restriction enzymes that are used for cloning and subsequently confirm by DNA sequencing.
6. Inoculate the freshly transformed BL21 (DE3) strain harboring the desired construct for further expression studies (*see* **Note 5**).

3.3 Expression of Soluble PBPs

Before purification of any protein in its active form, it is necessary to optimize its expression. In the aforementioned expression systems, very high levels of gene expression can be achieved by increasing the concentration of IPTG and increasing the post-induction period. However, sometimes a majority of the protein might remain as inclusion bodies. Moreover, IPTG is toxic to cells and is expensive too. So, to achieve a balance between cell survival and protein production, we have to optimize several parameters such as the IPTG concentration, the growth phase at which IPTG is to be added, the post-induction temperature, and the period of post–induction.

For expression optimization:

1. Allow the culture to grow overnight in a kanamycin-supplemented LB broth.
2. Next morning, inoculate 10 mL of kanamycin-supplemented LB broth with 0.1% inoculum from the overnight culture.
3. Express the gene by varying any one of the four parameters at a time, i.e., IPTG concentration, incubation temperature, phase of induction, and period of post–induction (*see* **Note 6**).
4. To check the expression levels, harvest the cells by centrifuging at 5000 × *g* at 4 °C for 10 min.
5. Resuspend the cell pellets in 200 μL lysis buffer, and sonicate the above mixture (*see* **Note 2**).

6. Centrifuge the sonicated samples at 10,000 × *g* at 4 °C for 30 min. Insoluble cell components (membranes and cell debris) will be pelleted at the bottom of the tube (*see* **Note 1**).
7. Collect the supernatants containing the soluble proteins and transfer into a fresh tube. Estimate the protein concentration by Bradford protein estimation kit using BSA as the standard (*see* Subheading 3.1, **step** 7).
8. Run a 15% SDS-PAGE following the standard protocol, to determine the expression levels (*see* **Note** 7).
9. After finalizing the optimum expression conditions, scale up the volume depending on the yield obtained and the concentration of protein desired.

3.4 Purification of Soluble PBPs

3.4.1 Coupling of Ampicillin into Activated CH-Sepharose 4B

1. Suspend 2 g of freeze-dried activated CH-Sepharose 4B in 0.001 N HCl, and wash it with the same (*see* **Note 8**).
2. Dissolve 90 mg ampicillin in the coupling buffer, and allow it to mix with the resin for 2 h at room temperature with gentle shaking.
3. Wash away the excess ampicillin with 500 mL of coupling buffer.
4. Block the remaining active site groups by incubating the ampicillin-coupled resin in the blocking buffer for 1 h.
5. Finally, wash the ampicillin-coupled resin thrice alternatively, with wash buffer A and buffer B.

3.4.2 Purification Through Ampicillin-Affinity Chromatography

1. Mix the cell lysate with ampicillin-coupled activated CH-Sepharose 4B, and incubate for 1 h at 30 °C.
2. Wash the resin with 1 L of wash buffer C followed by 1 L of wash buffer D, using a sintered glass funnel.
3. Elute the PBPs bound to the resin with the elution buffer with a flow rate of 1 mL/min. Monitor the elution profile by measuring the absorbance of the eluents at 280 nm [11].
4. Immediately pool the purified protein fractions, and dialyze for 12 h with three changes against dialysis buffer (*see* **Note 9**).
5. Estimate the concentration of the purified protein using Bradford assay.
6. Store the protein on ice at 0 °C for immediate use or at −80 °C (with 50% glycerol) for long-term storage (*see* **Note 10**).

A schematic diagram of the purification process is illustrated in Fig. 3.

3.4.3 Ni-NTA Affinity Chromatography

1. In a chromatography system, wash the HisTrap affinity (Ni-NTA) column with degassed Milli-Q water to remove the ethanol. The washing is complete after the conductivity and the

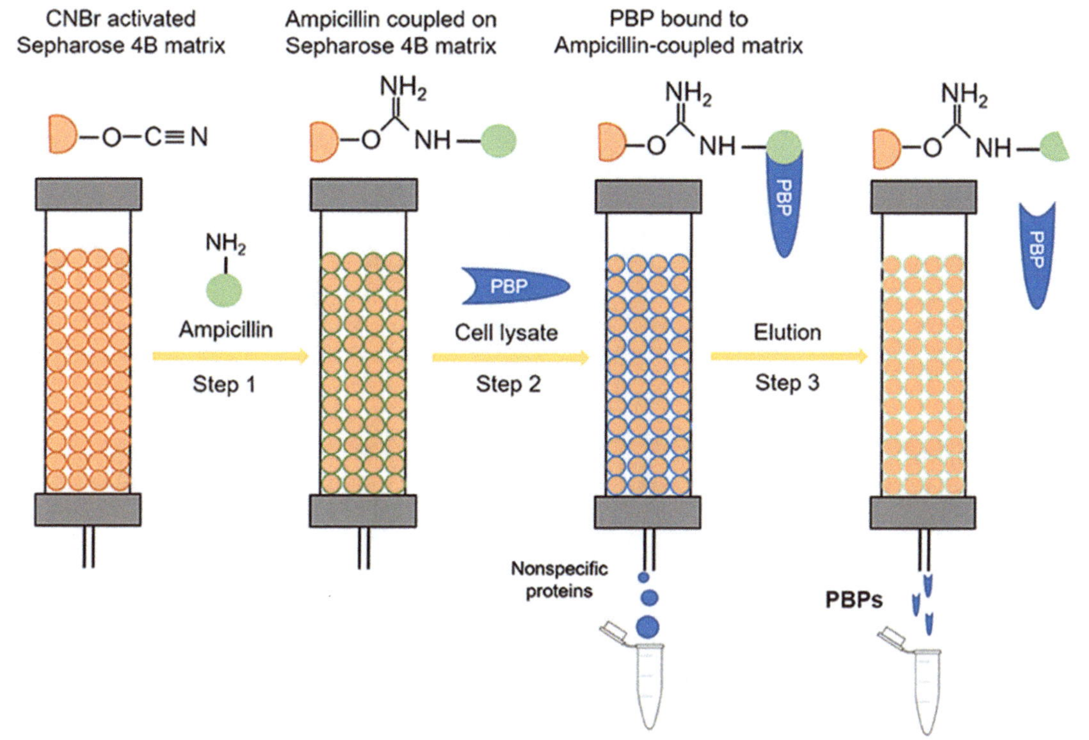

Fig. 3 Schematic of purification of PBPs using ampicillin-affinity chromatography

protein concentration levels reach zero (*see* **Note 11**). The purification can also be done without a chromatography system. In that case, one has to use a glass column with Ni-NTA resin and follow the protocol from **step 2**.

2. Equilibrate the column with the five column volumes (CV) of binding buffer. Stop the equilibration process once the conductivity becomes constant.
3. Load the cell lysate onto the pre-equilibrated column. Choose the sample loop according to the volume of your cell lysate (*see* **Note 12**).
4. Wash the column with five CV of the binding buffer.
5. Elute the proteins from the column with the elution buffer after making a gradient between 25 and 500 mM imidazole (*see* **Note 13**).
6. Dialyze the fractions collected with three changes of the dialysis buffer for a duration of 12 h.
7. Estimate the concentration of the purified protein, and store the protein properly (*see* **Note 10**).

3.5 BOCILLIN Binding Assay

This assay is done to confirm whether the purified soluble proteins are enzymatically active. The PBPs are allowed to bind to its

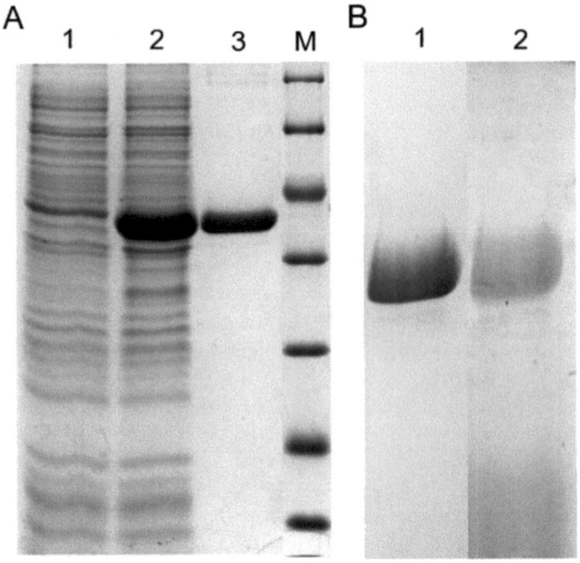

Fig. 4 (**a**) Purification of PBP, 15% SDS-PAGE: Lane 1, sonicated supernatant of uninduced sample; Lane 2, sonicated supernatant of induced sample; Lane 3, purified soluble PBP; M, protein molecular weight marker. (**b**) BOCILLIN binding; Lane 1, purified PBP stained with Coomassie Brilliant Blue; Lane 2, purified PBP labelled with BOCILLIN

substrate, BOCILLIN. A distinct band of BOCILLIN-bound PBP at 488 nm reveals that the purified protein is active.

1. Dissolve the purified PBP (5 μg) in 10 mM potassium phosphate buffer, pH 7.4, and label it with 50 μg/mL of BOCILLIN.
2. Incubate the mixture in the dark at 35 °C for 30 min.
3. Run the samples through SDS-PAGE following the standard protocol.
4. To visualize the labelled PBPs, wash the gel twice with deionized water, and scan immediately through a Typhoon imager at the excitation and emission wavelengths of 488 and 526 nm, respectively (*see* **Note 14**). The gel can also be visualized using a UV transilluminator, though a Gel Doc system having a 470 nm filter would be preferred.
5. After staining and destaining the gel, visualize the purified protein bands, and document the gel image.

Representative images of SDS-PAGE and BOCILLIN binding of a purified PBP are shown in Fig. 4.

3.6 Kinetic Analysis of the Interaction of PBPs with BOCILLIN

When a PBP enzyme (E) interacts with its substrate (S) (either peptide or beta-lactam), the kinetic parameters of the interaction are calculated based on the following reaction:

$$E + S \underset{k_{-1}}{\overset{k_1}{\leftrightarrow}} E \cdot S \xrightarrow{k_2} E - S \xrightarrow{k_3} E + P$$

where "$E - S$" is the covalent acyl-enzyme complex, "P" denotes the product released, and $K = (k_{-1} + k_2)/k_1$, "$E{\cdot}S$" is the non-covalent Michaelis complex.

3.6.1 Determination of the Acylation Rate Constant

The efficiency of PBP binding to penicillin is determined by the rate of acyl-enzyme complex formation. It is determined by the second-order rate constant k_2/K at subsaturating concentrations of the substrate (beta-lactam antibiotics) [12]. The acylation rate k_2/K is obtained from the time course study of the enzyme-substrate complex formation with BOCILLIN [13]. The second-order rate constant is determined by calculating the pseudo-first-order rate constant k_a using the equation given below:

$$k_a = -\ln\{1 - ([\text{PBP}_D]/[\text{PBP}_0])\}/t$$

In the equation, "PBP_D" represents the density (experimentally derived) of the fluorophore associated with PBP-bound BOCILLIN at time t for a particular BOCILLIN concentration D, and "PBP_0" shows the density at which the enzyme was saturated with BOCILLIN. Different k_a values obtained are plotted against the corresponding BOCILLIN concentrations to determine k_2/K (*see* **Note 15**).

1. Prepare four different concentrations of the BOCILLIN substrate (25, 50, 75, and 100 μM) in 10 mM potassium phosphate buffer, pH 7.4, in separate vials (*see* **Note 16**).
2. Preincubate the vials at 35 °C for 10 min.
3. Add the purified PBP to the BOCILLIN substrate and incubate in the dark (*see* **Note 17**).
4. Aliquot the samples (15 μL) at specific time intervals (*see* **Note 18**).
5. Immediately after aliquoting the sample, stop the reaction by adding it to the denaturing buffer. Keep collecting samples at specific time intervals till the last time point.
6. Run the BOCILLIN-labelled PBP samples and analyze through SDS-PAGE. The sequence in which the samples are loaded should match the sequence in which they are collected.
7. Quantify the band intensities of the BOCILLIN-bound proteins by densitometric scanning using a Typhoon.
8. Analyze the data to determine the acylation rate constant (k_2/K) of the PBP.

3.6.2 Determination of the Deacylation Rate Constant

The deacylation of the BOCILLIN-PBP complex leads to the release of free PBP and the cleaved BOCILLIN. The deacylation reaction is described by the first-order rate constant k_3, which is determined from the following equation:

$$\ln(\mathrm{PBP}_t/\mathrm{PBP}_0) = -k_3 t$$

In this equation, PBP_t is the residual acyl-enzyme concentration at time t, and PBP_0 is the initial concentration of acyl-enzyme complex. The deacylation rate constant (k_3) of the soluble PBP is determined by quantifying the fluorescence of enzyme-bound BOCILLIN (the acyl-enzyme complex) that decreases over time and is determined using the above equation [12, 14].

1. Incubate the PBP with BOCILLIN (50 μM) at 35 °C for 30 min in 10 mM potassium phosphate buffer, pH 7.4 (*see* **Notes 15** and **19**).
2. At t_0, add an excess of penicillin G (3 mM), and immediately aliquot out 15 μL and mix it with a denaturing buffer to stop the reaction (*see* **Note 20**).
3. Similarly, keep aliquoting 15 μL samples at various time intervals.
4. Run all the collected samples in a sequence through SDS-PAGE.
5. Measure the fluorescent intensity of the BOCILLIN remained bound to the PBP, at different time points by densitometric scanning.
6. Analyze the data to determine the deacylation rate constant (k_3) of the PBP.

3.7 Kinetic Analysis of DD-Carboxypeptidase Activity of PBP

The DD-CPase activity of the soluble PBP is determined using the artificial peptide, N_α,N_ε-diacetyl-L-Lys-D-Ala-D-Ala, and the peptidoglycan-mimetic pentapeptide substrate, L-Ala-γ-D-Glu-L-Lys-D-Ala-D-Ala. The kinetic parameters for the DD-CPase assay are deduced from the linear regression of the double reciprocal plot (Lineweaver-Burk plot):

$$\frac{1}{V} = \frac{K_m}{V_{max}}\frac{1}{[S]} + \frac{1}{V_{max}}$$

where reaction velocity (the reaction rate) is denoted by "V," Michaelis-Menten constant by "K_m," maximum reaction velocity by "V_{max}," and substrate concentration by "$[S]$." K_m (mM) is the concentration of substrate at which the half-maximal velocity is achieved. The kinetic data can be analyzed through the "Enzyme Kinetic Module" of SigmaPlot (Systat Software, San Jose, CA) to determine "K_m" and "k_{cat}" values [12].

1. In a 96-well plate, mix 1 μg of PBP with varying concentrations (2–10 mM) of the respective peptides, and adjust the reaction volume to 60 μL with 50 mM Tris–HCl, pH 8.5.
2. Incubate the reaction mixture for 30 min at 37 °C.

3. Add 10 μL of *o*-dianisidine dihydrochloride and 140 μL of the enzyme-coenzyme mixture to it. (Perform this step in the dark.) (*see* **Note 21**)
4. Incubate in dark conditions at 37 °C (*see* **Note 22**) [15].
5. Stop the reaction by adding 40 μL of 0.005 N HCl.
6. Estimate the free D-alanine generated, and compare against the standard D-alanine (1 mg/mL) solution by measuring the absorbance at 460 nm.
7. Analyze the data to determine the kinetic parameters, turnover number (k_{cat}), and substrate concentration at half the maximum velocity (K_m).

The reaction mechanism of the amount of free D-alanine estimation after DD-carboxypeptidation is shown in Fig. 5a. A representative image of a typical DD-CPase assay result is shown in Fig. 5b. It can be seen that little or no color is developed in the negative control well (absence of DD-CPase), whereas samples containing the DD-CPase enzyme showed a color development due to the release of D-alanine which is also seen in the positive control well containing D-alanine. A gradual increase in the color intensity is observed with an increase in the concentration of peptide substrate.

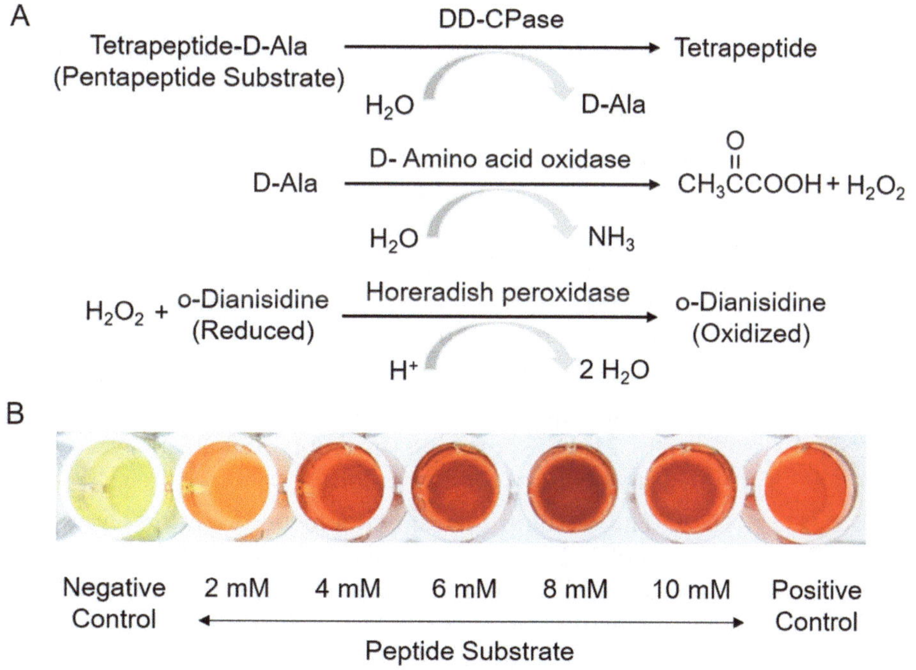

Fig. 5 **(a)** Reaction mechanism of amount of free D-alanine estimation after DD-carboxypeptidation. **(b)** Representative image of the DD-CPase assay

4 Notes

1. PMSF is a serine protease inhibitor used to protect the proteins from proteolysis during cell lysate preparation. The stock solution (made in isopropanol) is stable at room temperature for about 9 months. However, storing it at −20 °C is preferred.
2. All the centrifugation should be carried out at 0–4 °C.
3. For rupturing of cells using sonication: (1) Maintain the samples on ice to avoid heating of the samples which might denature the proteins. Larger sample volumes take longer time to lyse. In such cases, pause and incubate the sample on ice and repeat after some time. (2) For complete breakage of cells, the pulses should be given till the opaque bacterial suspension turns clear (or translucent). Ideally, a 15-s pulse followed by a 15-s break with an amplitude of 45% (it can range from 30% to 50%) is used. However, the parameters can be changed as per one's requirement. Ideally, the pellet obtained from 100 mL culture volume is resuspended in 2 mL lysis buffer and should lyse completely within 15 min. Sometimes, the solution turns white upon prolonged sonication which could be due to protein aggregation and should be avoided.
4. For competent cell preparation and transformation, one should be careful that (1) the total procedure is to be done on ice at 0 °C to avoid unwanted heat shock which may lower the efficiency of the competent cells, (2) no brake should be applied while centrifugation to avoid a tight cell pellet, and (3) while plating the transformed cell mixture, it should not be pressed hard on the agar surface to prevent cell rupturing.
5. It is better to freshly transform the expression plasmid into the expression host cells just before expression and purification.
6. During expression of the recombinant proteins, (1) IPTG used (for induction) should not exceed a final concentration of 1 mM, and (2) the OD_{600} of the culture should preferably be between 0.5 and 0.7, as the concentration of cell at the time of induction is important.
7. Load equal concentrations of proteins in all the lanes. Use BL21(DE3) carrying the empty pET-28a(+) vector as the negative control for expression.
8. The washing of resin should be done for several times with 0.001 N HCl, in order to remove the preservatives. The ampicillin-coupled matrix should not be reused, even for the same protein.
9. Dialysis should be done with constant stirring at 4 °C, and the buffer should be changed at least three times. The dialysis time with one change of buffer should not exceed 4–6 h to avoid

precipitation. For dialyzing 2 mL of protein solution, a minimum of 2 L of buffer is required.

10. PBPs are not very stable enzymes, and the experiments should be conducted with the freshly purified protein within a week. During this period, the protein should not be frozen and is to be maintained on ice at 0 °C. Prolonged storage on ice results in degradation of the protein and loss of activity. However, the proteins can be stored for longer durations at −80 °C (with 50% glycerol), but it is preferable to use the pure protein immediately as the glycerol may interfere with some experiments or else the glycerol has to be further removed.
11. The purification should be conducted at 4 °C. The matrix in the column is stored in 20% ethanol. Before starting the purification process, the ethanol should be removed completely. Buffers should be filtered and degassed or centrifuged at 10,000 × *g* for 20 min. Before loading the samples into the column, make sure it does not contain any particulate matter to avoid clogging. To ensure this, centrifuge the samples at 10,000 × *g* for 5 min before loading.
12. For better purification using Ni-NTA affinity chromatography: (1) In case of purification using Akta Prime, the sample should be passed slowly (0.2 mL/min) through the column for better binding efficiency. (2) Incubate the protein sample with the equilibrated resin for ~30 min for better binding. (3) Make sure that no traces of detergent are present in the sample. Otherwise, the column may be damaged due to the air bubbles created.
13. Before elution, the column should be washed with the binding buffer containing a lower concentration of imidazole (25–50 mM) which helps to wash off non-specifically bound proteins. During elution, the first peak is the flow through and does not have our desired protein. This peak is usually broad and flat. The next peak is that of our desired protein and should be monitored properly and collected. After the purification is over, wash the column with 5–10 CV of water followed by 5–10 CV of 20% ethanol. Finally, store the column in 20% ethanol at 4 °C.
14. The gel in which the BOCILLIN-treated samples are loaded should be protected from light. The setup should be kept in the dark during the duration of electrophoresis. Even after the gel is run, it should be protected from light before it is scanned.
15. Kinetic analysis: (1) Kinetic parameters should be determined within a couple of days after purification (post dialysis), to get the maximum efficiency of the enzymes. Do not use the protein if it starts precipitating. If needed run an SDS-PAGE to check whether the protein is intact or has started degrading. (2) The

molar concentration of protein should be measured after dialysis for the ease of calculating kinetic parameters. (3) For kinetic analysis the substrates should be prepared freshly.

16. The total reaction volume is calculated by multiplying the number of time points covered by the volume of sample aliquoted. For example, if there are ten time points and the aliquot volume is 15 μL, then the total is 150 μL. However, the total reaction volume in this case should be 160 μL, as it is advisable to keep 5–10 μL extra (due to pipetting error).
17. As soon as you add the protein sample, mix it with the substrate quickly, and take out the first aliquot. Do not aliquot before mixing. The time taken to execute this should not exceed more than a few seconds.
18. The acylation rates of different PBPs are different. It is better to set the first acylation reaction with a wide time range. Based on this result, the next reaction can be set up with a narrow time range, and the time intervals can be set up accordingly. Ideally, start with 30 s and end at around 1 h.
19. For deacylation, the PBP was incubated with BOCILLIN for 30 min. However, the duration of incubation should be decided based on the results obtained from the acylation experiment. The time point where the maximum BOCILLIN binding is seen in the acylation reaction should be selected.
20. In this experiment, penicillin G has been used for the deacylation reaction. However, this experiment can be performed using other β-lactam substrates as well.
21. Upon addition of the reagents, a white precipitate is visible. Quickly mix the ingredients properly in the wells before incubating it.
22. The duration of incubation depends on the activity of the DD-CPase enzyme. If it is very active, then the color development will be fast, compared to a weak enzyme. Therefore, keep observing and measuring the OD every 5–10 min till you see the color development. Usually, the results are visible within 5–30 min. Prolonged incubation results in precipitation.

Acknowledgment

The authors are thankful to the past and present members of the Molecular Microbiology Laboratory, Department of Biotechnology, IIT Kharagpur, with special thanks to Mr. Gaurav Kumar, Dr. Satya Deo Pandey, Dr. Chiranjit Chowdhury, Dr. Mouparna Dutta, and Dr. Debasish Kar. SP thanks IIT Kharagpur for her graduate fellowship.

References

1. Ghosh AS, Chowdhury C, Nelson DE (2008) Physiological functions of D-alanine carboxypeptidases in *Escherichia coli*. Trends Microbiol 16(7):309–317
2. Ghuysen JM (1991) Serine beta-lactamases and penicillin-binding proteins. Annu Rev Microbiol 45:37–67
3. Cayo R, Rodriguez MC, Espinal P et al (2011) Analysis of genes encoding penicillin-binding proteins in clinical isolates of *Acinetobacter baumannii*. Antimicrob Agents Chemother 55(12):5907–5913
4. Holtje JV (1998) Growth of the stress-bearing and shape-maintaining murein sacculus of *Escherichia coli*. Microbiol Mol Biol Rev 62 (1):181–203
5. Zhao G, Meier TI, Kahl SD et al (1999) BOCILLIN FL, a sensitive and commercially available reagent for detection of penicillin-binding proteins. Antimicrob Agents Chemother 43(5):1124–1128
6. Nicholas RA, Strominger JL (1988) Site-directed mutants of a soluble form of penicillin-binding protein 5 from *Escherichia coli* and their catalytic properties. J Biol Chem 263(4):2034–2040
7. Hochuli E, Dobeli H, Schacher A (1987) New metal chelate adsorbent selective for proteins and peptides containing neighbouring histidine residues. J Chromatogr 411:177–184
8. Studier FW, Moffatt BA (1986) Use of bacteriophage T7 RNA polymerase to direct selective high-level expression of cloned genes. J Mol Biol 189(1):113–130
9. Petersen TN, Brunak S, von Heijne G, Nielsen H (2011) SignalP 4.0: discriminating signal peptides from transmembrane regions. Nat Methods 8(10):785–786
10. Gautier R, Douguet D, Antonny B et al (2008) HELIQUEST: a web server to screen sequences with specific alpha-helical properties. Bioinformatics 24(18):2101–2102
11. Zhang W, Shi Q, Meroueh SO et al (2007) Catalytic mechanism of penicillin-binding protein 5 of *Escherichia coli*. Biochemistry 46 (35):10113–10121
12. Chowdhury C, Nayak TR, Young KD et al (2010) A weak DD-carboxypeptidase activity explains the inability of PBP 6 to substitute for PBP 5 in maintaining normal cell shape in *Escherichia coli*. FEMS Microbiol Lett 303 (1):76–83
13. Chambers HF, Sachdeva MJ, Hackbarth CJ (1994) Kinetics of penicillin binding to penicillin-binding proteins of *Staphylococcus aureus*. Biochem J 301(Pt 1):139–144
14. Di Guilmi AM, Mouz N, Petillot Y et al (2000) Deacylation kinetics analysis of *Streptococcus pneumoniae* penicillin-binding protein 2x mutants resistant to beta-lactam antibiotics using electrospray ionization-mass spectrometry. Anal Biochem 284(2):240–246
15. Frere JM, Leyh-Bouille M, Ghuysen JM et al (1976) Exocellular DD-carboxypeptidases-transpeptidases from *Streptomyces*. Methods Enzymol 145:610–636

Chapter 21

Extraction and Visualization of Capsular Polysaccharide from *Acinetobacter baumannii*

Kyle A. Tipton and Philip N. Rather

Abstract

Methods to determine the presence of a bacterial capsule typically involve the use of stains specific for capsule, the exclusion of a dye by the capsule or by electron microscopy. However, these procedures require equipment that may not be readily available to all researchers. Here we describe a method for extraction, visualization, and quantification of extracellular capsular polysaccharide from *Acinetobacter baumannii* with commonly used reagents and equipment.

Key words *Acinetobacter baumannii*, Capsule, Polysaccharide, Polyacrylamide gel electrophoresis, Alcian blue staining

1 Introduction

Polysaccharide capsule is an extracellular structure produced by many pathogenic and nonpathogenic bacteria [1]. Extracellular capsule is commonly composed of repeating saccharide subunits, which are joined through varied glycosidic bonds with organic or nonorganic substitutions yielding diverse structures [1]. Capsule aids bacterial cells in many processes including resistance to host defenses and phagocytosis, biofilm formation, and virulence [2–4]. Capsule biosynthetic loci and the capsular polysaccharide produced by different species of *Acinetobacter baumannii* can vary substantially [5]. The expression of capsule can vary from between strains, and in *A. baumannii* ATCC 17978, Geisinger and Isberg have shown that capsule production can be stimulated by the presence of sub-inhibitory concentrations of antibiotics which target translation [6]. Moreover, capsule expression can vary between individual cells of the same strain as recently described in *A. baumannii* strain AB5075 [7, 8]. *A. baumannii* strain AB5075 has been found to produce a capsule composed of repeating trisaccharide subunits as product of the KL25 biosynthetic locus [9]. In this chapter, a modified capsule extraction and

Indranil Biswas and Philip N. Rather (eds.), *Acinetobacter baumannii: Methods and Protocols*, Methods in Molecular Biology, vol. 1946, https://doi.org/10.1007/978-1-4939-9118-1_21, © Springer Science+Business Media, LLC, part of Springer Nature 2019

quantification protocol is described that is based on procedures utilized by Geising and Isberg, which were derived from previous work with *Salmonella* [10].

2 Materials

Prepare all solutions with distilled water prior to daily use.

2.1 Capsule Extraction

1. Lysis buffer: 2.4 mL 1 M Tris–HCl; pH 8, with 0.0002 g $CaCl_2$ and 0.081 g $MgCl_2$; and vortex to mix. Bring to volume 40 mL with distilled water (*see* **Note 1**).
2. 20 μg/mL DNase.
3. 20 μg/mL RNase.
4. 3 mg/mL lysozyme.
5. 10% sodium dodecyl sulfate (SDS) solution.
6. Proteinase K solution: Using lysis buffer prepared in 1, add proteinase K to 2 mg/mL.
7. 75% Ethanol. Store at −20 °C.

2.2 SDS PAGE

1. 4–12% Bis–Tris gel.
2. SDS running buffer.
3. SDS sample load buffer with reducing agent.
4. Molecular weight marker.

2.3 Alcian Blue Polysaccharide Staining

1. 20 mM sodium acetate, pH 4.75.
2. Stain base solution: 40% ethanol, 60% 20 mM sodium acetate, pH 4.75.
3. 0.1% w/v Alcian blue in stain base solution.

3 Methods

3.1 Capsule Extraction

1. Streak bacterial strains to be analyzed on 0.5× LB agar plates. Incubate plates at 37 °C for approximately 18 h.
2. Scrape cells from plate surface with plastic loop and suspend cells in 2 mL nutrient broth. Normalize optical density of suspensions (*see* **Note 2**).
3. Harvest cells from 1 mL of each suspension by centrifugation. Aspirate remaining supernatant from cell pellets.
4. Resuspend cells in 200 μL lysis buffer without proteinase K. Incubate suspensions at 37 °C for 60 min (*see* **Note 3**). Vortex suspensions after incubation.

5. Cycle extractions through three freeze/thaw cycles at −80 and 37 °C.
6. Add 1 μL DNase and RNase to extractions. Incubate at 37 °C for 30 min.
7. Add 10 μL 10% SDS to extractions and incubate at 37 °C for 30 min.
8. Boil extractions at 100 °C in boiling water bath for 10 min. Allow extractions to cool to room temperature after boiling.
9. Add 150 μL lysis buffer with proteinase K. Incubate at 60 °C for 60 min.
10. Centrifuge extractions at maximum speed for 2 min at room temperature.
11. Aspirate supernatant of each extraction to a clean microcentrifuge tube. Add 400 μL cold 75% ethanol to each extraction. Mix by inversion and incubate extractions at −20 °C overnight.
12. Centrifuge overnight precipitations at maximum speed for 30 min at 4 °C.
13. Decant supernatant from capsule pellets (*see* **Note 4**).
14. Centrifuge briefly at maximum speed for 2 min at 4 °C. Aspirate supernatant from pellets (*see* **Note 5**).
15. Once tubes are dried of residual solution, suspend pellets in sample load buffer (30 μL total volume).

3.2 SDS PAGE Separation

1. Boil samples for 5 min at 100 °C in boiling water bath.
2. Cool samples on ice then load aliquots into a 4–12% Bis–Tris gel.
3. Run gel at 150 V until dye has reached the bottom of the gel.

3.3 Alcian Blue Polysaccharide Staining

1. Following electrophoresis, trim wells from gel. Rinse gel in 200 mL distilled water for 10 min at room temperature with gentle shaking. Repeat five times (*see* **Note 6**).
2. Stain gel with Alcian blue polysaccharide stain for 60 min at room temperature with gentle shaking.
3. Destain gel in stain base solution for 60 min or until background staining is sufficiently minimized (*see* **Note 7**).

3.4 Imaging and Quantification

1. Following destaining procedure, gel images are captured with a digital camera (Fig. 1).
2. Digital images are analyzed with ImageJ software [11] (*see* **Note 8**).
3. Capsule is quantified by comparison to wild-type opaque variant (*see* **Note 9**).

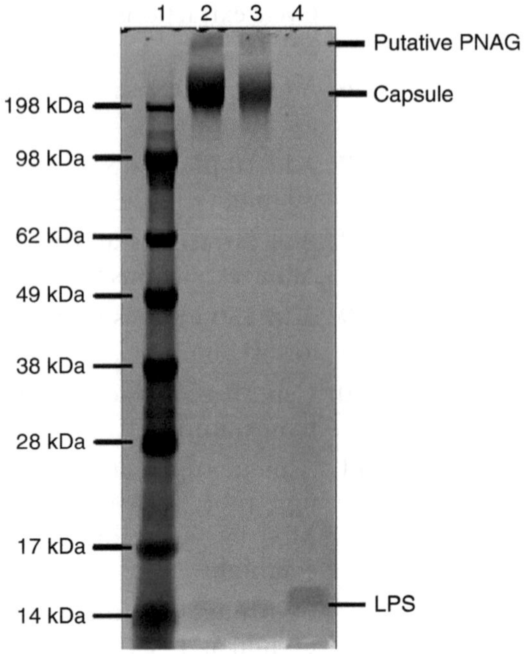

Fig. 1 Gel image of capsule polysaccharide purified from *A. baumannii* strain AB5075. Polysaccharide was purified from *A. baumannii* strain AB5075 VIR-O opaque and AV-T translucent variants and unencapsulated *A. baumannii* Δ*wzc* mutant. Extractions were visualized by polyacrylamide gel electrophoresis. Lanes: (*1*) MW standard; (*2*) *A. baumannii* strain AB5075 VIR-O opaque variant; (*3*) *A. baumannii* strain AB5075 AV-T translucent variant; (*4*) *A. baumannii* Δ*wzc* mutant. Molecular weights are listed to the left of the gel image. Designation of putative bands on gel are shown at the right and based on previous studies in *A. baumannii* [6]

4 Notes

1. Make lysis buffer fresh for each experimental trial. Results of extraction and quantification become unreliable in lysis buffer that has been stored at room temperature for more than a few hours.
2. Typically, cell suspensions are normalized to OD_{600} ~ 0.65. However, the density can be adjusted to suit experimental needs.
3. Begin lysis of cells on the same day as cells are scraped from plate. Storage of cell pellets over time (even at −80 °C) will yield unreliable results from capsule extraction and quantification.
4. Capsule pellets may be very small and nearly invisible.

5. Capsule pellets may be loose upon aspiration. If pellet is loose, leave 20–30 μL of remaining solution in tube. Dry pellets by incubating open tubes at 70 °C in a heat block.
6. Wash step is repeated for a total of six washes, in 60 min and 1200 mL distilled water.
7. Destaining procedure may require longer incubation than 60 min. Gels can be destained overnight in stain base solution.
8. Image analysis can be performed with any software package that allows for densitometric measurement.
9. A standard curve of dilutions from a solution with known capsule concentration could be used for absolute quantification of capsule production.

Acknowledgment

This work is supported by the following awards to P.N.R., VA Merit Award I01 BX001725 and Research Career Scientist Award IK6BX004470, both from the Department of Veterans Affairs, and R01AI072219 from the National Institutes of Health.

References

1. Roberts I (1996) The biochemistry and genetics of capsular polysaccharide production in bacteria. Annu Rev Microbiol 50:285–315
2. Lee K-J, Kim J-A, Hwang W, Park S-J, Lee K-H (2013) Role of capsular polysaccharide (CPS) in biofilm formation and regulation of CPS production by quorum-sensing in *Vibrio vulnificus*. Mol Microbiol 90(4):841–857
3. Russo T, Luke N, Beanan J, Olson R, Sauberan S, MacDonald U, Schultz L, Umland T, Campagnari A (2010) The K1 capsular polysaccharide of *Acinetobacter baumannii* strain 307-0294 is a major virulence factor. Infect Immun 78(9):3993–4000
4. Weber B, Harding C, Feldman M (2016) Pathogenic *Acinetobacter*: from the cell surface to infinity and beyond. J Bacteriol 198 (6):880–887
5. Kenyon J, Hall R (2013) Variation in the complex carbohydrate biosynthesis loci of *Acinetobacter baumannii* genomes. PLoS One 8(4): e62160
6. Geisinger E, Isberg R (2015) Antibiotic modulation of capsular exopolysaccharide and virulence in *Acinetobacter baumannii*. PLoS Pathog 11(2):e1004691. https://doi.org/10.1371/journal.ppat.1004691
7. Tipton K, Dimitrova D, Rather P (2015) Phase-variable control of multiple phenotypes in *Acinetobacter baumannii* strain AB5075. J Bacteriol 197(15):2593–2599
8. Chin CY, Tipton KA, Farokhyfar M, Burd EM, Weiss DS, Rather PN (2018) A high-frequency phenotypic switch links bacterial virulence and environmental survival in Acinetobacter baumannii. Nat Microbiol 3(5):563–569. https://doi.org/10.1038/s41564-018-0151-5
9. Senchencova S, Shashkov A, Popova A, Shneider M, Arbatsky N, Miroshnikov K, Volozhantsev N, Knirel Y (2015) Structure elucidation of the capsular polysaccharide of *Acinetobacter baumannii* having the KL25 capsule biosynthesis locus. Carbohydr Res 408:8–11
10. Hitchcock P, Brown T (1983) Morphological heterogeneity among *Salmonella* lipopolysaccharide chemotypes in silver-stained polyacrylamide gels. J Bacteriol 154(1):269–277
11. Schneider C, Rasband W, Eliceiri K (2012) NIH Image to ImageJ: 25 years of image analysis. Nat Methods 9:671–675

Chapter 22

Isolation of Lipid Cell Envelope Components from *Acinetobacter baumannii*

Matthew J. Powers, Carmen M. Herrera, Ashley T. Tucker, Bryan W. Davies, and M. Stephen Trent

Abstract

With the increasing occurrence of antibiotic resistance among *Acinetobacter* sp., the race is on for researchers to not only isolate resistant isolates but also utilize basic and applied microbiological techniques to study mechanisms of resistance. For many antibiotics, the limit of efficacy against Gram-negative bacteria is dependent on its ability to permeate the outer membrane and access its target. As such, it is critical that researchers be able to isolate and analyze the lipid components of the cell envelope from any number of *Acinetobacter* sp. that are either resistant or sensitive to antibiotics of interest. The following chapter provides in-depth protocols to confirm the presence or absence of lipooligosaccharide (LOS) in *Acinetobacter* sp., isolate lipid A, and glycerophospholipids and analyze them using qualitative (mass spectrometry) and semiquantitative (thin-layer chromatography) methods.

Key words Lipopolysaccharides, LPS, Lipid A, Thin-layer chromatography, Mass spectrometry, MALDI-TOF

1 Introduction

Acinetobacter baumannii is an increasingly relevant nosocomial pathogen associated with staggeringly high rates of antibiotic resistance [1]. Of note is its ability to gain moderate to high resistance to the polymyxin class of antibiotics, which include the last resort drugs polymyxin E (colistin) and polymyxin B. These antibiotics preferentially target the outer membrane of Gram-negative bacteria due to its net negative charge. The outer membrane consists of either lipopolysaccharide (LPS) or lipooligosaccharide (LOS), which differ only in the length of the O-antigen component (Fig. 1). *Acinetobacter* sp. synthesizes LOS which is what we will refer to in this chapter; however, most methods do not differ regarding LPS or LOS analysis.

The lipid anchor of LOS is lipid A, a hydrophobic glycolipid whose biosynthetic pathway is conserved among almost all Gram-

Indranil Biswas and Philip N. Rather (eds.), *Acinetobacter baumannii: Methods and Protocols*, Methods in Molecular Biology, vol. 1946, https://doi.org/10.1007/978-1-4939-9118-1_22, © Springer Science+Business Media, LLC, part of Springer Nature 2019

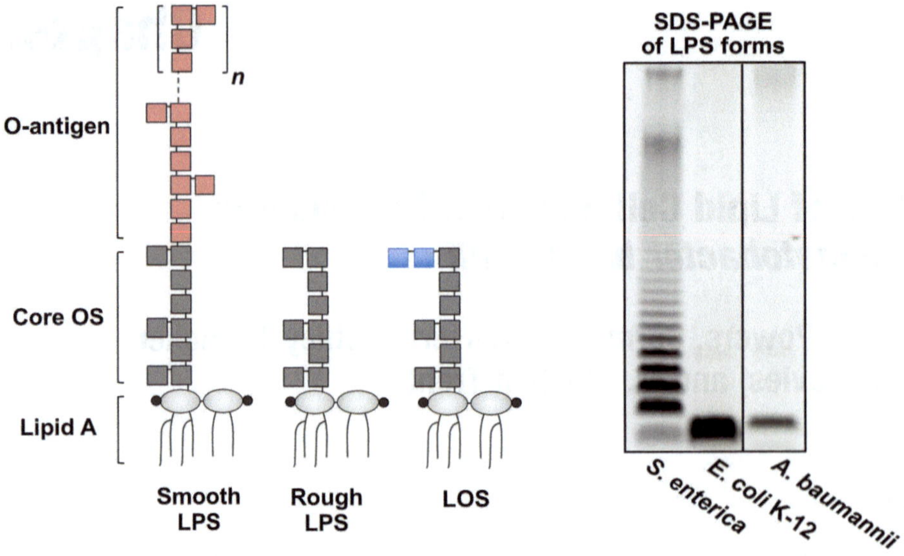

Fig. 1 Different glycoforms of lipopolysaccharide. Lipopolysaccharide is ubiquitous in Gram-negative bacteria; however, there is a wide range of diversity in the sugar composition of the core and O-antigen. Smooth LPS, depicted on the left, is the canonical, full-length LPS structure, complete with an intact core oligosaccharide (core OS) and repeating polysaccharide (O-antigen). This repeating polysaccharide can change even between strains of the same species; this is referred to as a serotype [26]. A truncation in the LPS structure due to mutations in the biosynthetic machinery of O-antigen results in rough LPS [27]. Some organisms, including *A. baumannii*, have lipooligosaccharide (LOS), which contains core OS and several additional sugars. Examples of how each of these glycoforms run (in the order discussed above) on a standard SDS-PAGE gel and stained is depicted on the right. Due to the repeating nature of the O-antigen, strains that have intact LPS exhibit a characteristic banding pattern. Each band in this case corresponds to the addition of a repeating oligomer unit. Strains with rough LPS or LOS (*E. coli* K-12 and *A. baumannii*, respectively) present one glycoform on the surface. When separated by size, this runs close to the bottom of an SDS-PAGE gel as shown above

negative bacterial species except for members of the *Spirochaetes* phyla [2, 3]. Despite this conservation, there is a high degree of variation in core structure, O-antigen serotypes, and conditional enzymatic modification of the lipid molecule [4]. *A. baumannii* can modify its lipid A with either galactosamine or phosphoethanolamine to reduce the negative charge of the molecule and increase its resistance to polymyxins [5–7] (Fig. 2). More strikingly, *A. baumannii* is one of the three organisms that has the ability to tolerate loss of its LOS, which is a drastic change to the outer membrane that is accompanied by an unprecedented resistance to polymyxin drugs [8–12]. Taken together, these characteristics have the ability to isolate and analyze lipid species from *A. baumannii* of critical importance both for basic and applied microbiology.

There are multiple methods to extract LOS from the bacterial surface, with the most common being hot phenol-water extraction introduced by Westphal et al. [13]. From intact LOS, lipid A is

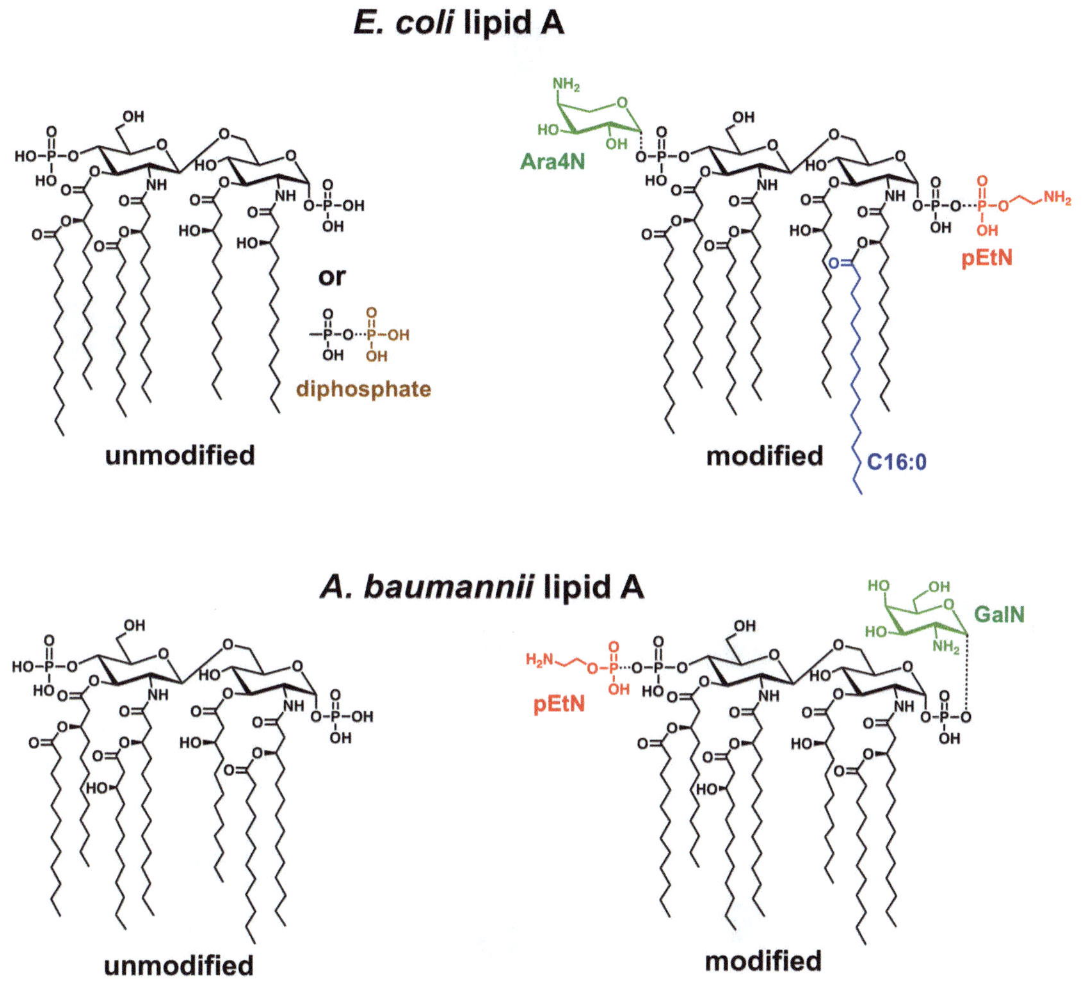

Fig. 2 Modified variants of lipid A. The canonical lipid A species is a hexacylated glucosamine disaccharide with phosphates at the 1 and 4′ moieties. Bacteria can modify lipid A to alter both its hydrophobicity and net-negative charge. The unmodified version of *Escherichia coli* (top left) can be modified (top right) at the 1 and 4′ phosphates with 4-amino-4-deoxy-L-arabinose (Ara4N) or phosphoethanolamine (pEtN) [28–30]. Additionally, the lipid A species can be heptaacylated by the addition of a secondary palmitate (C16:0) [31]. The unmodified lipid A species for *A. baumannii* differs from *E. coli* in that it is constitutively heptaacylated and has a 3-hydroxyacyl chain in an acyloxyacyl linkage at the 2′ position (bottom left) [32]. It can be modified via the addition of pEtN or galactosamine (GalN) at the 1 and 4′ phosphates [5, 33]

released by mild acid hydrolysis which breaks the linkage between the distal glucosamine and the first Kdo (3-deoxy-D-*manno*-oct-2-ulosonic acid). This method of extraction is burdened by both the amount of time and extensive generation of hazardous waste, primarily phenol.

Our laboratory expanded on the methods proposed by Odegaard [14] and Caroff [15] which uses Bligh-Dyer solvent extractions and mild acid hydrolysis to extract lipid A. Bligh-Dyer extractions typically are used for whole lipid species isolation from

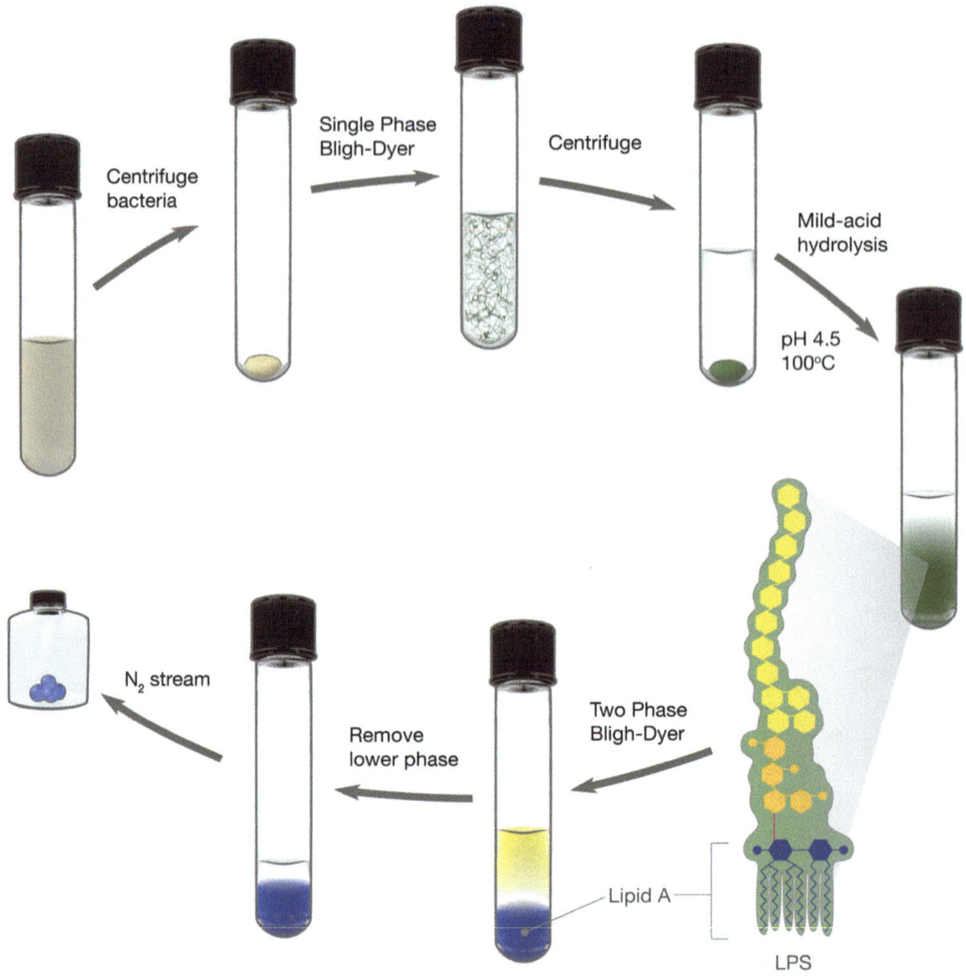

Fig. 3 Schematic of a two-phase Bligh-Dyer to isolate lipid A species. Lipids can be isolated directly from a bacterial culture using a two-phase Bligh-Dyer. Cells are pelleted and resuspended in a single-phase Bligh-Dyer. This lyses the bacterial cells and releases cellular material. Glycerophospholipids and prenyl lipids remain soluble after centrifugation. Lipid A, due to the core sugars and O-antigen, pellets with proteinaceous material. The core can be cleaved from the distal glucosamine ring of lipid A by mild acid hydrolysis, and free lipid A can be separated by a two-phase Bligh-Dyer. Ultimately, this leaves free lipid A molecules in the lower phase. This phase can be dried down and used for downstream analyses

a variety of samples [16] (Fig. 3). A Bligh-Dyer isolation of lipid A involves a series of solvent extractions with chloroform, methanol, and water in varying ratios to ultimately extract lipid A into the organic phase. This method is rapid, relatively unbiased, and minimizes accumulation of phenol waste.

Once isolated, lipid A can be analyzed qualitatively and semiquantitatively via thin-layer chromatography (TLC) coupled with densitometry and mass spectrometry (MS). In tandem, these approaches allow for robust identification and elucidation of lipid A structures. Advances in mass spectrometry techniques have yielded numerous methods for assessing lipid A structures that have been previously explained in detail [17, 18]. The ability to phenotypically assess the lipid A and glycerophospholipid profiles of *A. baumannii* strains is imperative for determining whether resistance to critical antibiotics is mediated by modifications to lipid species or through the absence of lipid A entirely. With the increased rate of identification of antibiotic-resistant *Acinetobacter* sp., it is increasingly relevant for researchers to be able to isolate and study cell envelope components. These phenotypes can provide not only routes to discovering novel modifications or regulatory systems through basic research but also provide pertinent epidemiological data for the prevalence and mechanism of polymyxin-resistant *A. baumannii* isolates.

This chapter will include methods for both the confirmation of presence or absence of LOS in *Acinetobacter* sp. and the extraction of lipid A and glycerophospholipids from *A. baumannii* for TLC or mass spectrometry analysis.

2 Materials

Prepare all solutions using ultrapure water, and store all reagents at room temperature unless indicated otherwise. Measure all solvents in a glass graduated cylinder, and store in glass solvent bottles with Teflon-lined caps. Amber bottles are recommended for long-term storage of chloroform solutions [19]. Teflon centrifuge tubes and rotary evaporator flasks should be rinsed with methanol and chloroform prior to use. Follow appropriate regulations for disposal of solvents and radioactive waste.

2.1 Detection of LOS Species

1. ProQ-Emerald 300 Lipopolysaccharide Gel Staining Kit (Invitrogen).
2. 4–12% gradient gel, 1 mm thickness (*see* **Note 1**).
3. Protein loading buffer supplemented with 4% β-mercaptoethanol.
4. E-test strips for vancomycin and polymyxin B.
5. Lysogeny broth: 10 g of tryptone, 5 g yeast extract, and 10 g sodium chloride.
6. Proteinase K (20 mg/mL).

2.2 ^{32}P-Labeled Lipid A Isolation

1. Lysogeny broth: Subheading 2.1, **item 5**.
2. Orbital shaker.
3. Spectrophotometer.
4. Inorganic ^{32}P (1 mCi in 100 μL).
5. Glass centrifuge tubes with Teflon-lined caps.
6. Fixed angle clinical centrifuge.

7. 1× phosphate-buffered saline (PBS): 8 g sodium chloride, 0.2 g potassium chloride, 1.42 g disodium phosphate, and 0.24 g monopotassium phosphate in 1 L of distilled water.
8. Chloroform.
9. Methanol.
10. Single-phase Bligh-Dyer mixture: chloroform, methanol, aqueous (1:2:0.8, v/v). To generate a 380 mL solution, combine 100 mL of chloroform, 200 mL of methanol, and 80 mL of water in a solvent bottle. Cap bottle, shake to mix, vent in fume hood, and seal.
11. Hydrolysis buffer: 50 mM sodium acetate pH 4.5, 1% sodium dodecyl sulfate (SDS). For a 250 mL solution, add 1.03 g of sodium acetate and 2.5 g of SDS to 200 mL of water. Once in solution, adjust the pH to 4.5. Raise the solution to a final volume of 250 mL in a graduated cylinder.
12. Pre-equilibrated two-phase Bligh-Dyer mixture: chloroform, methanol, aqueous (2:2:1.8, v/v). To generate a 580 mL solution, combine 200 mL of chloroform, 200 mL of methanol, and 180 mL of water to a solvent bottle. Cap bottle, shake to mix, vent in fume hood, and seal. Let sealed solution equilibrate overnight in a fume hood.
13. Bath sonicator.
14. Water bath.
15. Nitrogen gas.

2.3 Visualization of ^{32}P-Labeled Lipid A Species via Thin-Layer Chromatography

1. TLC tank.
2. 20 × 20 cm silica gel 60 TLC plates.
3. 3MM Chr chromatography paper (23 cm × 100 m).
4. TLC solvent system: chloroform, pyridine, 88% formic acid, aqueous (50:50:16:5, v/v). To generate a 242 mL solution, combine 100 mL of chloroform, 100 mL of pyridine, 32 mL of 88% formic acid, and 10 mL of water in a solvent bottle. Cap bottle, shake to mix, vent in fume hood, and seal.
5. Chloroform-methanol (4:1, v/v). To generate a 125 mL solution, combine 100 mL of chloroform and 25 mL of methanol in a solvent bottle. Cap bottle, shake to mix, vent in fume hood, and seal.
6. Scintillation counter and vials.
7. Scintillation cocktail.
8. Plastic wrap.
9. Phosphorimager screen.
10. Imaging analysis software.

2.4 ^{32}P-Labeled Phospholipid Isolation

1. Lysogeny broth: Subheading 2.1, **item 5**.
2. Orbital shaker.
3. Spectrophotometer.
4. Inorganic ^{32}P (1 mCi in 100 μL).
5. Glass centrifuge tubes with Teflon-lined caps.
6. Fixed angle clinical centrifuge.
7. 1× PBS: Subheading 2.2, **item** 7.
8. Chloroform.
9. Methanol.
10. Single-phase Bligh-Dyer mixture: Subheading 2.2, **item 10**.
11. Pre-equilibrated two-phase Bligh-Dyer mixture: Subheading 2.2, **item 12**.
12. Nitrogen gas.

2.5 Visualization of ^{32}P-Labeled Phospholipid Species via 1 or 2D Thin-Layer Chromatography

1. TLC tank.
2. 20 × 20 cm silica gel 60 TLC plates.
3. 3MM Chr chromatography paper (23 cm × 100 m).
4. TLC solvent system A: chloroform, methanol, aqueous (60:25:4, v/v). To generate a 267 mL solution, add 180 mL of chloroform, 75 mL of methanol, and 12 mL of water to a solvent bottle. Cap bottle, shake to mix, vent in fume hood, and seal.
5. TLC solvent system B: chloroform, methanol, acetic acid, water (80:12:15:4, v/v). To generate a 250 mL solution, add 180 mL of chloroform, 27 mL of methanol, 34 mL of acetic acid, and 9 mL of water to a solvent bottle. Cap bottle, shake to mix, vent in fume hood, and seal.
6. Chloroform-methanol (4:1, v/v): Subheading 2.3, **item 5**.
7. Scintillation counter and vials.
8. Scintillation cocktail.
9. Plastic wrap.
10. Phosphorimager screen.
11. Imaging analysis software.

2.6 Lipid A Isolation for Mass Spectrometry Analysis

1. Lysogeny broth: Subheading 2.1, **item 5**.
2. Orbital shaker.
3. Spectrophotometer.
4. Tabletop centrifuge and associated rotors.
5. 1× PBS: Subheading 2.2, **item** 7.
6. 250 mL Teflon centrifuge tubes.

7. Chloroform.
8. Methanol.
9. Single-phase Bligh-Dyer mixture: Subheading 2.2, **item 10**.
10. Hydrolysis buffer: Subheading 2.2, **item 11**.
11. Pre-equilibrated two-phase Bligh-Dyer mixture: Subheading 2.2, **item 12**.
12. Probe tip sonicator.
13. Water bath.
14. Rotary evaporator and flasks.

2.7 Matrix-Assisted Laser Desorption/Ionization (MALDI) Analysis of Lipid A

1. Chloroform-methanol (4:1, v/v): Subheading 2.3, **item 5**.
2. Nitrogen tank.
3. MALDI plate (polished).
4. Calibrant mixture (varies by MALDI manufacturer).
5. ATT matrix mixture: prepare a mixture of saturated 6-aza-2-thiothymine (ATT) in 50% acetonitrile and saturated tribasic ammonium citrate (20:1, v/v) by vortexing 20 μL of stock ATT solution with 1 μL of stock tribasic ammonium citrate solution. Store at 4 °C. The stock ATT solution is prepared by adding 10 mg of ATT to a 1:1 solution of water (500 μL) and acetonitrile (500 μL) to yield a supersaturated solution. The stock saturated solution of tribasic ammonium citrate is prepared by adding 1 mg of the salt to 500 μL of water. The precipitate should be readily apparent. Vortex all stock solutions just before use, and only the saturated supernatant is used.
6. MALDI-TOF capable of acquiring linear spectra in the negative- and positive-ion modes.

3 Methods

3.1 Detection of LOS Species

3.1.1 Phenotyping by Antibiotic Resistance

1. Start an overnight culture of your *Acinetobacter* sp. strain of interest in LB.
2. The following day, inoculate a culture to an OD_{600} of 0.05 in LB media. Grow with shaking (220 rpm) at 37 °C to an OD_{600} of ~1.0.
3. Spread 150 μL of culture on an LB plate using glass beads.
4. Let dry for 5–10 min (*see* **Note 2**).
5. Gently lay the E-strip on the plate, ensuring that no air bubbles are present.
6. Incubate at 37 °C overnight.

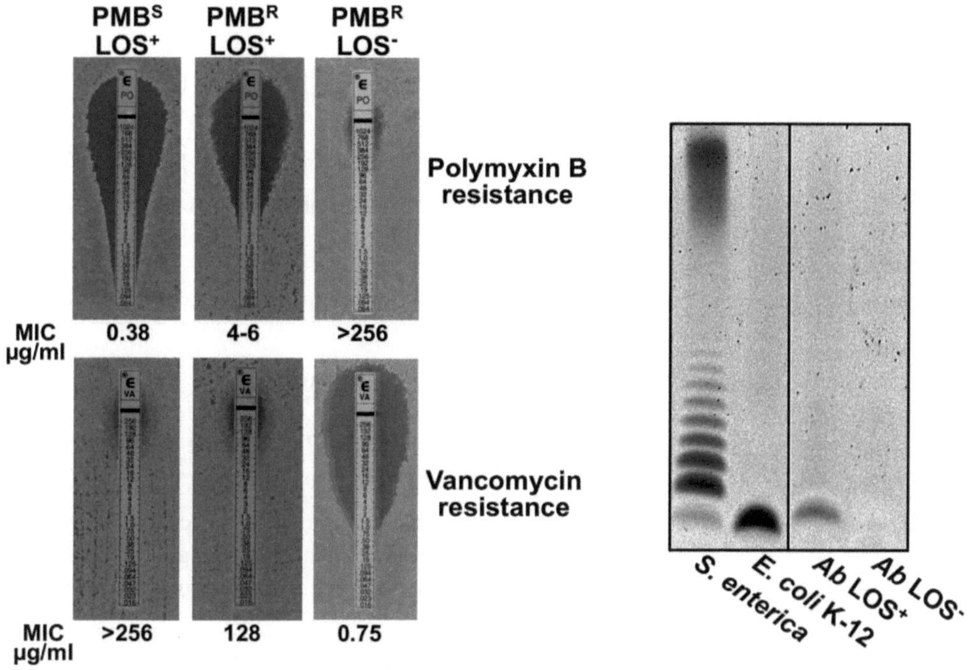

Fig. 4 Modifications to lipid A can increase the resistance to polymyxin antibiotics. Polymyxin B preferentially targets Gram-negative outer membranes due to the net-negative charge of lipid A, resulting in high susceptibility to this class of antibiotics. Modifications to lipid A (as shown in Fig. 2) can increase resistance. In this case, the integrity of the outer membrane as assessed by permeability of the outer membrane to the bulky, hydrophobic antibiotic vancomycin stays relatively the same. When *A. baumannii* lacks LOS, resistance to polymyxin B increases drastically with a concurrent increase in sensitivity to vancomycin. It is vital to assess clinical isolates in such a manner to indicate whether the presence or absence of LOS may play a role in resistance levels. On the right is a standard SDS-PAGE gel stained for LOS. This clearly shows that LOS-deficient *A. baumannii* lack stainable LOS

7. The following morning, read the MIC where the zone of inhibition intersects with the E-strip (*see* **Note 3**) (Fig. 4).

3.1.2 Phenotyping by SDS-PAGE Analysis of LOS

1. Start an overnight culture of your *Acinetobacter* sp. strain of interest in LB.
2. The following day, inoculate a culture to an OD_{600} of 0.05 in LB media. Grow with shaking (220 rpm) at 37 °C to an OD_{600} of ~1.0.
3. Measure the OD_{600} of the culture and normalize 1 mL of cells to an OD_{600} of 1.0.
4. Pellet the cells via centrifugation at 10,000 × *g* for 2 min. Wash the pellet with 1 mL of 1× PBS. Repeat the washing step for a total of two washes.
5. Resuspend the pellet in 100 µL of 1× loading buffer.
6. Boil sample for 15 min and let cool to at or below 55 °C.

7. Add proteinase K to a 25 μg mL^{-1}. Incubate at 55 °C for 1 h or overnight (*see* **Note 4**).
8. Boil sample for 5 min and load 10 μL/well on an SDS-PAGE gel.
9. Run samples for 1 h at 150 V (*see* **Note 5**).
10. Stain and image gel per the protocol of the ProQ Emerald 300 Kit (*see* **Note 6**) (Fig. 4).

3.2 ^{32}P-Labeled Lipid A Isolation

1. Inoculate 5 mL of LB using a single colony. Grow overnight at 37 °C.
2. Following overnight growth, measure the OD_{600}, and use the culture to inoculate 10 mL of LB at a starting OD_{600} of 0.05. Add 5 μCi/mL of inorganic ^{32}P (*see* **Note** 7). Grow cells to an OD_{600} of 0.8–1.0 (*see* **Note 8**).
3. Harvest the cells in glass centrifuge tubes with Teflon-lined caps using a fixed angle clinical centrifuge at 1500 × *g* for 10 min. Discard the radioactive waste in an appropriate radioactive waste container.
4. Wash the cell pellet with 5 mL of 1× PBS. Centrifuge for 10 min at 1500 × *g*. Discard the supernatant (*see* **Notes 9** and **10**).
5. Resuspend the cells in 5 mL of single-phase Bligh-Dyer mixture (Fig. 3).
6. Vortex and incubate at room temperature for 20 min to ensure complete cell lysis.
7. Centrifuge in a clinical centrifuge at 1500 × *g* for 20 min. Gently pour off the supernatant, which contains glycerophospholipids and isoprenyl lipids (*see* **Note 11**).

3.2.1 For Glycerophospholipid Isolation, Skip to Subheading 3.4, Step 8

1. Resuspend the LPS pellet in 1.8 mL of hydrolysis buffer via vortexing.
2. Sonicate the sample in a bath sonicator until the pellet is equally dispersed.
3. Incubate the sample for 30 min in a boiling water bath to fragment the intact molecule at the first Kdo sugar and distal glucosamine (*see* **Note 12**) (Figs. 1 and 3).
4. Remove the sample and allow to cool at room temperature for 5–10 min.
5. Convert the solution into a two-phase Bligh-Dyer mixture by the addition of 2 mL of chloroform and 2 mL of methanol. The final mixture at this point will be 2:2:1.8, v/v (*see* **Note 13**).
6. Vortex and separate phases using a fixed angle clinical at 1500 × *g* for 10 min. Extract the lower phase into a clean glass centrifuge tube using a glass pipet. Keep the upper phase!

7. To the remaining upper phase, add 2 mL of pre-equilibrated lower phase (*see* **Note 14**).
8. Vortex and centrifuge at 1500 × *g* for 10 min. Remove the lower phase and combine with the lower phase from **step 13**.
9. To wash the pooled lower phase, add 7.6 mL of pre-equilibrated upper phase (*see* **Note 14**).
10. Vortex and centrifuge at 1500 × *g* for 10 min. Remove the lower phase to a clean glass tube and dry under a stream of nitrogen (*see* **Note 15**). Dried samples can be stored at −20 °C for further use (*see* **Note 16**).

3.3 Visualization of ^{32}P-Labeled Lipid A Species via Thin-Layer Chromatography

1. Use a TLC tank that will accommodate 20 × 20 cm plates for TLC analysis. Line the back of the TLC tank with ~39 cm of chromatography paper (*see* **Note 17**).
2. Prepare the TLC system by adding 200 mL of the TLC solvent system to the tank. Allow the tank to pre-equilibrate for ~3 h (*see* **Note 18**).
3. Dissolve the ^{32}P-labeled sample in 500 μL of chloroform-methanol (4:1, v/v). Vortex.
4. Add 5 μL of sample to a scintillation vial containing 5 mL of scintillation cocktail. Vortex.
5. Count in a scintillation counter and calculate the total CPMs of the sample (*see* **Note 19**).
6. Spot 10,000–20,000 CPM/sample on the TLC plate and allow the silica to dry (*see* **Notes 20** and **21**).
7. Place the TLC plate into the pre-equilibrated tank. Once the solvent front reaches the top of the TLC plate (~ 2.5–3 h), remove and dry. A blow-dryer can be used to speed up the drying process (~15 min, hot air) (*see* **Note 22**).
8. Wrap in plastic wrap and expose to a phosphorimager screen overnight.
9. The next morning, scan the screen to obtain the image (*see* **Note 23**) (Fig. 5), and quantify lipid A species using an imaging analysis software.

3.4 ^{32}P-Labeled Phospholipid Isolation

1. Inoculate 5 mL of LB using a single colony. Grow overnight at 37 °C.
2. Following overnight growth, measure the OD_{600}, and use the culture to inoculate 10 mL of LB at a starting OD_{600} of 0.05. Add 5 μCi/mL of inorganic ^{32}P. Grow cells to an OD_{600} of 0.8–1.0 (*see* **Notes 7** and **8**).
3. Harvest the cells in glass centrifuge tubes with Teflon-lined caps using a fixed angle clinical centrifuge at 1500 × *g* for

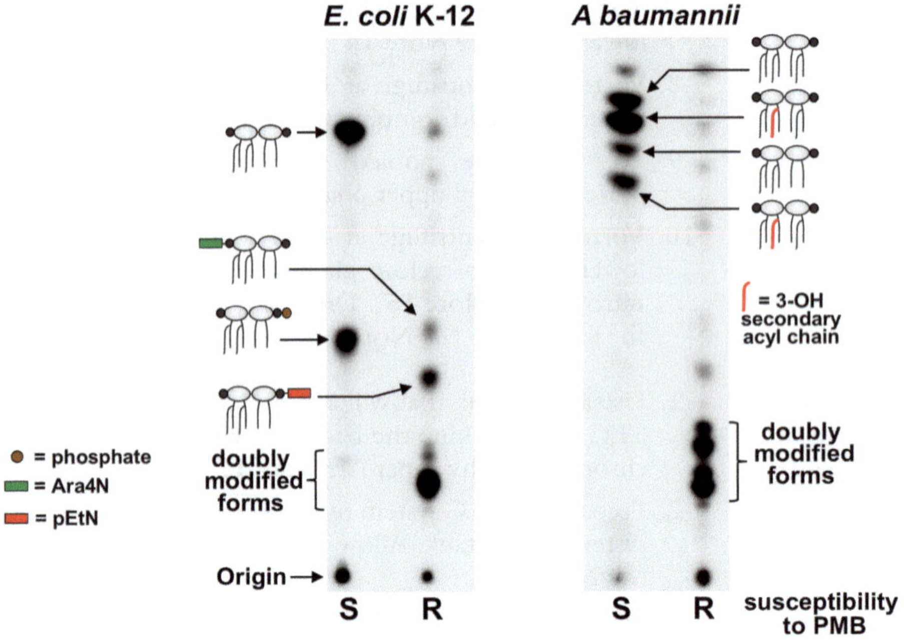

Fig. 5 Thin-layer chromatography (TLC) allows for differentiation of lipid A species. Single dimension TLC can separate lipid A species to determine whether lipid A modifications are responsible for alterations to the cell envelope and antibiotic resistance. The two left lanes are a PMB^S (indicated by S) and PMB^R (indicated by R) strain of *E. coli.* Variants of lipid A are indicated by a cartoon of their structure. Modifications that reduce hydrophobicity result in slower migration in the tank used—chloroform, pyridine, 88% formic acid, and aqueous (50:50:16:5 v/v). As such, highly modified species present in the R lane migrate much lower. *A. baumannii* has four main lipid A species. These are represented in the PMB^S lane as hepta- or hexaacylated, with or without a 3-hydroxyacyl chain in an acyloxyacyl linkage at the 2′ position. The PMB^R strain efficiently modifies each of these lipid A species with either GalN and/or pEtN

10 min. Discard the radioactive waste in an appropriate radioactive waste container.

4. Wash the cell pellet with 5 mL of 1× PBS. Centrifuge for 10 min at 1500 × *g*. Discard the supernatant.
5. Resuspend the cells in 5 mL of single-phase Bligh-Dyer mixture.
6. Vortex and incubate at room temperature for 20 min.
7. Centrifuge in a clinical centrifuge at 1500 × *g* for 20 min. Transfer the supernatant to a fresh tube, which contains glycerophospholipids and isoprenyl lipids.
8. Convert the solution to a two-phase Bligh-Dyer by the addition of 1.32 mL chloroform and 1.32 mL water to a final ratio of chloroform, methanol, aqueous (2:2:1.8, v/v).
9. Vortex and separate phases using a fixed angle clinical at 1500 × *g* for 10 min. Extract the lower phase into a clean glass centrifuge tube using a glass pipet.

10. To the remaining upper phase, add 2.64 mL of pre-equilibrated lower phase (*see* **Note 14**).
11. Vortex and separate phases using a fixed angle clinical at 1500 × *g* for 10 min. Extract the lower phase and pool with the previous extraction.
12. To wash the pooled lower phases, add 10 mL of pre-equilibrated upper phase.
13. Vortex and centrifuge at 1500 × *g* for 10 min. Remove the lower phase to a clean glass tube and dry under a stream of nitrogen (*see* **Note 15**). Dried samples can be stored at −20 °C for further use (*see* **Note 16**).

3.5 Visualization of ^{32}P-Labeled Phospholipid Species via 1D or 2D Thin-Layer Chromatography

1. Use a TLC tank that will accommodate 20 × 20 cm plates for TLC analysis. Line the back of the TLC tank with ~39 cm of chromatography paper (*see* **Note 17**).
2. Prepare the first tank by adding 200 mL of TLC solvent system A to the tank. Allow the tank to pre-equilibrate for ~3 h (*see* **Note 18**).
3. Dissolve the ^{32}P-labeled sample in 500 μL of chloroform-methanol (4:1, v/v). Vortex.
4. Add 5 μL of sample to a scintillation vial containing 5 mL of scintillation cocktail.
5. Count in a scintillation counter and calculate the total CPMs of the sample (*see* **Note 24**).
6. Spot 10,000–20,000 CPM/sample on the TLC plate and allow the silica to dry (*see* **Notes 20** and **21**).

   For 1D analysis, samples can be spotted evenly across the bottom of the plate.

   For 2D analysis, spot one sample per plate in the bottom left corner.
7. Place the TLC plate into the pre-equilibrated tank. Once the solvent front reaches the top of the TLC plate, remove plate for drying. For 1D analysis, the plate may be dried using a stream of hot air for 15 min. However, for 2D analysis the plate must be dried using cool air to prevent degradation of lipids (*see* **Note 22**).

   For 1D analysis, stop here and follow ***steps*** *8 and 9 of Subheading 3.3.*
8. While the first tank is running, prepare a second tank with TLC solvent system B (second dimension).
9. After the plate has dried, rotate the plate 90° and place in the TLC tank (*see* **Note 25**). Run until the solvent front reaches the top of the TLC plate. Remove and dry using a blow-dryer (15 min, hot air).

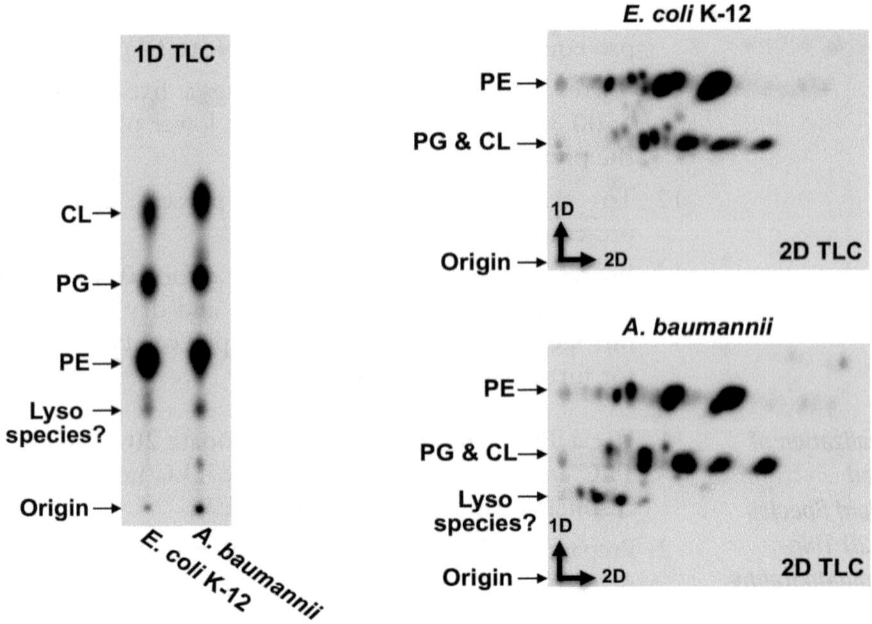

Fig. 6 TLC analysis of glycerophospholipids. (*Left*) 1D TLC can separate the three major glycerophospholipid species of *A. baumannii* (and other bacteria): phosphatidylethanolamine (PE), phosphatidylglycerol (PG), and cardiolipin (CL) as well as potential lysospecies. This is a representative TLC tank of chloroform, methanol, and acetic acid (65:25:5, v/v) loaded with 20,000 CPM/sample; however, an endless number of systems are available depending on the separation required. For reference, refer to Avanti Polar Lipids tech support (https:/avantilipids.com/tech-support/analytical-procedures/tlc-solvent-systems/) to determine which TLC tank suits your analysis. Note that *A. baumannii* glycerophospholipids have different acyl chain lengths resulting in faster migration in this tank. (*Right*) 2D TLC provides greater separation of glycerophospholipids and allows for identification of minor species that are initially masked in the 1D analysis. Shown here are two representative 2D TLC images for *E. coli* (top) and *A. baumannii* (bottom), each loaded with 100,000 CPM. Dimension 1 is chloroform, methanol, and aqueous (60:25:4, v/v), and dimension 2 is chloroform, methanol, acetic acid, and aqueous (80:12:15:4, v/v). It is important to note that this does not separate PG and CL in the first dimension

10. Wrap in plastic wrap and expose to a phosphorimager screen overnight.
11. The next morning, scan the screen to obtain the image (Fig. 6) (*see* **Note 23**).

3.6 Lipid A Isolation for Mass Spectrometry Analysis

1. Inoculate 5 mL of LB using a single colony. Grow overnight with shaking (220 rpm) at 37 °C.
2. Following overnight growth, measure the OD_{600}, and use the overnight culture to inoculate 100–250 mL of LB at a starting OD_{600} of 0.05. Grow the cells with shaking (220 rpm) at 37 °C until an OD_{600} of 0.8–1.0.
3. Harvest the cells via centrifugation at 10,000 × *g* for 10 min. Wash the pellet with 100 mL of 1× PBS.

4. Resuspend the pellet in 40 mL of 1× PBS, and divide into two 250 mL Teflon centrifuge tubes.
5. Add 25 mL of chloroform and 50 mL of methanol to each tube to create a single-phase Bligh-Dyer (chloroform, methanol, water, 1:2:0.8 v/v). Shake vigorously. Vent and reseal tube.
6. Incubate at room temperature for 20 min to ensure cell lysis. Shake vigorously, vent, and reseal tube.
7. Centrifuge the mixture at 2000 × *g* for 20 min. This yields an insoluble LPS pellet. Discard the supernatant (which contains glycerophospholipids and isoprenyl lipids).
8. Wash the LPS pellet with single-phase Bligh-Dyer mixture (1:2:0.8, v/v) (*see* **Note 26**).
9. Centrifuge at 2000 × *g* for 20 min. Discard the supernatant.
10. Resuspend the LPS pellet in 27 mL of hydrolysis buffer.
11. Sonicate the sample using a probe tip sonicator at a constant duty cycle for 20 s at 50% output. Repeat sonication of the sample two times (20 s per burst).
12. Incubate the sample for 30 min in a boiling water bath. Remove the sample and allow to cool to room temperature.
13. Convert the solution to a two-phase Bligh-Dyer mixture by adding 30 mL of chloroform and 30 mL of methanol, yielding a chloroform, methanol, water (2:2:1.8, v/v) mixture. Mix by vigorously shaking the tube, vent, and reseal (*see* **Note 13**).
14. Centrifuge the sample for 10 min at 2000 × *g*. Extract the lower phase into a clean Teflon centrifuge tube using a glass pipet.
15. To the remaining upper phase, add 30 mL of pre-equilibrated lower phase. Mix by vortexing (*see* **Note 14**).
16. Centrifuge at 2000 × *g* for 10 min. Extract the lower phase and pool it with the first extraction.
17. Wash the pooled lower phases (60 mL total) by adding 114 mL of pre-equilibrated upper phase. Mix by vortexing.
18. Centrifuge at 2000 × *g* for 10 min.
19. Remove the lower phase to a clean glass rotary evaporator flask and dry the sample.
20. Store the dried sample at −20 °C for MS analysis (*see* **Note 16**).

3.7 MALDI Analysis of Lipid A

1. Resuspend the dried lipid A sample in ~200 μL of chloroform-methanol (4:1, v/v).
2. Transfer the lipid A sample to a glass vial and dry under a stream of nitrogen.
3. Clean the MALDI plate with 100% methanol. Let air dry.

4. Prepare the MALDI plate (*see* **Note 27**) by adding the preferred calibrant mixture and matrix as specified by your instrument.

3.7.1 Two Methods of Sample Preparation Can Be Used: Thin Layer or Dried Droplet

Thin Layer

1. Spot 0.5 μL of ATT matrix (*see* **Note 28**) in a unique target region of the MALDI plate for the number of samples to be analyzed. Allow to air dry (*see* **Note 29**).
2. Spot 0.5 μL of sample onto the matrix spot. Dry at 37 °C for 5 min prior to acquisition of spectra (*see* **Note 30**).

Dried Droplet

1. Mix 0.5 μL of ATT matrix and 0.5 μL of sample. Spot onto MALDI plate. Dry at 37 °C for 5 min prior to acquisition of spectra (*see* **Note 30**).
2. For intact lipid A analysis, acquire spectra in linear negative mode (Fig. 7).
3. To determine approximate location of modifications, spectra can be collected in the linear positive mode which preferentially fragments the molecule at the β-1,6 linkage [20] (*see* **Note 31**).

4 Notes

1. If alterations to the core sugar structure are expected, a tricine gel can be used to achieve better separation [21].
2. If plates are wet, dry in a biosafety hood for 10–15 min.
3. If confluent lawns are not achieved, increase the concentration of cells plated. For LOS-deficient *Acinetobacter* sp., confluent lawns may only be achieved if an overnight culture is plated.
4. Increasing proteinase K concentration can be necessary to completely remove protein contamination depending on the strain being tested.
5. Run the gel until the blue dye front reaches the bottom of the gel. For separation of higher molecular weight species, the dye front can be run off.
6. Silver staining can also be used if preferred [22].
7. This concentration of ^{32}P is 2× higher than that used for *Escherichia coli* or *Salmonella enterica*. If low labeling is still achieved, the concentration can be increased or the media changed to a minimal media. G56 minimal media supplemented with glucose and a low molarity of phosphate (<0.5 mM) has been used successfully in the Trent Lab. Any minimal media that supports growth of your organism with low phosphate should increase labeling efficiency, as it increases the incorporation of ^{32}P.

A

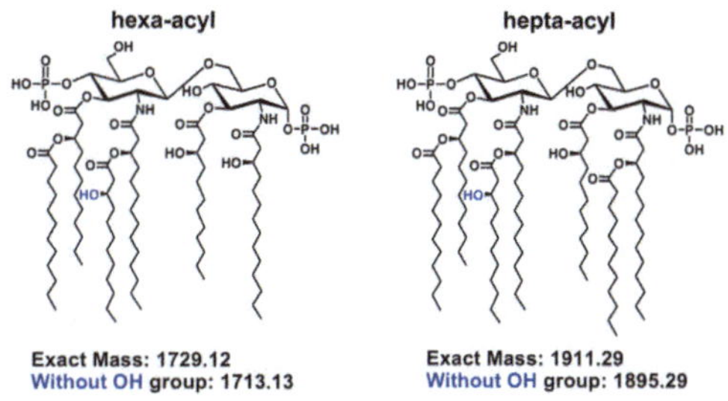

B

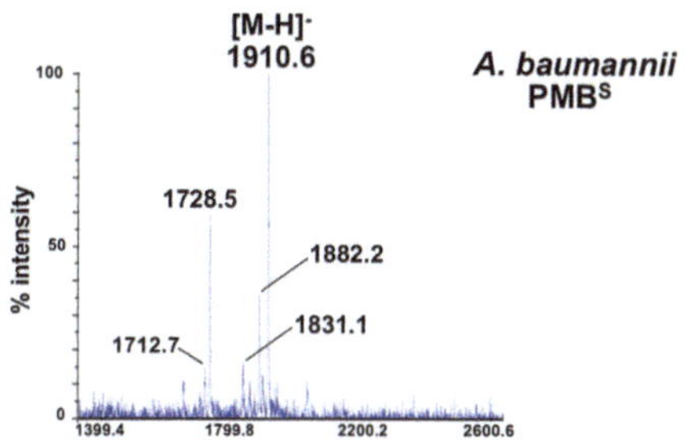

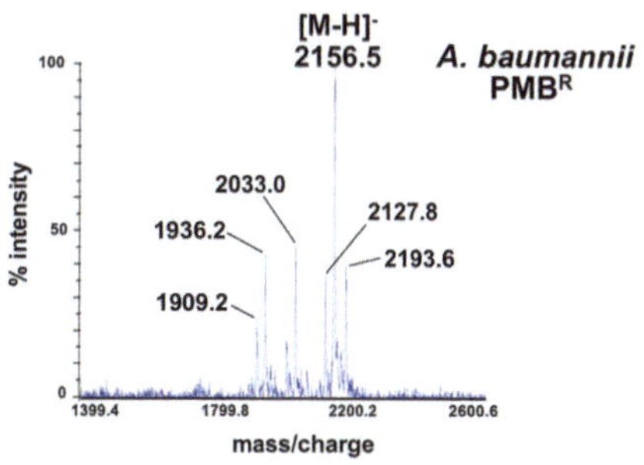

C

[M-H]⁻	No. acyl chains	Additional hydroxy group	pEtN	GalN
1712.7	6	No	No	No
1728.5	6	Yes	No	No
1831.1*	7	Yes	No	No
1882.2, 1909.2, 1910.6, 1936.2	7	Yes	No	No
2033.0	7	Yes	Yes, 1	No
2127.8, 2156.5	7	Yes	Yes, 2	No
2193.6	7	Yes	Yes, 1	Yes, 1

*Loss of phosphate at 1-position

Fig. 7 Mass spectrometric analysis of lipid A can confirm TLC phenotyping. (**a**) The structure and molecular masses of the four main lipid A species in *A. baumannii.* (**b**) MALDI-TOF MS of lipid A isolated from a sensitive strain (top) or resistant strain (bottom). The main species detected with an *m/z* of 1910.6 corresponds to heptaacylated lipid A with a hydroxyacyl chain (refer to table in **c**). The three other main species and their corresponding masses are also detected by this method. The peak with an *m/z* if 1831.1 is due to hydrolysis of the phosphate at the 1 position. This phosphate is the most labile and is typically an artifact in mass spectrometric analysis of lipid A. The MS profile of the polymyxin B-resistant strain (bottom) shows a predominant peak of an *m/z* equal to 2156.5, which corresponds to the addition of two phosphoethanolamines. Additional detected peaks can be assigned to other modified (and unmodified) variants of lipid A

8. Many LOS-deficient strains grow very slowly relative to wild-type *A. baumannii.* If necessary, cultures can be labeled overnight.
9. LOS pellets from LOS-deficient strains are significantly smaller and fragile. If the pellet is accidentally disrupted, repeat the centrifugation step.
10. If contaminating glycerophospholipids are detected in the final analysis, an additional wash step can be added here.
11. Isoprenyl lipids are a minor lipid species and are notoriously difficult to isolate. If these are your target lipid, please refer to [23].
12. For organisms that modify core sugars, hydrolysis is not efficient. In these cases, boiling steps can be increased to 1 h.

13. When lipid A has either reduced hydrophobicity through deacylation or phosphate modifications, the standard Bligh-Dyer extraction can be less efficient. To enhance lipid isolation, an acidic Bligh-Dyer can be used. In this case, the two-phase Bligh-Dyer final mixture should be chloroform-methanol (0.1 M HCl, 2:2:1.8 v/v).
14. Using a pre-equilibrated phase enhances the efficiency of isolation; however, the Trent Lab has noticed that in many cases this is not necessary. If desired, the pre-equilibration step can be avoided, and the addition of reagents can be done directly to the Bligh-Dyer itself to constitute the appropriate mixture.
15. If samples are taking a long time to dry under nitrogen, the addition of chloroform-methanol (4:1, v/v) can increase efficiency.
16. Samples should always be stored dry to help preserve sample integrity.
17. If fewer samples are being run, 10×20 cm plates can be used.
18. Tanks can be set up the night before if the seal is strong. Alternatively, tanks can be rapidly prepped by saturating the chromatography paper during the addition of the solvent system.
19. LOS-deficient samples will have negligible CPM reads. Load the entire sample.
20. Depending on the concentration of the sample, it may be necessary to dry down and resuspend in a smaller amount of chloroform-methanol (4:1, v/v).
21. Generally aim to spot 5–10 μL of sample.
22. Never use hot air prior to running the TLC plate.
23. Provided all samples are on the same TLC plate, this method is semiquantitative using densitometry.
24. CPM values for phospholipid preps are always higher than that of lipid A.
25. The corner which was originally spotted with sample should always be placed in the solvent.
26. An additional wash step can improve purity of lipid A.
27. Use a polished MALDI plate for the best results.
28. While this is the Trent Lab's preferred matrix for lipid A analysis, other matrices can be tested for differential ionization. Alternative matrices that may be worth testing are 5-chloro-2-mercaptobenzothiazole (CMBT), 2,5-dihydroxybenzoic acid (DHB) [24], and β-carboline, 9*H*-pyrido[3,4-*b*]indole (NRM) [25], although it is important to note that this list is not extensive.

29. Matrix is dried when it appears crystallized.
30. Depending on the spectra obtained, it may be necessary to adjust the concentration of the sample prior to spotting the target plate.
31. Positive ion mode collection is only required when the approximate localization of lipid A modifications is desired.

Acknowledgments

This work was supported by the National Institutes of Health (NIH) Grants AI064184, AI76322, AI119879, and AI129940 to M.S.T. and NSF 2017238867 M.J.P. We also acknowledge ARO grant W911NF-16-1-0146 to B.W.D., NIH grant AI125337 to B. W.D., and NIH F32 GM116523 to A.T.T.

References

1. Perez F, Hujer AM, Hujer KM et al (2007) Global challenge of multidrug-resistant Acinetobacter baumannii. Antimicrob Agents Chemother 51:3471–3484
2. Haake DA, Zückert WR (2015) The leptospiral outer membrane. Curr Top Microbiol Immunol 387:187–221
3. Whitfield C, Trent MS (2014) Biosynthesis and export of bacterial lipopolysaccharides. Annu Rev Biochem 83:99–128
4. Needham BD, Trent MS (2013) Fortifying the barrier: the impact of lipid A remodelling on bacterial pathogenesis. Nat Rev Microbiol 11:467–481
5. Arroyo LA, Herrera CM, Fernandez L et al (2011) The pmrCAB operon mediates polymyxin resistance in Acinetobacter baumannii ATCC 17978 and clinical isolates through phosphoethanolamine modification of lipid A. Antimicrob Agents Chemother 55:3743–3751
6. Chin C-Y, Gregg KA, Napier BA et al (2015) A PmrB-regulated deacetylase required for lipid A modification and polymyxin resistance in Acinetobacter baumannii. Antimicrob Agents Chemother 59:7911–7914
7. Boll JM, Tucker AT, Klein DR et al (2015) Reinforcing lipid A acylation on the cell surface of Acinetobacter baumannii promotes cationic antimicrobial peptide resistance and desiccation survival. MBio 6:e00478–e00415
8. Boll JM, Crofts AA, Peters K et al (2016) A penicillin-binding protein inhibits selection of colistin-resistant, lipooligosaccharide-deficient Acinetobacter baumannii. Proc Natl Acad Sci U S A 113:E6228–E6237
9. Peng D, Hong W, Choudhury BP et al (2005) Moraxella catarrhalis bacterium without endotoxin, a potential vaccine candidate. Infect Immun 73:7569–7577
10. Moffatt JH, Harper M, Harrison P et al (2010) Colistin resistance in Acinetobacter baumannii is mediated by complete loss of lipopolysaccharide production. Antimicrob Agents Chemother 54:4971–4977
11. Powers MJ, Trent MS (2018) Expanding the paradigm for the outer membrane: Acinetobacter baumannii in the absence of endotoxin. Mol Microbiol 107:47–56
12. Steeghs L, den HR, den BA et al (1998) Meningitis bacterium is viable without endotoxin. Nature 392:449–450
13. Westphal O, Westphal O, Jann K (1965) Bacterial lipopolysaccharide-extraction with phenol water and further application of procedure. Methods Carbohydr Chem 1:83–91
14. Odegaard TJ, Kaltashov IA, Cotter RJ et al (1997) Shortened hydroxyacyl chains on lipid A of Escherichia coli cells expressing a foreign UDP-N-acetylglucosamine O-acyltransferase. J Biol Chem 272:19688–19696
15. Caroff M, Deprun C, Karibian D et al (1991) Analysis of unmodified endotoxin preparations by 252Cf plasma desorption mass spectrometry. Determination of molecular masses of the constituent native lipopolysaccharides. J Biol Chem 266:18543–18549

16. Bligh EG, Dyer WJ (1959) A rapid method of total lipid extraction and purification. Can J Biochem Physiol 37:911–917
17. Hankins JV, Madsen JA, Needham BD et al (2013) The outer membrane of Gram-negative bacteria: lipid A isolation and characterization. Methods Mol Biol Clifton NJ 966:239–258
18. Henderson JC, O'Brien JP, Brodbelt JS et al (2013) Isolation and chemical characterization of lipid A from gram-negative bacteria. J Vis Exp (79):e50623
19. Garrett TA, Raetz CRH, Son JD et al (2011) Non-enzymatically derived minor lipids found in Escherichia coli lipid extracts. Biochim Biophys Acta 1811:827–837
20. Cotter RJ, Honovich J, Qureshi N et al (1987) Structural determination of lipid a from gram negative bacteria using laser desorption mass spectrometry. Biol Mass Spectrom 14:591–598
21. Heinrichs DE, Yethon JA, Amor PA et al (1998) The assembly system for the outer core portion of R1- and R4-type lipopolysaccharides of Escherichia coli the R1 core-specific β-glucosyltransferase provides a novel attachment site for O-polysaccharides. J Biol Chem 273:29497–29505
22. Hitchcock PJ, Brown TM (1983) Morphological heterogeneity among Salmonella lipopolysaccharide chemotypes in silver-stained polyacrylamide gels. J Bacteriol 154:269–277
23. Trent MS, Ribeiro AA, Doerrler WT et al (2001) Accumulation of a polyisoprene-linked amino sugar in polymyxin-resistant Salmonella typhimurium and Escherichia coli structural characterization and transfer to lipid A in the periplasm. J Biol Chem 276:43132–43144
24. Zhou P, Altman E, Perry MB et al (2010) Study of matrix additives for sensitive analysis of lipid A by matrix-assisted laser desorption ionization mass spectrometry. Appl Environ Microbiol 76:3437–3443
25. Scott AJ, Flinders B, Cappell J et al (2016) Norharmane matrix enhances detection of endotoxin by MALDI-MS for simultaneous profiling of pathogen, host and vector systems. Pathog Dis 74:ftw097
26. Raetz CRH, Whitfield C (2002) Lipopolysaccharide endotoxins. Annu Rev Biochem 71:635–700
27. Stevenson G, Neal B, Liu D et al (1994) Structure of the O antigen of Escherichia coli K-12 and the sequence of its rfb gene cluster. J Bacteriol 176:4144–4156
28. Trent MS, Ribeiro AA, Lin S et al (2001) An inner membrane enzyme in Salmonella and Escherichia coli that transfers 4-amino-4-deoxy-L-arabinose to lipid A: induction on polymyxin-resistant mutants and role of a novel lipid-linked donor. J Biol Chem 276:43122–43131
29. Herrera CM, Hankins JV, Trent MS (2010) Activation of PmrA inhibits LpxT-dependent phosphorylation of lipid A promoting resistance to antimicrobial peptides. Mol Microbiol 76:1444–1460
30. Lee H, Hsu F-F, Turk J et al (2004) The PmrA-regulated pmrC gene mediates phosphoethanolamine modification of lipid A and polymyxin resistance in Salmonella enterica. J Bacteriol 186:4124–4133
31. Bishop RE, Gibbons HS, Guina T et al (2000) Transfer of palmitate from phospholipids to lipid A in outer membranes of gram-negative bacteria. EMBO J 19:5071–5080
32. Gibbons HS, Lin S, Cotter RJ et al (2000) Oxygen requirement for the biosynthesis of the S-2-hydroxymyristate moiety in Salmonella typhimurium lipid A. Function of LpxO, A new Fe2+/alpha-ketoglutarate-dependent dioxygenase homologue. J Biol Chem 275:32940–32949
33. Pelletier MR, Casella LG, Jones JW et al (2013) Unique structural modifications are present in the lipopolysaccharide from colistin-resistant strains of Acinetobacter baumannii. Antimicrob Agents Chemother 57:4831–4840

Chapter 23

Methods for Detecting *N*-Acyl Homoserine Lactone Production in *Acinetobacter baumannii*

Aimee R. Paulk Tierney and Philip N. Rather

Abstract

Acinetobacter baumannii cell-cell signaling is mediated by production and release of *N*-acyl homoserine lactones (AHLs), primarily *N*-(3-hydroxydodecanoyl)-L-homoserine lactone (3-hydroxy-C_{12}-HSL). The secretion of this signal can be readily assessed via simple plate-based or thin-layer chromatography-based assays utilizing an *Agrobacterium tumefaciens traG-lacZ* containing biosensor. In this chapter, we describe methods to assay for secreted *N*-acyl homoserine lactone production in *Acinetobacter baumannii*.

Key words *Acinetobacter baumannii*, *N*-Acyl homoserine lactone, Quorum sensing, Biosensor, *Agrobacterium tumefaciens*, Thin-layer chromatography

1 Introduction

Similar to other Gram-negative bacteria, *Acinetobacter baumannii* communicates via production and secretion of *N*-acyl homoserine lactone (AHL) autoinducers utilizing a LuxI-/LuxR-type system [1]. In this system, an autoinducer synthase produces the AHL signal, which typically diffuses across the cell membrane, reenters the cell at high concentrations, and then binds and activates a transcriptional regulatory protein to mediate global changes in gene expression [2]. In *A. baumannii*, the autoinducer synthase and transcriptional regulator proteins are termed AbaI and AbaR, respectively [1].

Secretion levels of AHL signal can be qualitatively assessed by plate-based methods that utilize bacterial biosensor strains containing a reporter gene that is transcriptionally activated in the presence of AHL signals. Typical reporters include luminescence (*lux*), β-galactosidase (*lacZ*), or *Chromobacterium violaceum* purple pigment violacein [3, 4]. Here, we demonstrate detection of *A. baumannii* AHL signals using an *Agrobacterium tumefaciens* biosensor containing a plasmid-localized *traG-lacZ* fusion (pZLR4). The host biosensor strain does not make AHL signals,

Indranil Biswas and Philip N. Rather (eds.), *Acinetobacter baumannii: Methods and Protocols*, Methods in Molecular Biology, vol. 1946, https://doi.org/10.1007/978-1-4939-9118-1_23,

but exogenous AHL from other bacteria will bind to TraR, coactivating *traG* and inducing production of β-galactosidase [4, 5]. Strains of interest can be surveyed for AHL secretion by (a) cross-streaking the *A. tumefaciens* biosensor against the strain of interest on a LB plate containing 5-bromo-4-chloro-3-indolyl-β-D-galactopyranoside (X-gal); (b) direct plating of bacterial culture, cell-free supernatants, or patched colonies onto a soft agar lawn containing the *A. tumefaciens* biosensor and X-gal; or (c) separating extracts on reversed-phase C_{18} thin-layer chromatography plates and overlaying the dried plates with a soft agar lawn containing the *A. tumefaciens* biosensor.

2 Materials

Prepare all solutions with distilled water prior to daily use. Store reagents at room temperature unless otherwise indicated.

1. Media: Luria broth (LB)—10 g tryptone, 5 g yeast extract, 5 g sodium chloride per liter.
2. Soft agar. Luria broth containing 0.7% agar.
3. Gentamicin stock solution 20 mg/mL in water.
4. Petri dishes and tubes for bacterial culture.
5. X-gal (5-bromo-4-chloro-3-indolyl-β-D-galactopyranoside) solution at 20 μg/mL. Store at −20 °C.
6. Reversed-phase C-18 thin-layer chromatography (TLC) plates.
7. Glass chromatography chamber for developing TLC plates.
8. 60% methanol.

3 Methods

3.1 Cross-Streak Assay

1. Inoculate 2 mL LB containing gentamicin 30 μg/mL with the *Agrobacterium tumefaciens* biosensor strain. Grow overnight at 28 °C, stationary (*see* **Note 1**).
2. Prepare a LB agar plate (15 g agar/L) with 80 μL of 20 μg/mL X-gal per 20 mL (*see* **Note 2**). Dry the plate inverted with the top off at 37 °C for at least 1 h.
3. While holding the plate at an angle, add 30 μL of the *A. tumefaciens* biosensor strain ($OD_{600} = 0.2$) and allow the culture to run down the surface of the plate to form a line of cells. Streak a loop of *A. baumannii* strain to be tested for AHL production next to the biosensor strain. Alternatively, spot 5 μL of the culture to be tested for signal production adjacent to the biosensor strain and allow the spot to soak into the agar.

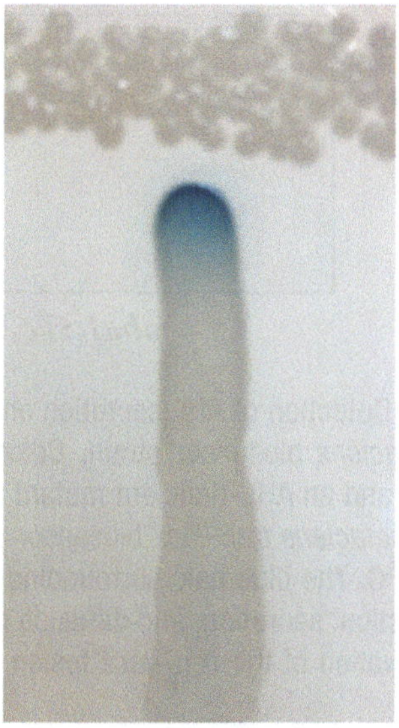

Fig. 1 Cross-streak of *Agrobacterium tumefaciens* indicator strain and *Acinetobacter baumannii*. *A. baumannii* (horizontal line, top of plate) was streaked perpendicular to the *A. tumefaciens* biosensor (vertical line). Blue pigmentation of the biosensor strain near the point of intersection to *A. baumannii* indicates production, secretion, and diffusion of AHL through the medium

4. Incubate plates at 28 °C for 12 h or until a blue color develops in the biosensor strain.

Blue pigmentation of the *A. tumefaciens* biosensor near the strain(s) of interest indicates production of the AHL signal (Fig. 1).

3.2 Soft Agar Assay

1. Inoculate 2 mL LB containing gentamicin (30 μg/mL) with the *Agrobacterium tumefaciens* biosensor strain. Grow overnight at 28 °C, stationary (*see* **Note 1**).
2. Prepare LB soft agar (0.7 g agar/L). Cool to 45 °C and add 80 μL of 20 μg/mL X-gal per 20 mL melted agar and 250 μL of the *A. tumefaciens* culture (OD $A_{600} = 0.2$) per 20 mL melted agar. Mix and pour plate immediately (*see* **Notes 2** and **3**).
3. Allow plate to set and then dry inverted at 37 °C for 30–45 min.
4. Plate or patch bacterial cultures, colonies, or supernatants as desired. Incubate plates at 28 °C. Strains of interest that are positive for AHL secretion will produce a blue halo (Fig. 2).

Fig. 2 Detection of AHL secretion on a soft agar lawn containing *Agrobacterium tumefaciens* biosensor strain. Colonies of an *A. baumannii* wild-type control (right) and an AHL-deficient mutant (left) were patched onto a soft agar lawn of *A. tumefaciens traG-lacZ* biosensor and X-gal. The plate was incubated for 24 h at 28 °C. The blue halo surrounding the wild-type *A. baumannii* patch indicates production, secretion, and diffusion of AHL through the medium, where it results in activation of the *traG-lacZ* fusion in the biosensor strain

3.3 Thin-Layer Chromatography (TLC) Assay

1. Spot samples to be tested on a reversed-phase C_{18} TLC plate. Spots should be approximately 2–3 cm from the bottom of the TLC plate.
2. Allow spots to dry completely.
3. Incubate the TLC plate in a glass chamber with 50 mL of 60% methanol and allow the solvent to completely ascend to the top of the plate.
4. Allow the plate to sit at room temperature for 60 min in a fume hood to allow the methanol to evaporate.
5. Seal all four sides of the TLC plate with tape in a manner that creates a lip of at least 1/2 in. on all sides. Laboratory tape works well for this (Fig. 3A).
6. Prepare 50 mL of LB soft agar (0.7 g agar/L). Cool to 45 °C and add 160 μL of 20 μg/mL X-gal and 400 μL of the *A. tumefaciens* culture. Mix and immediately pour the agar solution over the TLC plate to completely cover the surface. Allow the agar to solidify for 30 min (*see* **Note 4**).
7. Remove the tape border and place the TLC plate on top of an 8 in. × 8 in. glass Pyrex dish that contains 100 mL of water in the bottom, Fig. 3B (*see* **Note 5**). Cover the plate with an identical 8 in. × 8 in. glass Pyrex dish (Fig. 3C). Seal the two dishes by using Saran wrap. Place the sealed Pyrex dishes in a 28 °C incubator and incubate for 24–48 h until blue spots appear indicating activation of the biosensor.
8. Air-dry the TLC plate until the soft agar lawn has become completely dehydrated. This can be done at room temperature or in a 37 °C incubator (*see* **Note 6**).

A. B.

C.

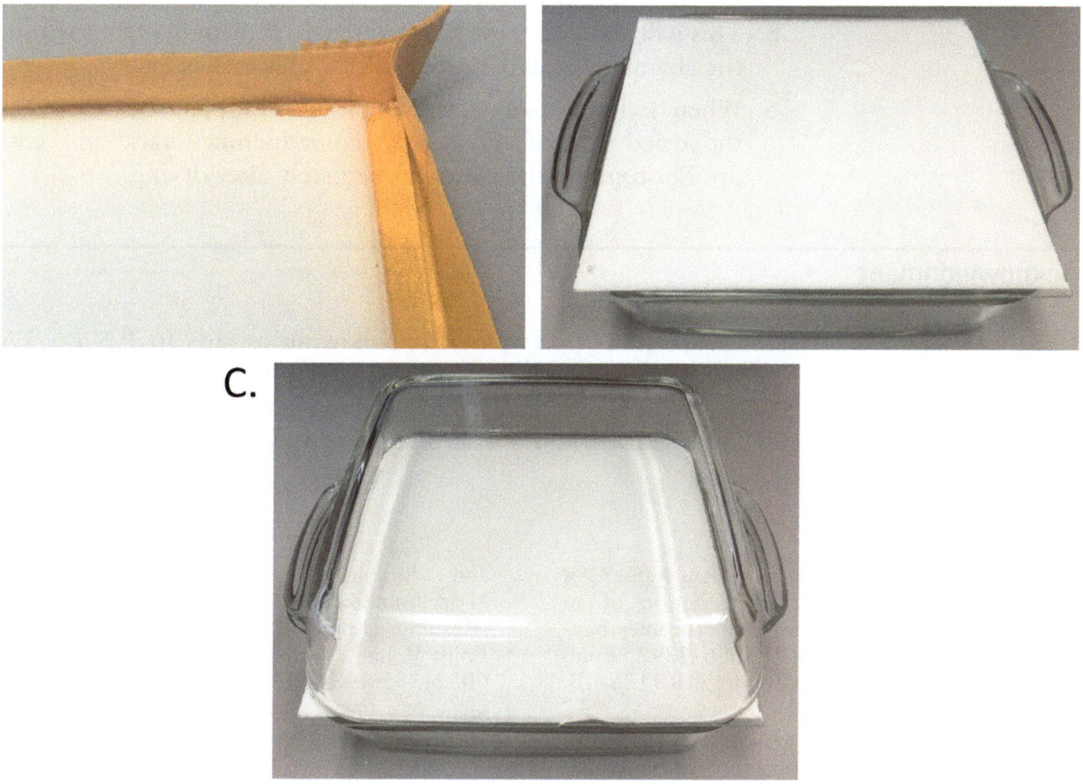

Fig. 3 Methods for incubating TLC plates with a soft agar overlay. This figure demonstrates how an enclosure can be generated with glass Pyrex dishes that can be used to incubate TLC plates in an incubator. Panel **A** demonstrates how tape can be used to create an agar enclosure for the TLC plate. Panels **B** and **C** demonstrate the use of Pyrex dishes to create a humidified enclosure

4 Notes

1. The *A. tumefaciens* with pZLR4 should be grown at 28 °C as the replication of this plasmid is temperature sensitive.
2. The concentration of X-gal can be adjusted to suit experimental needs.
3. The volume of *A. tumefaciens* culture added varies by culture density. The indicated volume of 250 μL per 20 mL agar is based on a culture at an OD_{600} of at least 0.2. It is generally better to err on the side of using a higher concentration as too low concentration will result in a sparse lawn with poor halo visualization. Troubleshooting the volume of culture needed may be required to suit experimental needs.
4. When the soft agar is added to the TLC plate, it is not necessary to have it at an even thickness in all areas of the plate. Once the

plate is done incubating and the agar overlay is dried, all changes in thickness will even out and be invisible.

5. This will allow the soft agar overlay to remain hydrated within the chamber created by the sealed Pyrex dishes.
6. When drying the soft agar lawn on the TLC plate, be aware that the dried agar/TLC matrix can sometime crack and curl up. Photograph the plates immediately after drying.

Acknowledgment

This work is supported by the following awards to P.N.R., VA Merit Award I01 BX001725 and Research Career Scientist Award IK6BX004470, both from the Department of Veterans Affairs, and R01AI072219 from the National Institutes of Health.

References

1. Niu C, Clemmer KM, Bonomo RA, Rather PN (2008) Isolation and characterization of an autoinducer synthase from Acinetobacter baumannii. J Bacteriol 190(9):3386–3392. Epub 19 Feb 2008. https://doi.org/10.1128/JB.01929-07
2. Bhargava N, Sharma P, Capalash N (2010) Quorum sensing in Acinetobacter: an emerging pathogen. Crit Rev Microbiol 36(4):349–360. Epub 18 Sept 2010. https://doi.org/10.3109/1040841X.2010.512269
3. McClean KH, Winson MK, Fish L, Taylor A, Chhabra SR, Camara M, Daykin M, Lamb JH, Swift S, Bycroft BW, Stewart GS, Williams P (1997) Quorum sensing and Chromobacterium violaceum: exploitation of violacein production and inhibition for the detection of N-acylhomoserine lactones. Microbiology (Reading, England) 143(Pt 12):3703–3711. Epub 9 Jan 1998. https://doi.org/10.1099/00221287-143-12-3703
4. Shaw PD, Ping G, Daly SL, Cha C, Cronan JE Jr, Rinehart KL, Farrand SK (1997) Detecting and characterizing N-acyl-homoserine lactone signal molecules by thin-layer chromatography. Proc Natl Acad Sci U S A 94(12):6036–6041. Epub 10 June 1997
5. Cook DM, Li PL, Ruchaud F, Padden S, Farrand SK (1997) Ti plasmid conjugation is independent of vir: reconstitution of the tra functions from pTiC58 as a binary system. J Bacteriol 179(4):1291–1297. Epub 1 Feb 1997

Chapter 24

Isolation and Characterization of the Acinetobactin and Baumannoferrin Siderophores Produced by *Acinetobacter baumannii*

William F. Penwell and Luis A. Actis

Abstract

Siderophores are high-affinity iron chelators produced and used by bacteria to prosper under iron-limiting conditions they normally encounter in the environment and hosts. In this chapter, we describe the isolation and purification of the siderophores acinetobactin and baumannoferrin produced by the bacterial pathogen *Acinetobacter baumannii* using XAD-7 batch adsorption and high-pressure liquid chromatography (HPLC). We also describe chemical tests and biological assays used to detect the presence of catechol and hydroxamate siderophores in culture supernatants, XAD-7 extracts, and HPLC fractions.

Key words Catechol and hydroxamate siderophores, XAD-7 adsorption, High-pressure liquid chromatography, Siderophore detection, Siderophore utilization assays

1 Introduction

Globally, *Acinetobacter baumannii* is an important human pathogen that is associated with hospital- and community-acquired infections [1–3]. This nosocomial pathogen typically affects immunocompromised patients and causes a multitude of diseases, which include pneumonia, bacteremia, urinary tract infections, meningitis, and wound-associated infections [4–7]. To successfully cause an infection and persist in the host, *A. baumannii* requires iron [8, 9]. This essential micronutrient is involved in numerous biological processes including electron transport, photosynthesis, tricarboxylic acid cycle-mediated metabolic processes, deoxynucleotide biosynthesis, and protection from free radicals [10–12]. Despite the importance of iron, in aerobic environments at neutral pH, iron occurs in an insoluble, oxidized, ferric form, which limits the availability of free iron significantly below the concentration required for optimal growth [12–15]. Similar restrictions are found within the human host; free iron is sequestered by

Indranil Biswas and Philip N. Rather (eds.), *Acinetobacter baumannii: Methods and Protocols*, Methods in Molecular Biology, vol. 1946, https://doi.org/10.1007/978-1-4939-9118-1_24,

high-affinity iron-binding proteins, which ultimately make this essential micronutrient unavailable for pathogens, like *A. baumannii*, to utilize for growth [16, 17].

To overcome iron limitations in both the environment and the host, bacteria can produce high-affinity iron acquisition systems to scavenge this essential micronutrient. The strategies found in bacteria include (1) direct interaction with host iron-binding proteins, such as transferring and lactoferrin, (2) uptake of heme, and (3) expression of high-affinity siderophore-mediated iron acquisition systems [10, 16]. The human pathogen *A. baumannii* utilizes the latter to obtain iron from the host during infections. Siderophore-mediated iron acquisition systems in *A. baumannii* have been shown to be required to cause a successful infection in the host using various infections models, which included in vivo infections assays using *Galleria mellonella* and a mouse sepsis model [9]. To date, three different siderophore uptake systems have been described in *A. baumannii* [8, 18–23].

One siderophore system that has been described in *A. baumannii* is fimsbactins A–F, which is a catechol-hydroxamate siderophore [8, 20]. The fimsbactin-mediated system seems to be unique to the *A. baumannii* ATCC 17978 clinical isolate, which was determined after bioinformatic examination of 50 different clinical isolates [22]. Analysis of the *A. baumannii* ATCC 17978 genome revealed a 26-kbp gene cluster that contains the genes needed to code for biosynthesis, secretion, and utilization of this siderophore. Another siderophore produced and utilized by *A. baumannii* strains is acinetobactin, which is the best described siderophore-mediated iron acquisition system in this human pathogen [18, 19, 22, 24]. Acinetobactin is an iron chelator that contains both a catechol and a hydroxamate functional group. Structurally, acinetobactin is composed of *N*-hydroxyhistamine, L-threonine, and 2,3-dihydroxybenzoic acid (DHBA) [25]. The 19 genes involved in biosynthesis, secretion, and utilization of this siderophore are found in two separate genetic loci [18, 19, 24]. Bioinformatic examination of 50 clinical isolates revealed that acinetobactin is ubiquitously found in the majority of the *A. baumannii* isolates [22]. Also found in the majority of the clinical isolates is the chromosomal locus coding for the baumannoferrin A and baumannoferrin B iron acquisition system [23]. The structure of this hydroxamate siderophore was recently resolved, which showed that these siderophores are composed of citrate 1,3-diaminopropane, 2,4-diaminobutyrate, decanoic acid, and α-ketoglutarate and that baumannoferrin A and B differ by a double bond in the decanoic chain [23].

Interestingly, the bioinformatic examination of genomes from 50 clinical isolates shows that the majority of strains contain both the acinetobactin and baumannoferrin gene clusters. Therefore, in this chapter, we focus on the major methods used for the isolation and analysis of these two *A. baumannii* siderophores. Specifically, this chapter includes protocols we have used for the isolation of these two siderophores using XAD-7 batch adsorption and high-pressure liquid chromatography (HPLC) and their detection using chemical tests and siderophore utilization bioassays.

2 Materials

Prepare all solutions using ultrapure Milli-Q water and analytical grade reagents. Unless indicated, prepare and store all reagents and solutions at room temperature. Unless indicated, all water volumes represent final volumes of the reagents described below. Overnight cultures were incubated between 12 and 14 h under the conditions described above.

2.1 Succinate Medium

1. 100× $MgCl_2$ and $CaCl_2$ stock solution: Dissolve 1.0 g of $MgCl_2{\cdot}6H_2O$ and 1.5 g of $CaCl_2$ in 100 mL final volume with water. Autoclave the solution and, once cooled, add 10 mL to 1 L sodium succinate medium to get a final concentration of 1×.
2. 15-μM $FeCl_3{\cdot}6H_2O$ stock solution: Dissolve 0.42 mg of $FeCl_3{\cdot}6H_2O$ in 100 mL 0.1 N HCl. Once dissolved, filter sterilize with a 0.22 μm filter. Add $FeCl_3{\cdot}6H_2O$ to sodium succinate medium to get a final concentration of 0.15 μM.
3. Sodium succinate medium: Dissolve 12.1 g of Tris, 3.7 g of KCl, 1.1 g of NH_4Cl, 0.272 g of KH_2PO_4, 0.142 g of Na_2SO_4, 5 g of NaCl, and 5 g of sodium succinate. Add water to a volume of 500 mL and adjust pH to 7.4 with NaOH. Add water up to 1 L. Autoclave the solution and store it at room temperature. To prepare complete sodium succinate medium, add $MgCl_2$ and $CaCl_2$ and $FeCl_3{\cdot}6H_2O$ stock solutions as indicated below.

2.2 Luria Bertani (LB) Agar and Broth Culture Media

To prepare LB agar, add 40 g of LB agar powder to 1 L of water. Heat to a boil while stirring to completely dissolve the powder. Autoclave to sterilize. To prepare LB broth, add 25 g of LB broth powder to 1 L of water. Stir to completely dissolve powder. Autoclave to sterilize. Iron-limiting agar plates were made by adding 2,2′-dipyridyl to a final concentration between 150 and 200 μM to LB agar kept at 50 °C.

2.3 10 mM 2,2′-Dipyridyl Stock Solution

Dissolve 156.18 mg of 2,2′-dipyridyl in water. Filter sterilize with a 0.22-μm filter, transfer it to an ambler flask, and keep the solution at 4 °C.

2.4 XAD-7 Adsorption Resin

XAD-7, which was extracted with methanol, acetonitrile, and diethyl ether using a Soxhlet extractor to remove organic impurities, is stored in 50% methanol. Prior to use, extensively wash the resin with water using a sidearm vacuum filtering flask, a Büchner funnel, and a glass fiber filter.

2.5 Chrome Azurol S Reagent

1. Solutions needed for chrome azurol S (CAS) reagent.
   (a) 2 mM stock solution of the chrome azurol S (CAS) dye: Dissolve 0.121 g of the dye in 100 mL water.
   (b) 1 mM stock solution of $FeCl_3$: Dissolve 0.0136 g of $FeCl_3 \cdot 6H_2O$ in 50 mL of 10 mM HCl.
   (c) Piperazine buffer solution: Dissolve 4.307 g of piperazine in 30 mL water. Once completely dissolved, add 6.75 mL of 12 N HCl. Adjust pH to 5.6 with 1 N HCl (*see* **Note 1**).
   (d) Hexadecyltrimethylammonium bromide (HDTMA) solution: Dissolve 0.219 g of HDTMA in 50 mL of water.
   (e) CAS shuttle solution: Dissolve 2.18 g of 5-sulfosalicylic acid in 50 mL of water. Stir until completely dissolved and then store in an amber glass bottle.
2. CAS assay solution.
   (a) In a volumetric flask, mix 1.5 mL of 1 mM $FeCl_3$ stock solution with 7.5 mL of the CAS dye stock solution.
   (b) Slowly add 30 mL of the piperazine solution over a 5-min period.
   (c) Bring the final volume to 100 mL with water and store in an amber plastic bottle.

2.6 Ferric Perchlorate Reagent

Weight 0.115 g of $Fe(ClO_4)_3$ and transfer to a glass bottle. In a chemical hood, add 50 mL of 0.1 M $HClO_4$. Stir until powder is completely dissolved and transfer to a glass bottle.

2.7 Arnow's Reagent

1. Reagent A: Add 2.5 mL 12 N HCl to 47.5 mL water.
2. Reagent B: Dissolve 10 g of $NaNO_3$ and 10 g of Na_2MoO_4 in 100 mL of water.
3. Reagent C: Dissolve 2 g NaOH 50 mL water.

2.8 Acetonitrile/TFA HPLC Column Equilibration Solution

HPLC columns used for siderophore purification were equilibrated with 17% acetonitrile containing 0.1% trifluoroacetic acid (TFA) prior to loading siderophore extracts.

2.9 Cleaning of Glassware

All glassware used for siderophore isolation is washed with Nochromix cleaning solution followed by a wash with 0.5 M HCl and a final extensive rinse with water. Clean flasks used for bacterial growth are rinsed with water and autoclaved before adding complete sodium succinate medium, which is then autoclaved for sterilization.

3 Methods

3.1 Bacterial Culturing and Preparation

1. From a frozen glycerol stock kept at −80 °C, streak *A. baumannii* strains onto LB agar plates and grow cultures overnight at 37 °C (*see* **Note 2**).
2. Pick an isolated colony from the LB agar plate and inoculate 100 mL of complete succinate medium. Grow cells in a shaking incubator overnight at 37 °C (*see* **Note 3**).
3. Perform a 1:100 dilution from overnight culture into 1500 mL of fresh complete succinate medium in a 3-L Fernbach culture flask. Grow overnight in shaking incubator at 37 °C (*see* **Note 4**).

3.2 Isolation of Acinetobactin and Baumannoferrin Using XAD-7 Resin

1. After incubation, transfer the culture to 500-mL Nalgene bottles and pellet the cells by centrifugation at 10,000 × *g* for 10 min at 4 °C. Pour the culture supernatant into a 5-L amber glass jug (*see* **Note 5**).
2. Add 20 g (wet weight) of water-rinsed XAD-7 resin and shake the suspension slowly for 2.5 h at room temperature to allow for absorption of siderophores to the resin.
3. Collect the XAD-7 resin by filtration with a sidearm vacuum filtering flask and a Büchner funnel on a glass fiber filter.
4. Transfer the Büchner funnel with the XAD-7 resin to a clean 1-L sidearm vacuum filtering flask. Store the filtered culture supernatant at 4 °C to be tested for siderophore activity (*see* **Note 6**).
5. Wash the XAD-7 resin with adsorbed siderophore three times with 500 mL of water each time.
6. Elute the adsorbed siderophore with a total volume of 300 mL of methanol using 50-mL aliquots of alcohol (*see* **Note** 7).
7. Dry the 300-mL extract with a rotary evaporator. Once completed, resuspend the dried material with 3 mL of methanol using several small aliquots of solvent. Store methanol extracts in acid-washed glass tubers at −80 °C until further use (*see* **Notes 8** and **9**).

3.3 Siderophore Utilization Assay

1. Inoculate indicator strain in a glass test tube with 5 mL of LB broth containing appropriate antibiotics and incubate overnight at 37 °C in shaking incubator (*see* **Note 10**).
2. Prepare iron-limiting agar by supplementing 20 mL of sterile LB agar kept at 50 °C with 0.45 mL of the 2,2′-dipyridyl stock solution (225 μM final concentration) using a sterile glass tube. Invert the mix ten times without making bubbles to completely mix.

3. Add 200 μL of an overnight culture of the bacterial indicator strain to the iron-limiting agar medium kept at 50 °C. Mix the medium and cells without making bubbles and immediately pour into a Petri dish. Let the medium solidify at room temperature.
4. Dry the plates at 37 °C for 10–15 min and use them immediately for siderophore utilization assays.
5. Place 10 μL of "test" or control solution onto a 7-mm sterile filter disk. Once dried, place the disk on the surface of the iron-limiting agar plate inoculated with the appropriate indicator strain. Test solutions can range from cell-free culture supernatants to fractions containing purified siderophores. Sterile water and 10 mM of $FeCl_3$ are used as negative and positive controls, respectively. Incubate the plates side-up at 37 °C for at least 30 h (*see* **Notes 11–13**).

3.4 Chrome Azurol S Liquid Assay

1. To detect for the presence of a siderophore, add 1 mL of CAS assay solution to an equal volume of cell-free culture supernatants, XAD-7 purified siderophore solutions, or HPLC fractions and mix. For a negative control assay, mix CAS assay solution with equal volume of culture media that were used for cell growth or water in the case of testing purified siderophore dissolved in water.
2. Add 10 μL of shuttle solution, mix, and incubate at room temperature for at least 30 min.
3. If iron-binding activity is present, the CAS assay solution changes color from blue to orange within the incubation period (Fig. 3) (*see* **Note 14**).

3.5 Ferric Perchlorate Assay

1. To detect for the presence of hydroxamate compounds, dry under a stream of nitrogen gas 10–50 μL of siderophore purified from *A. baumannii* AYE culture supernatants using XAD-7.
2. Dissolve the dried siderophore sample with water to a final volume of 0.5 mL using a vortex mixer for 1 min.
3. Once dissolved, add 0.5 mL of $Fe(ClO_4)_3$ reagent and mix with a vortex. Incubate the mix at room temperature for 5 min.
4. Measure the absorbance of each sample at 450 nm using a solution containing equal volumes of water and the $Fe(ClO_4)_3$ reagent as a blank (*see* **Note 15**).

3.6 Arnow's Test

1. To detect for the presence of catechol compounds, transfer 1 mL of either cell-free culture supernatant or siderophore solution to glass test tube.

2. Add 1 mL of Arnow's reagent A to the test tube and mix via vortexing.
3. After mixing, add 1 mL of Arnow's reagent B to the test tube and mix. If catechols are present, the mixture will produce a light-yellow color at this point.
4. After mixing, add 1 mL of Arnow's reagent C to the test tube and mix. If catechols are present, the solution will become red to pink in color.
5. Incubate the samples at room temperature for 10 min, and measure the absorbance at a wavelength of 510 nm using a blank made with water or sterile culture medium mixed with equal volumes of reagents A, B, and C.

3.7 Acinetobactin HPLC Purification and Analysis

1. For HPLC purification and analysis of acinetobactin, either XAD-7 purified siderophore samples or cell-free culture supernatants can be used (*see* **Notes 16** and **17**).
2. Equilibrate a C-8 reversed-phase column (5 μm, 250 mm × 4.6 mm) with acetonitrile/TFA HPLC column equilibration solution with a flow rate of 1 mL/min for 1 h before use.
3. Use water and acetonitrile as the mobile phases for liquid chromatography. Add TFA to both water and acetonitrile at 0.13% and 0.1%, respectively. In the case of using the Agilent 1100 LC HPLC instrument, mobile phase degassing is not necessary since the instrument contains an in-line vacuum degasser.
4. Fractionate cell-free culture supernatants or methanol extracts containing acinetobactin using the following gradient series: 17% acetonitrile for 5 min and then from 17% to 30% within 30 min. Hold the 30% acetonitrile elution for 15 min using a 0.5 mL/min flow rate of mobile phases. The presence of acinetobactin is detected with the visible wavelength detector set at 317 nm. The purification of acinetobactin and DHBA by HPLC under the conditions described under these conditions is shown in Fig. 1.
5. Use Arnow's test, CAS reagent, and siderophore utilization assays to test the presence of DHBA and acinetobactin in the isolated fractions.

3.8 Baumannoferrin HPLC Purification and Analysis

1. For HPLC purification and analysis of baumannoferrin, use XAD-7 purified siderophore samples dissolved in methanol. First, dry a small volume (1–3 mL) of a siderophore-containing methanol extract, and resuspend it in 150 μL of water.

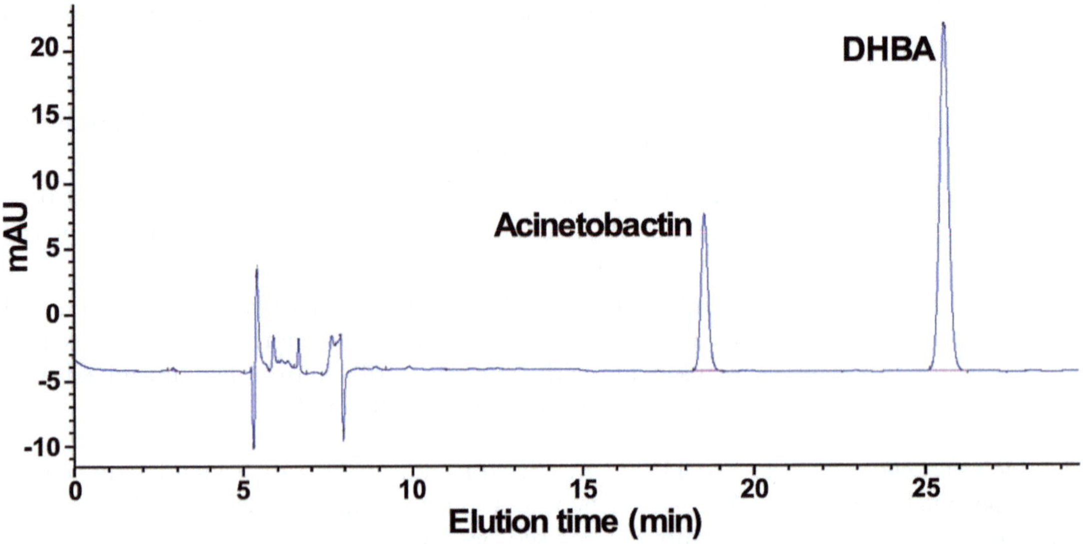

Fig. 1 HPLC profile from cell-free succinate culture supernatant fractionated using the conditions described in Subheading 3.7. The peaks eluted at ~19 and ~26 min correspond to acinetobactin and DHBA, respectively

2. Equilibrate a C-18 reversed-phase column (5 μm, 250 mm × 4.6 mm) with acetonitrile/TFA HPLC column equilibration solution with a flow rate of 1 mL/min for 1 h before use.
3. Use water and acetonitrile as the mobile phases for liquid chromatography. Add TFA to both water and acetonitrile at 0.1%. In the case of using the Agilent 1100 LC HPLC instrument, mobile phase degassing is not necessary since the instrument contains an in-line vacuum degasser.
4. Fractionate the methanol extract containing baumannoferrin using the following gradient series: 0% acetonitrile for 2 min and then 0–100% of acetonitrile within 25 min. Hold the 100% acetonitrile elution step for 2 min using a 1.5 mL/min flow rate of mobile phases. The presence of baumannoferrin is detected with the visible wavelength detector set at 254 nm. The purification of baumannoferrin A and B under these conditions is shown in Fig. 2.
5. Use the $Fe(ClO_4)_3$ and CAS reagents and siderophore utilization assays to test the presence of baumannoferrin in the isolated fractions. Fig. 3 shows the detection of siderophores present in the XAD-7 extract and the HPLC fraction eluted at 14.4 min during the analysis of baumannoferrin using the CAS reagent.

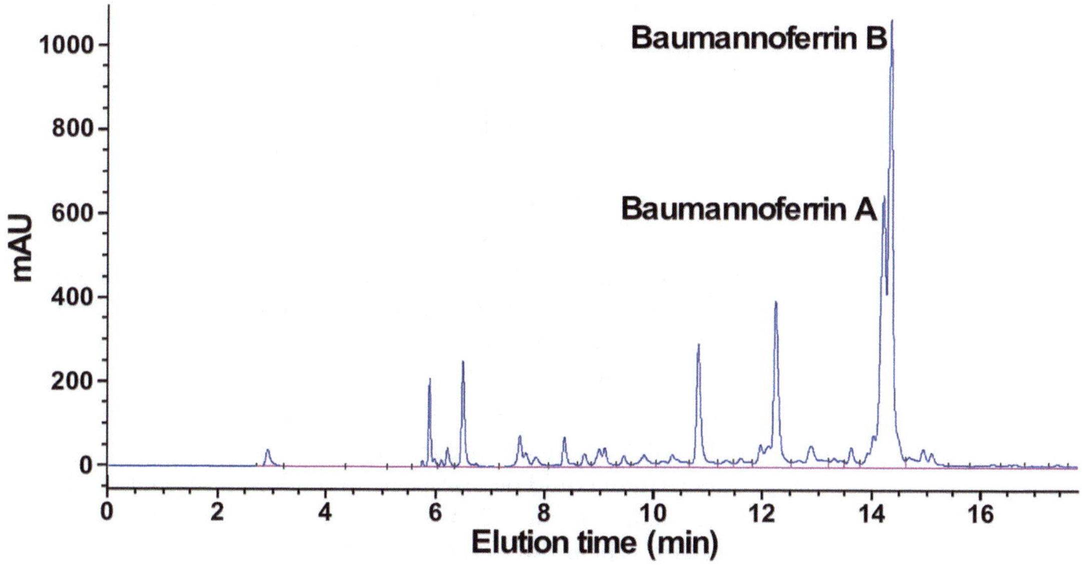

Fig. 2 HPLC profile of baumannoferrin A and B solution using the conditions described in Subheading 3.8. Fractions containing baumannoferrin A and B correspond to the absorption peaks eluted at 14.4 min

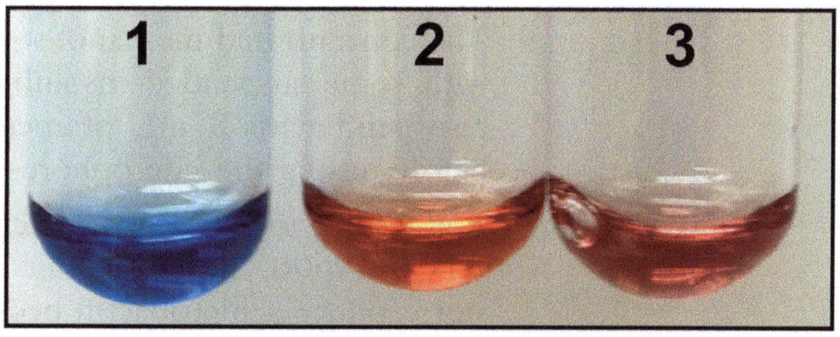

Fig. 3 Siderophore detection using the CAS reagent. The presence of baumannoferrin in XAD-7 extract (2) and the HPLC fraction eluted at 14.4 min (3) was tested using the CAS reagent using sterile complete succinate medium (1) as a blank

4 Notes

1. The pH of the piperazine buffer is extremely critical for the success of the CAS liquid assay. If the pH of this buffer is not close to 5.6, the final solution will not give a color change once a high-affinity iron-binding molecule is added to the reagent.
2. The *A. baumannii* ATCC 19606^T strain was used for the isolation of acinetobactin. This is because this strain only produces acinetobactin [24]. Although this strain does possess the baumannoferrin locus, it does not produce but uses this siderophore. For purification of baumannoferrin, the clinical

isolate AYE was used since it does not produce acinetobactin due to a frameshift mutation in *entA* [24].

3. Unless noted otherwise, shaking incubator is set at 200 rpm for all culturing conditions.
4. Generally, 9 L of complete sodium succinate medium is prepared each time to isolate large amounts of siderophores using XAD-7 batch adsorption.
5. For the isolation of acinetobactin, adjust the pH of the culture supernatant to 7.0 before the addition of XAD-7 resin to reduce the adsorption of DHBA, which is produced in excess during bacterial growth.
6. Store the filtered culture supernatant at 4 °C until it is tested for siderophore activity. If no siderophore activity is present, the supernatant could be discarded after adding 50 mL pure bleach and incubating for 2 h. However, if siderophore activity is present, repeat the XAD-7 adsorption step.
7. Elution of the siderophore from the XAD-7 resin produces an extract with a yellow tint.
8. Place the evaporation flask into a water bath set at 30–35 °C to accelerate methanol evaporation using a rotary evaporator.
9. The concentrated methanol solution could be brown in color with some suspended insoluble material once the extract is resuspend with 3 mL of methanol. Centrifuge the sample before testing the siderophore-containing solution.
10. For testing the presence of acinetobactin, the *A. baumannii* ATCC 19606^{T} BasD isogenic derivative, which uses but does not produce acinetobactin because of a mutation in the *basD* acinetobactin biosynthesis gene, is used as a reporter strain since it expresses all the functions needed for the transport and utilization of this siderophore. For testing the presence of baumannoferrin A and/or B, the *A. baumannii* ATCC 19606^{T} BauA isogenic derivative, which does not produce the acinetobactin outer membrane receptor, is used as a reported since this strain does not produce but uses these siderophores [24].
11. To prepare acinetobactin or baumannoferrin samples, remove 10 μL from the concentrated methanol solution, dry it under a stream of nitrogen gas, and dissolve it in 100 μL of water. Using concentrated siderophore methanol extracts is toxic to bacterial cells if directly applied to a plate containing the indicator strain.
12. This will ensure the filter disks will stick to the surface of the agar plate during incubation.

13. The test solutions diffuse away from the filter disks. Acinetobactin, baumannoferrin, and $FeCl_3$ (positive control) should promote the growth of the selected indicator strain around the filter disks, whereas water (negative control) should not.
14. The CAS liquid assay is a colorimetric assay, which when a siderophore is present, will remove iron from the dye complex resulting in visible color change from blue to orange. The color change can be measured by the absorbance of the solution at 630 nm. The relative siderophore amount (% siderophore units) can be determined by using the equation $[(A_r - A_s)/A_r] \times 100$. A_s is the absorbance of the solution with siderophore activity, and A_r is the reference (or blank) solution.
15. The ferric perchlorate assay is a colorimetric assay, which when the reagent is added to the solution containing an iron chelator will form complexes that can be orange to purple in color. The absorbance of the test solution is measured at a wavelength that is appropriate for the color formed. In the case of baumannoferrin, the solution turned a purple color once the ferric perchlorate reagent was added to purified siderophore.
16. Dry 20 μL of XAD-7 purified acinetobactin with nitrogen gas and dissolve with 500 μL of Milli-Q water. Transfer the acinetobactin samples to 11-mm amber glass crimp HPLC vials for HPLC analysis.
17. For cell-free culture supernatants, remove cells from an overnight culture by centrifugation for 5 min at 12,000 × *g* and filter the supernatant with a 0.45 μm cellulose acetate filter. After filtration, transfer 500 μL of the samples to an 11-mm amber glass crimp HPLC vial for HPLC analysis.

Acknowledgment

Funds from Miami University and Public Health grant AI44776 supported the work described in the reported protocols.

References

1. Perez F, Ponce-Terashima R, Adams MD et al (2011) Are we closing in on an "elusive enemy"? The current status of our battle with *Acinetobacter baumannii*. Virulence 2:86–90
2. Ong CWM, Lye DCB, Khoo KL et al (2009) Severe community-acquired *Acinetobacter baumannii* pneumonia: an emerging highly lethal infectious disease in the Asia-Pacific. Respirology 14:1200–1205
3. Sengstock DM, Thyagarajan R, Apalara J et al (2010) Multidrug-resistant *Acinetobacter baumannii*: an emerging pathogen among older adults in community hospitals and nursing homes. Clin Infect Dis 50:1611–1616
4. Peleg AY, Seifert H, Paterson DL (2008) *Acinetobacter baumannii*: emergence of a successful pathogen. Clin Microbiol Rev 21:538–582
5. Peleg AY, de Breij A, Adams MD et al (2012) The success of *Acinetobacter* species; genetic, metabolic and virulence attributes. PLoS One 7:e46984

6. Visca P, Seifert H, Towner KJ (2011) Acinetobacter infection—an emerging threat to human health. IUBMB Life 63:1048–1054
7. Bergogne-Bérézin E, Towner KJ (1996) *Acinetobacter* spp. as nosocomial pathogens: microbiological, clinical, and epidemiological features. Clin Microbiol Rev 9:148–165
8. Zimbler DL, Penwell WF, Gaddy JA et al (2009) Iron acquisition functions expressed by the human pathogen *Acinetobacter baumannii*. Biometals 22:23–32
9. Gaddy JA, Actis LA, Arivett BA (2012) Role of Acinetobactin-mediated iron acquisition functions in the interaction of *Acinetobacter baumannii* strain ATCC 19606^{T} with human lung epithelial cells, *Galleria mellonella* caterpillars, and mice. Infect Immun 80:1015–1024
10. Krewulak KD, Vogel HJ (2008) Structural biology of bacterial iron uptake. Biochim Biophys Acta Biomembr 1178:1781–1804
11. Crosa JH, Walsh CT (2002) Genetics and assembly line enzymology of siderophore biosynthesis in bacteria. Microbiol Mol Biol Rev 66:223–249
12. Andrews SC, Robinson AK, Rodríguez-Quiñones F (2003) Bacterial iron homeostasis. FEMS Microbiol Rev 27:215–237
13. Schaible UE, Kaufmann SHE (2004) Iron and microbial infection. Nat Rev Microbiol 2:946–953
14. Bullen JJ, Rogers HJ, Spalding PB et al (2006) Natural resistance, iron and infection: a challenge for clinical medicine. J Med Microbiol 55:251–258
15. Rohmer L, Hocquet D, Miller SI (2011) Are pathogenic bacteria just looking for food? Metabolism and microbial pathogenesis. Trends Microbiol 19:341–348
16. Skaar EP (2010) The battle for iron between bacterial pathogens and their vertebrate hosts. PLoS Pathog 6:1–2
17. Johnson EEE, Wessling-Resnick M (2012) Iron metabolism and the innate immune response to infection. Microbes Infect 14:207–216
18. Dorsey CW, Tomaras AP, Connerly PL et al (2004) The siderophore-mediated iron acquisition systems of *Acinetobacter baumannii* ATCC 19606 and *Vibrio anguillarum* 775 are structurally and functionally related. Microbiology 150:3657–3667
19. Mihara K, Tanabe T, Yamakawa Y, Funahashi T et al (2004) Identification and transcriptional organization of a gene cluster involved in biosynthesis and transport of acinetobactin, a siderophore produced by *Acinetobacter baumannii* ATCC 19606^{T}. Microbiology 150:2587–2597
20. Proschak A, Lubuta P, Grün P et al (2013) Structure and biosynthesis of Fimsbactins A-F, siderophores from *Acinetobacter baumannii* and *Acinetobacter baylyi*. Chembiochem 14:633–638
21. Eijkelkamp BA, Hassan KA, Paulsen IT et al (2011) Investigation of the human pathogen *Acinetobacter baumannii* under iron limiting conditions. BMC Genomics 12:126
22. Antunes LCS, Imperi F, Towner KJ et al (2011) Genome-assisted identification of putative iron-utilization genes in *Acinetobacter baumannii* and their distribution among a genotypically diverse collection of clinical isolates. Res Microbiol 162:279–284
23. Penwell WF, Degrace N, Tentarelli S et al (2015) Discovery and characterization of new hydroxamate siderophores, Baumannoferrin A and B, produced by *Acinetobacter baumannii*. Chembiochem 16:1896–1904
24. Penwell WF, Arivett BA, Actis LA The *Acinetobacter baumannii entA* gene located outside the acinetobactin cluster is critical for siderophore production, iron acquisition and virulence. PLoS One 5:e36493
25. Yamamoto S, Okujo N, Sakakibara Y (1994) Isolation and structure elucidation of acinetobactin, a novel siderophore from *Acinetobacter baumannii*. Arch Microbiol 162:249–254

Chapter 25

Skin and Soft Tissue Models for *Acinetobacter baumannii* Infection

Daniel V. Zurawski, Jaideep Banerjee, Yonas A. Alamneh, Jonathan P. Shearer, and Samandra T. Demons

Abstract

Multidrug-resistant *A. baumannii* are important Gram-negative pathogens causing persistent wound infections in both wounded and burned victims, which often result in secondary complications such as delayed wound healing, skin graft failure, and sometimes more serious outcomes such as sepsis and amputation. The choice of antibiotics to remediate these *A. baumannii* infections is becoming limited; and therefore, there has been a renewed interest in the research and development of new antibacterials targeting this pathogen. However, the evaluation of safety and efficacy is made more difficult by the lack of well-established in vivo models. This chapter describes established rodent and large animal models that have been used to investigate and develop treatments for *A. baumannii* skin and soft tissue infections.

Key words ESKAPE pathogens, Drug-resistant bacteria, Graft failure, Animal models, SSTI, Wound healing, Mice, Pigs, Murine, Porcine

1 Introduction

Acinetobacter baumannii is a Gram-negative bacterial pathogen that is one of the key nosocomial bacterial species responsible for antibiotic-resistant infections that is a burden to healthcare facilities worldwide [1–3]. *A. baumannii* can persist for months in a desiccated state and, therefore, can be readily transmitted in the hospital environment as well as in nursing homes [4, 5]. This is further complicated by the fact that during the last few years, strains of *A. baumannii* that are resistant to the majority of currently available antibiotics have emerged and have been defined as multidrug-resistant (MDR), extensively drug-resistant (XDR), and pandrug-resistant (PDR) [3, 6]. MDR *A. baumannii* has been increasingly linked to skin and soft tissue infections (SSTI) [7] and is an important pathogen causing persistent wound infections in burn victims, leading to loss of skin grafts and delayed wound healing [8–10]. The US military healthcare system has also recorded a

Indranil Biswas and Philip N. Rather (eds.), *Acinetobacter baumannii: Methods and Protocols*, Methods in Molecular Biology, vol. 1946, https://doi.org/10.1007/978-1-4939-9118-1_25,

significant increase in the incidence of MDR *A. baumannii* in combat-related wound infections in the last 10–15 years [11, 12]. The majority of those infections occurred during the conflict in Iraq and coincided with polytraumatic injuries, where osteomyelitis, amputation, and amputation revisions became the norm in addition to SSTI [13]. Further, there have been increases in the incidence of neonate sepsis caused by *A. baumannii* [14, 15], some of which could be linked to a wound infection at the cesarean section site [15, 16].

Unfortunately, many cases have left physicians with only colistin, a drug of last resort, to remediate *A. baumannii* infections; and therefore, the lack of treatment options has driven the consideration of narrow-spectrum approaches and the development of new drugs targeting this pathogen [17]. Eliminating such infections requires a deeper understanding of the factors that enable the pathogen to persist as well as the ability to design and create key tests for new antimicrobials targeted specifically against *A. baumannii*. However, the research to develop these new drugs could be limited by a lack of well-established in vivo models. This review describes the rodent and large animal *A. baumannii* infection SSTI models available along with the development of the first preclinical, mono-infection models for *A. baumannii* infection developed in our lab (*see* **Note 1**). These models could become standardized approaches for translational research, but there are, of course, limitations to these approaches as well.

One well-accepted limitation of rodent models, with regard to wound infection, is the dissimilarity in physiology and anatomy with the human skin. Rodents have an increased amount of hair coupled with thinner dermal and epidermal layers. Rodents also lack a layer of fat cells underneath the dermis, which is found in humans. Furthermore, rodents are generally recognized to heal through a combination of both contraction and reepithelialization [18]. In contrast, humans heal by reepithelialization only [18]. From an anatomic and physiologic standpoint, porcine skin bears more similarities to humans in terms of thickness, cellularity, elasticity, healing times, hair follicle distribution, vessel size, and orientation [19, 20]. The porcine mono-infection model is, therefore, highly relevant with regard to skin infection and wound healing and will be an increasingly important tool as several commercial and academic scientists are working on small-molecule and antibiotic alternatives that have specific activity against *A. baumannii* to include systemic and topical applications [21].

Another challenge with *A. baumannii* infection models is the choice of strain(s) for the model system. Numerous groups have developed wound models (Table 1) using different clinical isolates or ATCC laboratory strains; however, virulence is a key variable and can differ greatly between strains [22], which, in turn, can confound antibacterial evaluation. We found it was important to use a more

Table 1
***A. baumannii* wound models of infection**

SPECIES	TYPE OF WOUND	BACTERIA STRAIN	PURPOSE	REF
Mouse - Female BALB/c	**Partial thickness skin abrasion -** Scraped with no. 15 scalpel blade resulting in a wound measuring approx. 1.2 cm × 1.2 cm.	Bioluminescent multidrug resistant clinical isolate *A. baumannii* strain. No specific strain mentioned.	Investigate the use of ultraviolet C light for prevention of infection	[28]
Mouse - Female BALB/c (6-8 wks)	**Full thickness excisional** – 5 mm diameter.	AB0057	Test the capacity of a nitric oxide-releasing nanoparticle to treat infected wounds.	[29]
Mouse - Female BALB/c (6-8 wks)	**Full thickness excisional** – 6 mm diameter. (described in detail in this review)	AB5075	Test antibiotics in a mon-infection model.	[23]
Mouse – athymic nude mice	**Burn**– 12% full-thickness dorsal scald burn injury induced by immersion of the dorsal skin in a 90 °C water bath for 10 seconds	No specific strain mentioned.	Test the efficacy of a genetically engineered human tissue expressing human cathelicidin host defense proteins.	[30]
Mouse - Female BALB/c (6-7 wks)	**Burn -** brass block at 95°C at dorsal surface for 7 seconds, resulting in full-thickness, third-degree burns measuring approximately 1.2 cm× 1.2 cm.	Bioluminescent multidrug resistant clinical isolate *A. baumannii* strain. No specific strain mentioned.	Test the utility of antimicrobial blue light therapy	[31]
Mouse - Female CD-1	**Burn** - pressing a 1 g hot metal weight (diameter 11 mm, temperature 90'C) to the shaved dorsal skin for 2 min.	ATCC BAA-1805	Test the efficacy of a designer antibacterial peptide with that of imipenem and colistin	[32]

(continued)

Table 1
(continued)

Mouse – Female C57BL/6, 4-month-old	**Burn** - dorsal skin surface contact for 10 seconds with brass blocks at 95°C, resulting in a nonlethal 1 cm^2, full-thickness burn.	ATCC BAA 747	Test a wound disinfection method using high voltage, short pulsed electric fields	[33]
Mouse - Female BALB/c	**Burn -** applying a brass block at 95°C to the dorsal surface for 7 secs, resulting in full-thickness, third-degree burns measuring approx. 1 cm × 1.5 cm	Bioluminescent multidrug resistant clinical isolate *A. baumannii* strain	To investigate the use of ultraviolet C light for prevention of infection	[28]
Mouse – Male C57BL/6J (7–8 wks)	**Burn sepsis -** 12 % dorsal scald burn injury induced by submerging the back in hot water at 85°C for 8 seconds, resulting in a full-thickness burn.	No specific strain mentioned.	Test the efficacy of a novel antibacterial therapeutic technology using bacterial conjugation (carried within *Escherichia coli*)	[34]
Mouse – BALB/c	**Systemic infection**	ATCC 17978, HUMC1, HUMC4, HUMC5, HUMC6, HUMC12	Identify a recombinant vaccine immunogen that protected mice against lethal infection, and also induced protective antibodies when administered as passive immunization	[35]
Rat - Male obese Zucker	**Full thickness excisional -** surface area 4.6 cm^2	ATCC 19606	Test the topical delivery of nitric oxide (NO) through a wound dressing for its potential to reduce wound infections	[36]
Rat	**Full thickness excisional –** 1.5cm X 1.5 cm	T-10 and G7	Test the effectiveness of phage administration as an antibacterial remedy	[37]

(continued)

Table 1 (continued)

Rat - Sprague-Dawley (250–300 g)	**Burn** –skin contact with water at 100'C for 10 s, resulting in a full thickness burn	ATCC 19606	Compare the efficacy of different topical antibacterial agents	[38]
Pig – Female Yorkshire (25-35 kg)	**Burn -** electrically heated burn device with controlled pressure delivery at 150°C heated stylus for 50 seconds, resulting in 2"x2" inch full thickness burns.	ATCC 19606	To study polymicrobial wound biofilm and host interaction	[39]
Pig – Female Yorkshire (25-35 kg)	**Full thickness excisional** – 12 mm diameter (described in detail in this review).	AB5075	Mono-species to evaluate topical polymyxin B	[40]

Shaded rows are the models discussed in further detail in this review

virulent isolate and the same strain between all models for consistency when evaluating novel antibacterials. AB5075, an ST2 strain, better represents the isolates found in the clinic with increased antibiotic resistance and negative outcomes being observed in patients [22] (*see* **Note 2**). In this chapter, we will present our two excisional wound models in both murine and porcine species. These models rely on the increased and consistent virulence of AB5075 across animal systems and the use of cyclophosphamide to dampen the innate immune response [22] (*see* **Note 3**).

2 Materials

2.1 Murine Wound Infection Model

1. Female BALB/c mice (6–10 weeks of age) (14–20 g weight) (*see* **Note 4**).
2. 150 and 100 mg/kg cyclophosphamide in 0.9% sodium chloride injection solution (*see* **Note 5**) (final concentration—100 mg/mL).

3. 130 mg/kg ketamine.
4. 10 mg/kg xylazine.
5. 28 gauge syringe.
6. 0.05 mg/kg buprenorphine.
7. Electric razor.
8. Iodine solution.
9. Ethanol rinse solution.
10. 6.0 mm disposable skin biopsy punch.
11. 2.0×10^6 CFU/mL AB5075 *A. baumannii* cells in PBS suspension.
12. Lennox Luria-Bertani (LB) media.
13. Tegaderm roll, 3M Health Care.
14. Vetbond tissue adhesive.
15. Sterile water and food—dry rodent chow supplemented with DietGel Recovery (ClearH_2O, Portland, ME) (*see* **Note 6**).

2.2 Porcine Wound Infection Model

1. Female Yorkshire pigs (30–35 kg).
2. 12–20 mg/kg ketamine.
3. 2–4.4 mg/kg xylazine.
4. Isoflurane gas.
5. Sterile water and food—Laboratory Porcine Diet Grower 5084 (Purina LabDiet®, St. Louis, MO).
6. 25 mg/kg cyclophosphamide IV in 0.9% sodium chloride injection solution (final concentration—100 mg/mL).
7. Electric razor.
8. Chlorhexidine rinse.
9. 12 and 4 mm disposable skin biopsy punch.
10. Gauze.
11. 1.0×10^6 CFU/mL AB5075 *A. baumannii* cells in PBS suspension.
12. Lennox Luria-Bertani (LB) media.
13. Tegaderm roll, 3M Health Care.
14. Vetbond tissue adhesive.
15. 75 mcg/h fentanyl patch.
16. Bandaging tape (VetrapTM).

3 Methods

3.1 Murine Mono-species Wound Infection Model

Purpose: A *murine full-thickness, excisional wound model* was developed in which a diminutive inoculum of a clinically relevant, multidrug-resistant *A. baumannii* isolate was inoculated that could proliferate, form biofilms, and be effectively treated with antibiotics [23] (Figs. 1 and 2).

1. Female BALB/c mice were purchased from Charles River Laboratories (Frederick, MD). The mice used in these experiments were 6–10 weeks of age and weighed 14–20 g.
2. Cyclophosphamide (Baxter, Deerfield, IL) was dissolved in 0.9% sodium chloride injection solution (Hospira Inc., Lake Forest, IL) to obtain a final concentration of 100 mg/mL. Mice received 150 mg/kg (*see* **Note 5**) of body weight and 100 mg/kg cyclophosphamide via intraperitoneal (i.p.) injections, before wounding and infection, on days −4 and −1, respectively.
3. On day 0, the day of wounding and inoculation, mice were anesthetized with an injection of a mixed ketamine (130 mg/kg) and xylazine (10 mg/kg) solution i.p. with a 28-gauge syringe. Buprenorphine (0.05 mg/kg) was administered separately via intramuscular (i.m.) injection for pain management.
4. Hair was clipped with an electric razor from the cervical to mid-lumbar dorsum, and the skin was scrubbed with iodine solution followed by an ethanol rinse.
5. A 6.0 mm disposable skin biopsy punch was used to create a full-thickness skin defect overlying the thoracic spinal column and the adjacent musculature. The circular skin to be removed is lifted with forceps, and any subepithelial connective tissue is incised with iris scissors (*see* **Note 7**).
6. Aliquots of 50 μL containing 5.0×10^4 AB5075 cells in a PBS suspension were pipetted into the wound and allowed to absorb for 3 min.
7. A circular cutout (30 mm in diameter) of transparent dressing (Tegaderm roll; 3M Health Care, St. Paul, MN) was placed over the wound and secured with tissue adhesive (Vetbond; 3M Animal Care, St. Paul, MN) (*see* **Note 8**).
8. Beginning at 4 h post-inoculation, mice are treated with any antibiotic treatments once daily for a 6-day treatment period.
9. All mice received sterile food and water ad libitum, and dry rodent chow was supplemented with DietGel Recovery (ClearH_2O, Portland, ME) during 48 h following wounding.
10. On day 7, the transparent dressing was removed, the treatment was discontinued, and the wound was monitored for closure through day 25.

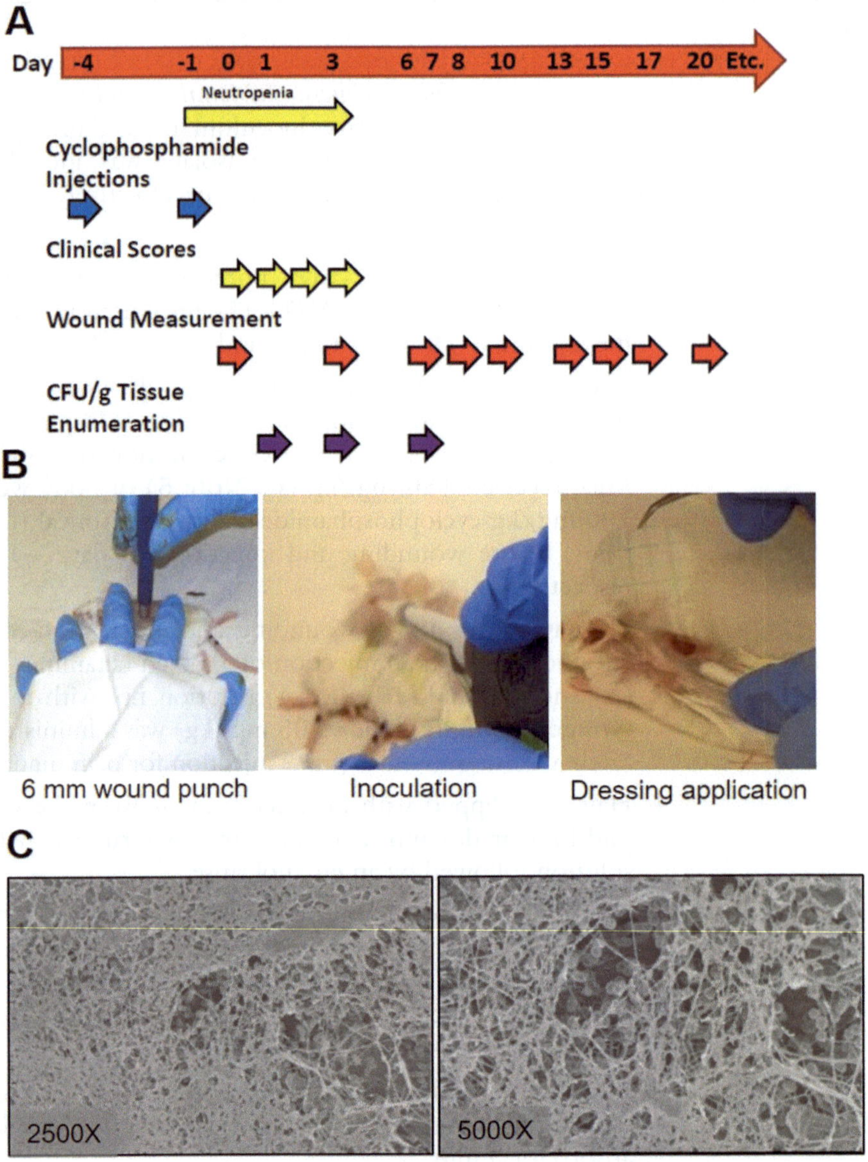

Fig. 1 Murine excisional *A. baumannii* wound model. (**a**, **b**) The murine dorsal wound puncture model uses 6–10-week-old female BALB/c mice that are treated with cyclophosphamide at 4 days prior to infection with (150 mg/kg) cyclophosphamide and again 1 day prior to infection (100 mg/kg). On the day of surgery, mice are weighed and anesthetized with ketamine/xylazine. Backs are shaved and sterilized with a povidone-iodine scrub followed by washing with 70% isopropanol. A full-thickness 6 mm diameter punch is taken from the back of the mouse. Once the wound is created, the mouse is administered a localized dose of buprenorphine and then inoculated with 5.0×10^4 cells of AB5075. To enclose the wound, a circular piece of Tegaderm™ dressing is applied with supplemental glue over the puncture with Vetbond, and the mice are allowed to recover on a heating pad. (**c**) Biofilm formation at 24 h post-infection analyzed using scanning electron microscope (left) 2500× zoom and (right) 5000× zoom

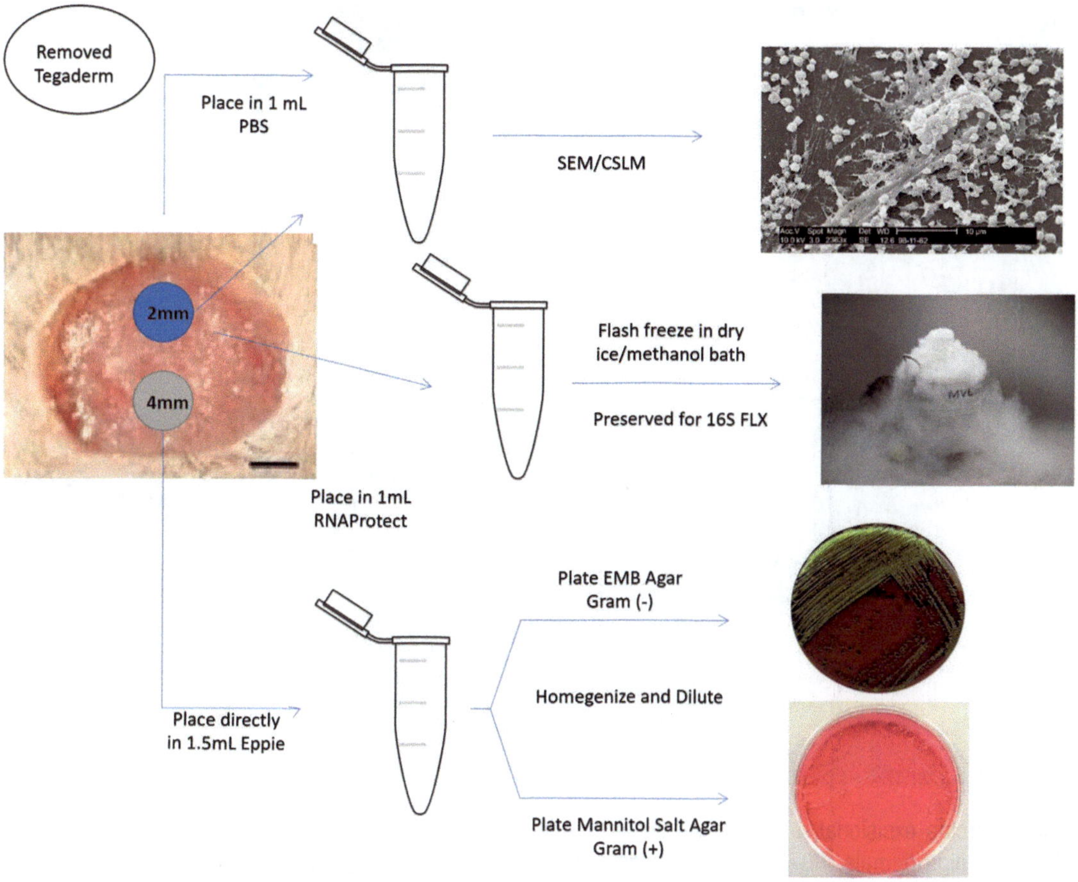

Fig. 2 Collection of biopsies from the murine wound

3.2 Porcine Mono-species Wound Infection Model

Purpose: An *excisional, mono-species infected porcine wound model* was developed, in which a diminutive inoculum (5.0 × 10^4 CFU) of a clinically relevant MDR *A. baumannii* isolate could proliferate, form biofilms, and be effectively treated with antibiotics (Figs. 3 and 4). This model can therefore simulate a skin and soft tissue wound infection preclinical model and can be used to assess novel antimicrobial compounds targeted specifically to *A. baumannii*.

1. Female Yorkshire pigs weighing 30–35 kg (Animal Biotech Industries, Doylestown, PA) were used. All pigs received measured amounts of Laboratory Porcine Diet Grower 5084 (Purina LabDiet®, St. Louis, MO) and water ad libitum.
2. Beginning on day −4 and at all subsequent time points (days 0, 1, 3, 7, and 10), pigs were anesthetized with ketamine 12–20 mg/kg and xylazine 2.0–4.4 mg/kg intramuscular (i.m.) injection, followed by tracheal intubation and anesthetic maintenance on 2.0–4.0% isoflurane gas.

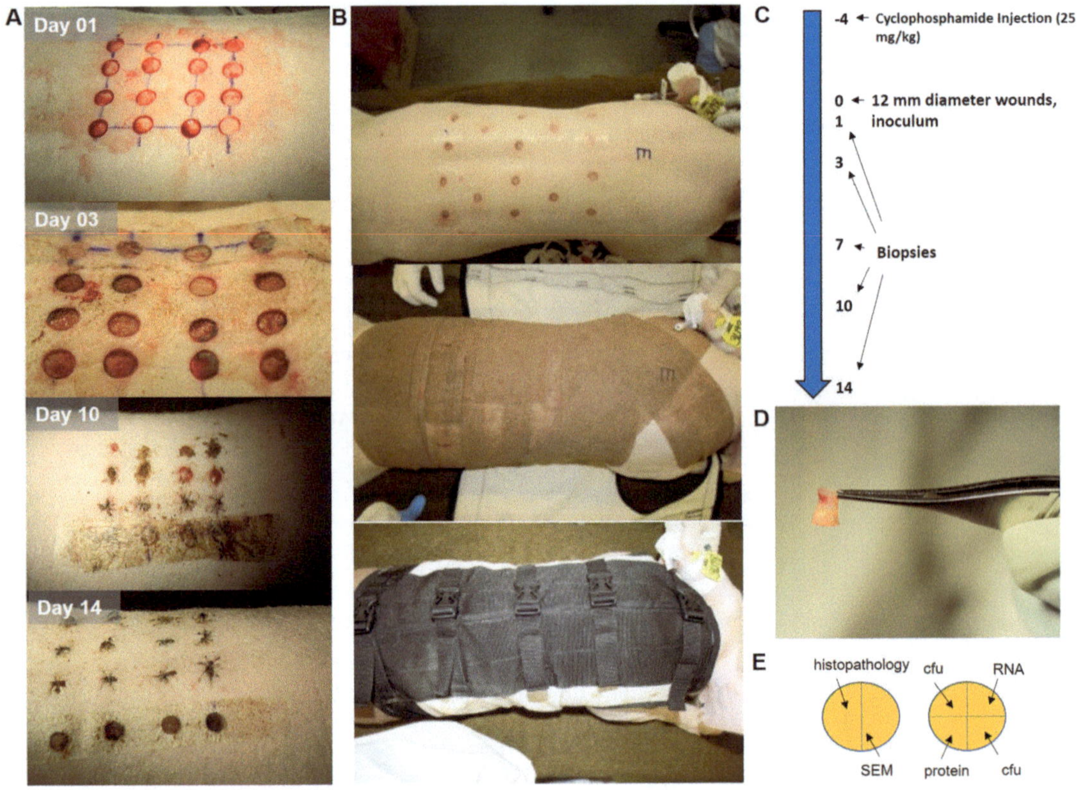

Fig. 3 Porcine excisional *A. baumannii* wound model. (**a**) The porcine dorsal wound puncture model uses Yorkshire pigs (female, 30–35 kg). The pigs are given a single 25 mg/kg intravenous cyclophosphamide injection at day −4. On day 0, anesthesia is induced, and 16 full-thickness dorsal skin wounds in a 4×4 grid formation are created using a 12 mm diameter biopsy punch. Each wound bed is inoculated with 5.0×10^4 CFU AB5075 and then covered with Tegaderm™ dressing. These wounds maintain a 1.0×10^6 AB5075 burden through day 10. (**b**) Progression from inoculated to dressed wounds, through bandaging, and vest application, (**c**) timeline of experiment, (**d**, **e**) 8 mm punches were used to collect biopsies for histopathology/RNA/protein/ELISA or CFU analysis

3. A 1.0–3.0 mL intravenous (IV) blood sample was taken for complete blood count (CBC) analyses.
4. On day −4, each pig received 25 mg/kg cyclophosphamide IV injection (*see* **Note 5**). The neutropenic agent cyclophosphamide (Baxter, Deerfield, IL) was dissolved in 0.9% sodium chloride injection solution (Hospira Inc., Lake Forest, IL) to obtain a final concentration of 100 mg/mL.
5. Pigs were fitted with a customized canvas vest for acclimation prior to wounding and bandaging.
6. On day 0, hair with an electric razor was clipped from the cervical to mid-lumbar dorsum, and the skin scrubbed with iodine solution followed by a chlorhexidine rinse.

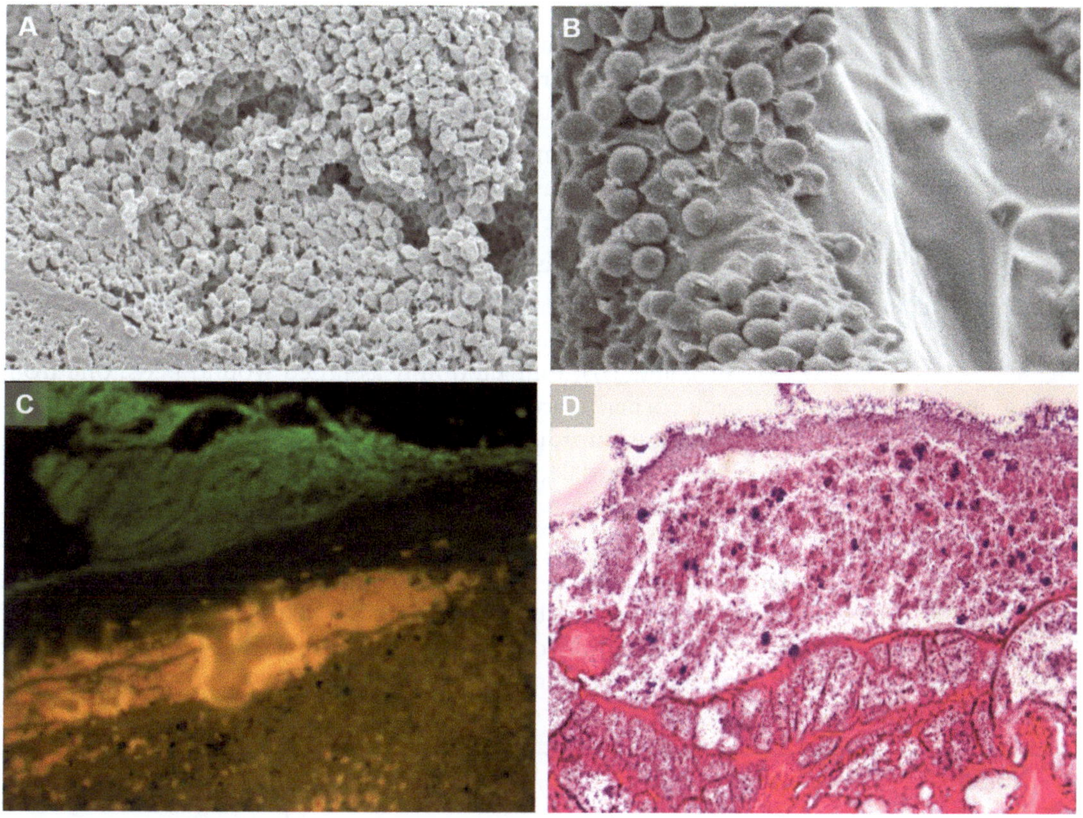

Fig. 4 Demonstration of *A. baumannii* infection in full-thickness excisional porcine wound. Scanning electron micrographs at (**a**) the wound bed and (**b**) the wound edge demonstrating biofilm formation. (**c**) Peptide nucleic acid fluorescent in situ hybridization of pig wound biopsy staining demonstrates the presence of bacteria (green stain). (**d**) Hematoxylin and eosin staining demonstrates the presence of bacteria (black dots) in the wound bed

7. A 12 mm disposable skin biopsy punch (Acuderm® Inc., Fort Lauderdale, FL) and surgical scissors were used to create 16 full-thickness skin wounds approximately 2 mm deep into the subcutaneous fat overlying thoracic and lumbar paraspinal musculature. Gauze is used to pack the wound to prevent excess bleeding until the inoculum is delivered (*see* **Note** 7).
8. On day 0, 50 μL containing 5.0×10^4 AB5075 cells in PBS suspension was pipetted into the wounds and allowed to absorb for 3 min. Bacteria were cultured in Lennox Luria-Bertani (LB) media (Becton, Dickinson and Co., Sparks, MD). 100 μL of AB5075 overnight culture was subcultured into 10 mL of LB and then grown at 37 °C and shaking at 250 rpm in a 250 mL Erlenmeyer flask. Cells were harvested when the culture grew to OD_{600} 0.7 (in mid-log growth phase). Cells were washed twice with sterile phosphate-buffered saline (PBS) and then resuspended in PBS at a concentration of 2.0×10^6 cells/mL.

9. After inoculation, a transparent dressing (Tegaderm™ Film 1622W, 3M Health Care, St. Paul, MN) was placed over each wound and secured with tissue adhesive (Vetbond™, 3M Animal Care, St. Paul, MN).
10. A 75 mcg/h fentanyl patch (Duragesic®, Janssen Pharmaceuticals, Inc., Titusville, NJ) was placed on the pig flank and replaced at subsequent sampling time points.
11. The pig torso was wrapped with bandaging tape (Vetrap™, 3M Animal Care, St. Paul, MN), and the canvas vest was reapplied.
12. Beginning on day 1 and at all subsequent time points (days 3, 7, and 10), four randomly assigned wound beds, one from each corner to account for anatomical variability, were sampled using a 4 mm disposable skin biopsy punch (Acuderm® Inc., Fort Lauderdale, FL). Biopsies were processed for histopathology, scanning electron microscopy, CFU analysis, ELISA, and qPCR.
13. A chlorhexidine-soaked gauze sponge was inserted into each sampled and debrided wound for 3 min to reduce bacterial load. Finally, wounds were closed with 3-0 polydioxanone suture material in a cruciate pattern.

4 Notes

1. *Topical vs. systemic treatments.* It should be noted that as companies and academic labs continue to develop new antibacterial treatments and toxicity is tested, many treatments, even at low doses, cannot be used systemically without causing adverse effects. This has spurred a secondary approach to use some of these treatments topically where the toxicity concern is less stringent. The benefit of the models presented is that both systemic and topical agents can be applied. It should also be noted that formulation is tremendously important. Understanding how a product performs as a gel, cream, or liquid and to what level the antibacterial reaches the wound bed can obviously affect outcomes. It is recommended that substantial testing in vitro with varying formulations be done before any animal experiments are considered.
2. *Bacterial strain selection.* For both models, any *A. baumannii* strain can likely be used, but the inoculum will need to be adjusted. For a strain more virulent than AB5075, cyclophosphamide may not be needed [26]. For less virulent strains, we found a log increase in the inoculum still resulted in a similar wound infection with some other strains (our unpublished data). However, this is certainly strain-dependent, and pilot

experiments will be needed to further assess the proper inoculating dose.

3. It should be noted that other agents can be used to dampen the innate immune response that still results in a prolonged infection. Two examples include the use of Ly-6G monoclonal antibody that blocks neutrophils specifically [24]. Another example is morphine that can dampen the overall inflammatory response [25]. The use of such agents depends on what is driving the experiment in the first place. For example, it may be better to use Ly-6G when components of the immune system are required for the antibacterial being tested such as immunomodulatory agents. Because cyclophosphamide has wide-ranging effects on the immune system, it may not be the best choice for that agent. In contrast, with standard antibiotic approaches that just attack the bacterium chemically, the use of cyclophosphamide might be more beneficial from a cost perspective.
4. *Mouse strain selection.* For the murine model, we routinely use BALB/c mice, but one should understand these are not the only mice that can be utilized when testing antibacterial agents. We initially selected BALB/c mice because they are skewed to a Th2 immune response, which appears to better suit Gram-negative infection that can be limited by the innate immune response (Th1 response) [27]. That said, we have also used male and female C57BL/6J mice, diabetic mice, and humanized mice with our protocol, each with varying outcomes and dependent on the agent being tested. However, female BALB/c mice have mild personalities and are easier to take care of, manipulate, and handle. They have less of tendency to bite when being held and less of a tendency to attack cage-mates or littermates during the duration of the experiment. It should also be noted that some strains of mice, such as A/J mice, are naturally susceptible to *A. baumannii* infection for various reasons and cyclophosphamide will not be needed [21]. The bottom line is that one should not feel limited by the mouse strain selection. Depending on the needs of the experiment, other strains of mice can certainly be used, but the wounding procedure and postsurgery care are the same.
5. For both of these models, the dosage of cyclophosphamide needs to be accurate. We explored both lesser and increased doses in pilot experiments with varying results. For example, when increased doses of cyclophosphamide were used, the animals remained neutropenic for a longer time, and the bacteria spread more easily into the bloodstream, and animals unfortunately perished. In contrast, when too little cyclophosphamide was used, the infection was cleared by the immune

system, and the wounds would close the same as uninfected wounds.

6. Enrichment and welfare for the animals is important. We have found that better care of the animals in both infection models results in better, more reproducible results. An example of this is the use of DietGel Recovery for the mice inside the cage after the procedure and inoculation. As the mice become infected, they become lethargic, and some animals have a tendency to nest rather than making repeated trips for food. Unfortunately, some animals can even succumb to infection under this stress, even though the infection itself is not a lethal dose. With the gel in the cage, the mice require less effort to obtain food, and this improves survivability and outcomes. Each mouse cage also includes a house for nesting and a toy. In the case of the porcine model, we enrich the animals' day-to-day life with toys and sometimes soothing music; therefore, at the time of the procedure, the animals are relaxed and easier to manipulate.
7. *Learning to make the wounds.* Making wounds with the punch biopsy instrument could be considered almost an art. Incisions made too deep result in wounds that significantly bleed and could overly harm the animal, and wounds made with the pressure being too light often did not result in an infection. Our staff are trained (and retrained) for many hours learning how to make the wounds with appropriate depth and with appropriate circular pressure. This included training with deceased animals before live animals were even attempted. Both models require a dedicated staff such as this. The murine model to effectively perform the procedure and treatment requires at least four people on the day of the surgery and at least two people for care and treatments, before and after the surgery. The porcine model requires even more dedicated staff as the animal is essentially treated as a human patient. We had at least two veterinarians monitoring and assisting with the pigs at all times.
8. For the murine wound model, single caging initially was required to prevent predation during the first few days of the experiment and after surgery and recovery. Again, as the animals become sick, they try to limit trips to food sources. Unfortunately, chewing on cage-mates' backs for food became a solution. Single caging prevented this during the initial stages of infection. After about a week (generally, after a few days of Tegaderm™ removal), animals can be housed together again as their immune systems fight the symptoms of infection, and animal health improves.

Acknowledgments

Our research has been supported by a number of grants from the Military Infectious Diseases Research and Development Program as well as supplemental funding from the Defense Health Agency. We would like to thank all the members of the Wound Infections Department, past and present, who have contributed to this work. We would also like to thank MAJ Matt Johnson of the NMRC and the Navy Dive team who supplied the custom vest for the pigs.

Material has been reviewed by the Walter Reed Army Institute of Research. There is no objection to its presentation and/or publication. The opinions or assertions contained herein are the private views of the author and are not to be construed as official or as reflecting true views of the Department of the Army or the Department of Defense. Research was conducted under approved animal protocols in an AAALAC-accredited facility in compliance with the Animal Welfare Act and other federal statutes and regulations relating to animals and experiments involving animals and adheres to principles stated in the *Guide for the Care and Use of Laboratory Animals*, NRC publication, 2011 edition.

References

1. Boucher HW, Talbot GH, Bradley JS, Edwards JE, Gilbert D, Rice LB, Scheld M, Spellberg B, Bartlett J (2009) Bad bugs, no drugs: no ESKAPE! An update from the Infectious Diseases Society of America. Clin Infect Dis 48 (1):1–12. https://doi.org/10.1086/595011
2. Talbot GH, Bradley J, Edwards JE Jr, Gilbert D, Scheld M, Bartlett JG, Antimicrobial Availability Task Force of the Infectious Diseases Society of America (2006) Bad bugs need drugs: an update on the development pipeline from the Antimicrobial Availability Task Force of the Infectious Diseases Society of America. Clin Infect Dis 42(5):657–668. https://doi.org/10.1086/499819
3. Peleg AY, Seifert H, Paterson DL (2008) Acinetobacter baumannii: emergence of a successful pathogen. Clin Microbiol Rev 21 (3):538–582. https://doi.org/10.1128/CMR.00058-07
4. Rastogi VK, Wallace L, Smith LS (2007) Disinfection of Acinetobacter baumannii-contaminated surfaces relevant to medical treatment facilities with ultraviolet C light. Mil Med 172(11):1166–1169
5. Gayoso CM, Mateos J, Mendez JA, Fernandez-Puente P, Rumbo C, Tomas M, Martinez de Ilarduya O, Bou G (2014) Molecular mechanisms involved in the response to desiccation stress and persistence in Acinetobacter baumannii. J Proteome Res 13(2):460–476. https://doi.org/10.1021/pr400603f
6. Durante-Mangoni E, Zarrilli R (2011) Global spread of drug-resistant Acinetobacter baumannii: molecular epidemiology and management of antimicrobial resistance. Future Microbiol 6(4):407–422. https://doi.org/10.2217/fmb.11.23
7. Ali A, Botha J, Tiruvoipati R (2014) Fatal skin and soft tissue infection of multidrug resistant Acinetobacter baumannii: a case report. Int J Surg Case Rep 5(8):532–536. https://doi.org/10.1016/j.ijscr.2014.04.019
8. Sharma M, Taneja N (2007) Burns, antimicrobial resistance & infection control. Indian J Med Res 126(6):505–507
9. Sharma BR (2007) Infection in patients with severe burns: causes and prevention thereof. Infect Dis Clin N Am 21(3):745–759, ix. https://doi.org/10.1016/j.idc.2007.06.003
10. Santucci SG, Gobara S, Santos CR, Fontana C, Levin AS (2003) Infections in a burn intensive care unit: experience of seven years. J Hosp Infect 53(1):6–13

11. Murray CK, Hospenthal DR (2008) Acinetobacter infection in the ICU. Crit Care Clin 24 (2):237–248, vii. https://doi.org/10.1016/j.ccc.2007.12.005
12. Scott P, Deye G, Srinivasan A, Murray C, Moran K, Hulten E, Fishbain J, Craft D, Riddell S, Lindler L, Mancuso J, Milstrey E, Bautista CT, Patel J, Ewell A, Hamilton T, Gaddy C, Tenney M, Christopher G, Petersen K, Endy T, Petruccelli B (2007) An outbreak of multidrug-resistant Acinetobacter baumannii-calcoaceticus complex infection in the US military health care system associated with military operations in Iraq. Clin Infect Dis 44(12):1577–1584. https://doi.org/10.1086/518170
13. Yun HC, Branstetter JG, Murray CK (2008) Osteomyelitis in military personnel wounded in Iraq and Afghanistan. J Trauma 64 (2 Suppl):S163–S168; discussion S168. https://doi.org/10.1097/TA.0b013e318160868c
14. Lee HY, Hsu SY, Hsu JF, Chen CL, Wang YH, Chiu CH (2018) Risk factors and molecular epidemiology of *Acinetobacter baumannii* bacteremia in neonates. J Microbiol Immunol Infect 51(3):367–376. https://doi.org/10.1016/j.jmii.2017.07.013
15. Investigators of the Delhi Neonatal Infection Study Collaboration (2016) Characterisation and antimicrobial resistance of sepsis pathogens in neonates born in tertiary care centres in Delhi, India: a cohort study. Lancet Glob Health 4(10):e752–e760. https://doi.org/10.1016/S2214-109X(16)30148-6
16. Ahmed S, Kawaguchiya M, Ghosh S, Paul SK, Urushibara N, Mahmud C, Nahar K, Hossain MA, Kobayashi N (2015) Drug resistance and molecular epidemiology of aerobic bacteria isolated from puerperal infections in Bangladesh. Microb Drug Resist 21 (3):297–306. https://doi.org/10.1089/mdr.2014.0219
17. Melander RJ, Zurawski DV, Melander C (2018) Narrow-spectrum antibacterial agents. Med Chem Commun 9:12–21
18. Chen L, Mirza R, Kwon Y, DiPietro LA, Koh TJ (2015) The murine excisional wound model: contraction revisited. Wound Repair Regen 23(6):874–877. https://doi.org/10.1111/wrr.12338
19. Greenhalgh DG (2005) Models of wound healing. J Burn Care Rehabil 26(4):293–305
20. Sullivan TP, Eaglstein WH, Davis SC, Mertz P (2001) The pig as a model for human wound healing. Wound Repair Regen 9(2):66–76
21. Wong D, Nielsen TB, Bonomo RA, Pantapalangkoor P, Luna B, Spellberg B (2017) Clinical and pathophysiological overview of Acinetobacter infections: a century of challenges. Clin Microbiol Rev 30 (1):409–447. https://doi.org/10.1128/CMR.00058-16
22. Jacobs AC, Thompson MG, Black CC, Kessler JL, Clark LP, McQueary CN, Gancz HY, Corey BW, Moon JK, Si Y, Owen MT, Hallock JD, Kwak YI, Summers A, Li CZ, Rasko DA, Penwell WF, Honnold CL, Wise MC, Waterman PE, Lesho EP, Stewart RL, Actis LA, Palys TJ, Craft DW, Zurawski DV (2014) AB5075, a highly virulent isolate of Acinetobacter baumannii, as a model strain for the evaluation of pathogenesis and antimicrobial treatments. mBio 5(3):e01076–e01014. https://doi.org/10.1128/mBio.01076-14
23. Thompson MG, Black CC, Pavlicek RL, Honnold CL, Wise MC, Alamneh YA, Moon JK, Kessler JL, Si Y, Williams R, Yildirim S, Kirkup BC Jr, Green RK, Hall ER, Palys TJ, Zurawski DV (2014) Validation of a novel murine wound model of Acinetobacter baumannii infection. Antimicrob Agents Chemother 58 (3):1332–1342. https://doi.org/10.1128/AAC.01944-13
24. Grguric-Smith LM, Lee HH, Gandhi JA, Brennan MB, DeLeon-Rodriguez CM, Coelho C, Han G, Martinez LR (2015) Neutropenia exacerbates infection by Acinetobacter baumannii clinical isolates in a murine wound model. Front Microbiol 6:1134. https://doi.org/10.3389/fmicb.2015.01134
25. Breslow JM, Monroy MA, Daly JM, Meissler JJ, Gaughan J, Adler MW, Eisenstein TK (2011) Morphine, but not trauma, sensitizes to systemic Acinetobacter baumannii infection. J Neuroimmune Pharmacol 6(4):551–565. https://doi.org/10.1007/s11481-011-9303-6
26. Jones CL, Clancy M, Honnold C, Singh S, Snesrud E, Onmus-Leone F, McGann P, Ong AC, Kwak Y, Waterman P, Zurawski DV, Clifford RJ, Lesho E (2015) Fatal outbreak of an emerging clone of extensively drug-resistant Acinetobacter baumannii with enhanced virulence. Clin Infect Dis 61(2):145–154. https://doi.org/10.1093/cid/civ225
27. Prabhakara R, Harro JM, Leid JG, Keegan AD, Prior ML, Shirtliff ME (2011) Suppression of the inflammatory immune response prevents the development of chronic biofilm infection due to methicillin-resistant Staphylococcus aureus. Infect Immun 79(12):5010–5018. https://doi.org/10.1128/IAI.05571-11

28. Dai T, Murray CK, Vrahas MS, Baer DG, Tegos GP, Hamblin MR (2012) Ultraviolet C light for Acinetobacter baumannii wound infections in mice: potential use for battlefield wound decontamination? J Trauma Acute Care Surg 73(3):661–667. https://doi.org/10.1097/TA.0b013e31825c149c
29. Han G, Martinez LR, Mihu MR, Friedman AJ, Friedman JM, Nosanchuk JD (2009) Nitric oxide releasing nanoparticles are therapeutic for Staphylococcus aureus abscesses in a murine model of infection. PLoS One 4(11):e7804. https://doi.org/10.1371/journal.pone.0007804
30. Thomas-Virnig CL, Centanni JM, Johnston CE, He LK, Schlosser SJ, Van Winkle KF, Chen R, Gibson AL, Szilagyi A, Li L, Shankar R, Allen-Hoffmann BL (2009) Inhibition of multidrug-resistant Acinetobacter baumannii by nonviral expression of hCAP-18 in a bioengineered human skin tissue. Mol Ther 17 (3):562–569. https://doi.org/10.1038/mt.2008.289
31. Zhang Y, Zhu Y, Gupta A, Huang Y, Murray CK, Vrahas MS, Sherwood ME, Baer DG, Hamblin MR, Dai T (2014) Antimicrobial blue light therapy for multidrug-resistant Acinetobacter baumannii infection in a mouse burn model: implications for prophylaxis and treatment of combat-related wound infections. J Infect Dis 209(12):1963–1971. https://doi.org/10.1093/infdis/jit842
32. Ostorhazi E, Rozgonyi F, Sztodola A, Harmos F, Kovalszky I, Szabo D, Knappe D, Hoffmann R, Cassone M, Wade JD, Bonomo RA, Otvos L Jr (2010) Preclinical advantages of intramuscularly administered peptide A3-APO over existing therapies in Acinetobacter baumannii wound infections. J Antimicrob Chemother 65(11):2416–2422. https://doi.org/10.1093/jac/dkq337
33. Golberg A, Broelsch GF, Vecchio D, Khan S, Hamblin MR, Austen WG Jr, Sheridan RL, Yarmush ML (2015) Pulsed electric fields for burn wound disinfection in a murine model. J Burn Care Res 36(1):7–13. https://doi.org/10.1097/BCR.0000000000000157
34. Shankar R, He LK, Szilagyi A, Muthu K, Gamelli RL, Filutowicz M, Wendt JL, Suzuki H, Dominguez M (2007) A novel antibacterial gene transfer treatment for multidrug-resistant Acinetobacter baumannii-induced burn sepsis. J Burn Care Res 28 (1):6–12. https://doi.org/10.1097/BCR.0b013e31802c8861
35. Luo G, Lin L, Ibrahim AS, Baquir B, Pantapalangkoor P, Bonomo RA, Doi Y, Adams MD, Russo TA, Spellberg B (2012) Active and passive immunization protects against lethal, extreme drug resistant-Acinetobacter baumannii infection. PLoS One 7(1):e29446. https://doi.org/10.1371/journal.pone.0029446
36. Neidrauer M, Ercan UK, Bhattacharyya A, Samuels J, Sedlak J, Trikha R, Barbee KA, Weingarten MS, Joshi SG (2014) Antimicrobial efficacy and wound-healing property of a topical ointment containing nitric-oxide-loaded zeolites. J Med Microbiol 63 (Pt 2):203–209. https://doi.org/10.1099/jmm.0.067322-0
37. Kusradze I, Karumidze N, Rigvava S, Dvalidze T, Katsitadze M, Amiranashvili I, Goderdzishvili M (2016) Characterization and testing the efficiency of Acinetobacter baumannii phage vB-GEC_Ab-M-G7 as an antibacterial agent. Front Microbiol 7:1590. https://doi.org/10.3389/fmicb.2016.01590
38. Selcuk CT, Durgun M, Ozalp B, Tekin A, Tekin R, Akcay C, Alabalik U (2012) Comparison of the antibacterial effect of silver sulfadiazine 1%, mupirocin 2%, Acticoat and octenidine dihydrochloride in a full-thickness rat burn model contaminated with multi drug resistant Acinetobacter baumannii. Burns 38 (8):1204–1209. https://doi.org/10.1016/j.burns.2012.04.009
39. Roy S, Elgharably H, Sinha M, Ganesh K, Chaney S, Mann E, Miller C, Khanna S, Bergdall VK, Powell HM, Cook CH, Gordillo GM, Wozniak DJ, Sen CK (2014) Mixed-species biofilm compromises wound healing by disrupting epidermal barrier function. J Pathol 233(4):331–343. https://doi.org/10.1002/path.4360
40. Zurawski DV, Black CC , Alamneh YA, Biggemann L, Banerjee J, Thompson MG, Wise MC, Honnold CL, Kim RK, Paranavitana C, Tyner SD, and Demons ST (2018) A porcine wound model of *Acinetobacter baumannii* infection. Adv. Wound Care. Published online Aug. 27. https://doi.org/10.1089/wound.2018.0786

Chapter 26

Assessing *Acinetobacter baumannii* Virulence and Persistence in a Murine Model of Lung Infection

Lauren D. Palmer, Erin R. Green, Jessica R. Sheldon, and Eric P. Skaar

Abstract

Acinetobacter baumannii is a Gram-negative opportunistic pathogen and a leading cause of ventilator-associated pneumonia. Murine models of *A. baumannii* lung infection allow researchers to experimentally assess *A. baumannii* virulence and host response. Intranasal administration of *A. baumannii* models acute lung infection. This chapter describes the methods to test *A. baumannii* virulence in a murine model of lung infection, including assessing the competitive index of a bacterial mutant and the associated inflammatory responses.

Key words Lung infection, Pneumonia, Murine model, *Acinetobacter baumannii*, Inflammation

1 Introduction

Acinetobacter baumannii is an important opportunistic pathogen that commonly infects critically ill patients in hospital settings. While this aerobic, Gram-negative coccobacillus was once largely considered an innocuous environmental organism, *A. baumannii* and related species have emerged in recent years as important threats to global public health. Driving its clinical significance is the rapid acquisition of multidrug resistance phenotypes, which have arisen at alarming rates [1, 2]. This has led the World Health Organization to prioritize multidrug-resistant *A. baumannii* among a group of bacterial pathogens that pose the greatest threat to human health [3].

A. baumannii causes a range of infections, including pneumonia, bacteremia, skin and soft tissue infections, as well as urinary tract infections [4, 5]. In hospital settings, these *A. baumannii* infections frequently occur when normal anatomical barriers are breached, such as in the case of ventilator-associated pneumonias, central line-associated bloodstream infections, surgical wound infections, and catheter-associated urinary tract infections. In particular, ventilator-associated pneumonias are among the most

Indranil Biswas and Philip N. Rather (eds.), *Acinetobacter baumannii: Methods and Protocols*, Methods in Molecular Biology, vol. 1946, https://doi.org/10.1007/978-1-4939-9118-1_26, © Springer Science+Business Media, LLC, part of Springer Nature 2019

frequent causes of hospital-acquired *A. baumannii* infections, where mortality rates approach 75% [6]. While less frequent, cases of community-acquired *A. baumannii* pneumonia have also been reported, with mortality rates as high as 64% [7]. Taken together, the challenges presented by this emerging pathogen underscore the importance of studying its biology and pathogenesis in animal infection models.

Over the last decade, there has been significant effort to characterize the mechanisms by which *A. baumannii* thrives in the host environment. This work has identified a number of virulence factors and systems required for optimal virulence, including protein secretion machinery [8–11], micronutrient acquisition and homeostasis systems [12–17], and proteins involved in the assembly and modification of outer membrane and capsular polysaccharide structures [18–20]. Additionally, a number of genome-wide transposon mutagenesis screens have identified *A. baumannii* genes required for infection in eukaryotic hosts, including those involved in micronutrient acquisition and capsule formation, amino acid and purine metabolism, biofilm formation, and oxidative and osmotic stress responses [21–23].

Studies examining the pathogenesis of *A. baumannii* have employed a number of eukaryotic host model organisms, including mouse, rat, guinea pig, rabbit, zebrafish, and insect larvae infection models [24–29]. Because of the low virulence of most *A. baumannii* strains in mice, some murine models of infection with this pathogen utilize mechanisms to suppress the host immune response by rendering mice neutropenic with antibody or cyclophosphamide treatments [27, 30–33]. While these model systems often result in more lethal infections, they confound the ability to study interactions between *A. baumannii* and the innate immune system. By contrast, the use of immunocompetent murine infection models has driven the discovery of several components of the host immune response to *A. baumannii* that are important for protection against this pathogen. These include detection by pattern recognition receptors [34–37], production of reactive oxygen species [38], and the release of antimicrobial proteins by immune cells [12].

Here, we describe a murine model of acute *A. baumannii* pneumonia in an immunocompetent host that can be utilized to characterize both bacterial factors important for *A. baumannii* pneumonia and the host immune response to this pathogen.

2 Materials

2.1 Lysogeny Broth (LB)

LB is used to culture *Acinetobacter* spp. Agar is added when solid medium is required. Mix the following in ddH_2O and autoclave:

1. Tryptone, 10 g/L.
2. Yeast extract, 5 g/L.
3. Sodium chloride, 10 g/L.
4. (*Optional*) agar, 15 g/L.

2.2 Phosphate-Buffered Saline (PBS)

Endotoxin-free PBS for Avertin stock can be purchased. To generate sterile PBS for bacterial suspensions and serial dilutions, mix the following in 800 mL ddH_2O:

1. 8 g sodium chloride.
2. 0.2 g potassium chloride.
3. 1.44 g sodium phosphate dibasic.
4. 0.24 g potassium phosphate monobasic.

Adjust the pH to 7.4 with HCl. Add ddH_2O to 1 L total volume and autoclave to sterilize.

2.3 Neutral Buffered Formalin

Neutral buffered formalin solution (10%) is used to fix lung tissue prior to paraffin embedding and sectioning. It can be purchased or generated by adding the following to 900 mL ddH_2O:

1. 4 g sodium dihydrogen orthophosphate (monohydrate).
2. 6.5 g disodium hydrogen orthophosphate (anhydrous).
3. 100 mL formalin (40% aqueous solution of formaldehyde).

2.4 Diluted Optimal Cutting Temperature (OCT) Compound (25%)

Diluted optimal cutting temperature compound is tracheally instilled in lungs to be frozen and sectioned. It can be purchased or generated by thoroughly mixing the following:

1. 75 mL ddH_2O.
2. 25 mL optimal cutting temperature (OCT) compound.

2.5 Mice

Specific pathogen-free age- and sex-matched 6–12-week-old C57BL/6, BALB/c mice (The Jackson Laboratory) or Swiss Webster mice (Charles River Laboratories). Mice should be allowed food and water ad libitum. All protocols and experimentation require approval from an institutional animal care and use committee.

2.6 Anesthesia and Euthanasia

2.6.1 Avertin Anesthesia

1. Avertin (2,2,2-tribromoethanol).
2. *tert*-Amyl alcohol (2-methyl-2-butanol).
3. Sterile, endotoxin-free PBS.

Avertin should always be protected from light [39]. Generate a 100% stock by dissolving 1 g Avertin (2,2,2-tribromoethanol) in 1 mL *tert*-Amyl alcohol (2-methyl-2-butanol) by vigorous vortexing. The 100% stock can be stored at 4 °C for approximately

1 month. Generate a 1.25% concentrated stock by diluting into sterile, endotoxin-free PBS pre-warmed to 37–42 °C. Ensure full dissolution of the stock by vigorous vortexing and/or rocking at room temperature. The working solution should be kept in the dark at 4 °C if it is not to be used immediately.

2.6.2 Euthanasia

CO_2 inhalation or ketamine/xylazine anesthesia with exsanguination is used for euthanasia. Ketamine/xylazine stock should be prepared in sterile PBS (or 0.9% saline) at 20 mg/mL ketamine and 2 mg/mL xylazine. For non-survival surgery, ketamine/xylazine should be administered at 150 mg/kg ketamine and 15 mg/kg xylazine, with additional as needed to maintain a surgical plane of anesthesia. Alternatively, overdose with ketamine/xylazine or pentobarbital can be used as approved by the institutional animal care and use committee. All drugs are administered intraperitoneally using a 0.5–1 mL insulin syringe with an attached 28 G needle.

2.7 Bacterial Suspension

1. LB.
2. Sterile PBS.
3. Sterile 96-well plate.
4. Sterile petri plates with agar media containing appropriate antibiotics.
5. Disposable 1-cm cuvettes or alternative method to measure optical density at 600 nm.

2.8 Intranasal Application

The following items should be prepared prior to the anesthesia procedure:

1. Isothermal pads pre-warmed to 37 °C.
2. Cotton swabs.
3. Puralube ointment.
4. P50–P200 pipettor and tips.

2.9 Measurement of Bacterial Load

1. 70% ethanol spray bottle.
2. Styrofoam block (e.g., from conical tube holders).
3. 21 G needles.
4. Forceps and scissors.
5. 50-mL conical tubes filled with ethanol and sterile water to sterilize surgical tools.
6. Sterile Bullet Blender lysis tubes and Bullet Blender Tissue Homogenizer (Next Advance Inc., Troy, NY); alternatively, sterile Whirl-Pak bags (e.g., Nasco, Fort Atkinson, WI; Fisher B00736WA) and plastic or metal rolling pin.
7. Sterile PBS (for serial dilutions of bacteria).

8. Sterile 96-well plates.
9. P10–P300 multichannel pipettor.
10. Petri dishes with agar media containing appropriate antibiotics.

2.10 Digestion of Lungs for Flow Cytometry

1. 1.5-mL tubes.
2. Dulbecco's Modified Eagle Medium (DMEM).
3. Fetal bovine serum (FBS).
4. Type I collagenase.
5. DNase.
6. 2× digestion medium: DMEM + 10% FBS containing 2 mg/mL collagenase type I and 0.2 mg/mL DNase, filter sterilized.
7. 70-μm filters.
8. 50-mL conical tubes.
9. Red blood cell (RBC) lysis buffer.

2.11 Tracheal Instillation for Bronchoalveolar Lavage Fluid Collection and Preparation of Formalin-Fixed Paraffin Embedded or Frozen Lung Tissue for Histopathology

1. 70% ethanol.
2. Styrofoam block (e.g., from conical tube holders).
3. 21 G needles.
4. Fine-tipped angled forceps and fine-tipped sharp scissors.
5. 50-mL conical tubes filled with ethanol and sterile water to sterilize surgical tools
6. 21 G blunt needle or tracheostomy tool.
7. Suture threads cut to 2″ lengths and one cut to 5″.
8. *Bronchoalveolar lavage fluid (BALF)*: sterile PBS, sterile 1 mL syringes.
9. *Formalin-fixed and paraffin-embedded lung tissue*: neutral buffered formalin solution (10%), 50-mL conical tubes containing 10 mL neutral buffered formalin.
10. *Frozen lung tissue*: 25% optimal cutting temperature compound (OCT) in ddH_2O, dry ice, small weigh boats (~4.5 cm).

3 Methods

3.1 Preparation of Acinetobacter spp. for Acute Lung Infection of Mice

1. Streak *Acinetobacter* sp. strain from −80 °C stock to LB agar with appropriate antibiotics for single colonies, and incubate overnight at 37 °C.
2. Inoculate 3 mL LB using 1–2 colonies *Acinetobacter* sp., and incubate with shaking 37 °C at 180 rpm for 12–16 h. If antibiotics must be used to maintain selection, appropriate control strains should be generated to allow identical growth conditions.

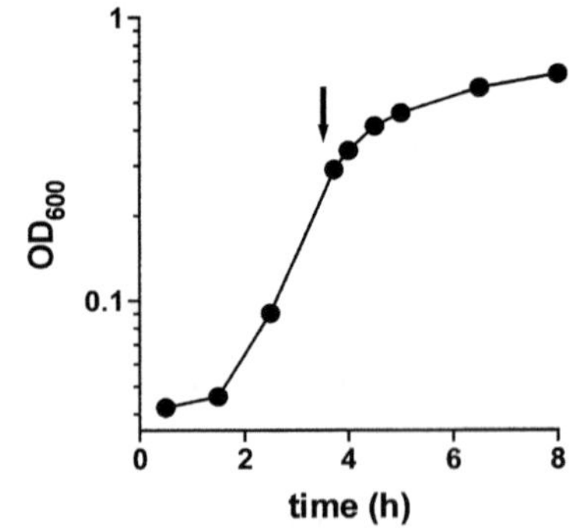

Fig. 1 *A. baumannii* growth curve. *A. baumannii* 17978 was grown in 10 mL LB, and bacterial growth was monitored by optical density at 600 nm (OD_{600}). The arrow indicates 3.5 h or mid-exponential phase when bacteria are typically used for intranasal inoculation

3. Inoculate 10 mL LB with 10 μL overnight culture in an appropriate tube (e.g., 50-mL conical tube with cap sealed loosely), and incubate at 37 °C with shaking at 180 rpm until mid-logarithmic growth phase (*see* example in Fig. 1; *see* **Note 1**). To achieve high-density suspensions (i.e., 1×10^9 colony-forming units), inoculate 50 mL LB with 50 μL overnight culture in a 250-mL Erlenmeyer flask, and incubate at 37 °C with shaking at 180 rpm.
4. Centrifuge at 5,900 × *g* in a fixed-rotor centrifuge for 6 min at 4 °C, discard the supernatant, and resuspend in an equal volume of ice-cold sterile PBS. Repeat centrifugation and resuspension.
5. Centrifuge at 5,900 × *g* in a fixed-rotor centrifuge for 6 min at 4 °C, discard the supernatant, and resuspend in 1 mL ice-cold PBS. Keep bacterial suspensions on ice.
6. Measure the culture optical density at 600 nm (OD_{600}). Quantify bacteria against a standard curve for the number of colony-forming units (CFU)/OD_{600} (*see* **Note 2**).
7. Dilute the bacteria to the desired density in ice-cold sterile PBS, and keep bacterial suspensions on ice. For a competitive infection, mix wild-type and mutant bacteria at 1:1 ratio following generation of individual strain suspensions at the appropriate density. Bacterial suspensions prepared in this way can be similarly used for systemic infections (*see* **Note 3**).
8. Following infection, transfer 10 μL of each inoculum to the top row of a 96-well plate containing 90 μL ice-cold sterile PBS in each well (i.e., to generate dilutions 10^{-1}–10^{-8}). Using a multichannel pipettor, plate 10 μL serial dilutions to LB agar plates. Use dried LB agar plates to maintain separation of the

plated dilution series. If the experiment is to calculate the competitive index of two strains, plate to media containing the appropriate antibiotics to differentiate between the strains (i.e., wild type and mutant grow on LB, while only the mutant grows on LB containing appropriate antibiotics) to confirm the plating scheme.

9. Incubate plates overnight at 37 °C and count total bacterial input to the lungs.

3.2 Mouse Infection via Intranasal Application

1. Neutropenia can be induced 24 h prior to anesthesia if desired (*see* **Note 4**). For Avertin anesthesia, all chemicals and stock solutions should be kept protected from light. Prepare an Avertin stock as described above, and administer 125–250 mg/kg to mice by intraperitoneal injection. Mice should be unresponsive to toe pinching, but breathing regularly.
2. Apply lubricating ointment to the eyes of the mouse using a sterile cotton swab.
3. Holding the mouse gently by the scruff, position the animal upright using a forefinger to maintain a straight trachea (Fig. 2). Confirm the mouse is breathing normally.
4. Cover a 37 °C heated isothermal pad with a paper towel and position it in the cage at an angle.
5. Using a P50–P200 pipettor, apply 30–50 μL bacterial suspension to the nostril as the animal inhales, breathing in the bacterial suspension (*see* **Note 5**).
6. Hold the mouse upright for 1 min before placing the mouse prone on the heated isothermal pad with the head elevated. Successful intranasal administration will lead to a change in breathing rate that should normalize before placing the mouse on the heated pad. If the animals are not sufficiently anesthetized, they may snort or cough up the inoculum, which reduces the likelihood of successful infection.
7. Monitor the animal until fully awake and remove the heated isothermal pad.

3.3 Measurement of Lung Bacterial Loads and Immune Response

1. Acute lung infection can cause mice to temporarily show signs of illness including loss of coat quality. Mice should be observed for clinical signs of illness including responsiveness. Any moribund animals should be euthanized humanely in accordance with an approved animal care protocol. Acute infection of wild-type mice can be observed for up to 48 h post-infection (hpi), at which point bacterial burdens diminish (Fig. 3) [40].
2. Euthanize mice by CO_2 inhalation and record body weight.

Fig. 2 Intranasal application of *Acinetobacter* suspension. Anesthetized mice are held upright for intranasal inoculation with bacterial suspension

3. Sterilize the Styrofoam board surgical surface by spraying down with 70% ethanol. Pin the mouse to the board using 21 G needles. Disinfect the mouse coat with 70% ethanol. Sterilize the surgical tools by keeping in 50-mL conical filled with ethanol, then dipping in sterile water prior to use. Expose the peritoneum by a vertical excision with blunt scissors and removal of the ventral skin. Expose the thoracic cavity by cutting through the peritoneum and diaphragm. Excise the lungs from the animal, and place in a Bullet Blender lysis tube filled with 0.7 mL ice-cold PBS or in a Whirl-Pak bag. If desired, measure the lung mass by weighing the lysis tube with and without the lung. If the samples will be used for protein analysis, add protease inhibitors to the PBS, and sterilize by filtration prior to filling the tubes.
4. Sterilizing tools in between each step, harvest additional organs such as the liver and spleen to assess bacterial dissemination. Spleens are placed in Bullet Blender lysis tubes with 0.7 mL PBS or Whirl-Pak bags. Livers can be divided into two 1.5-mL

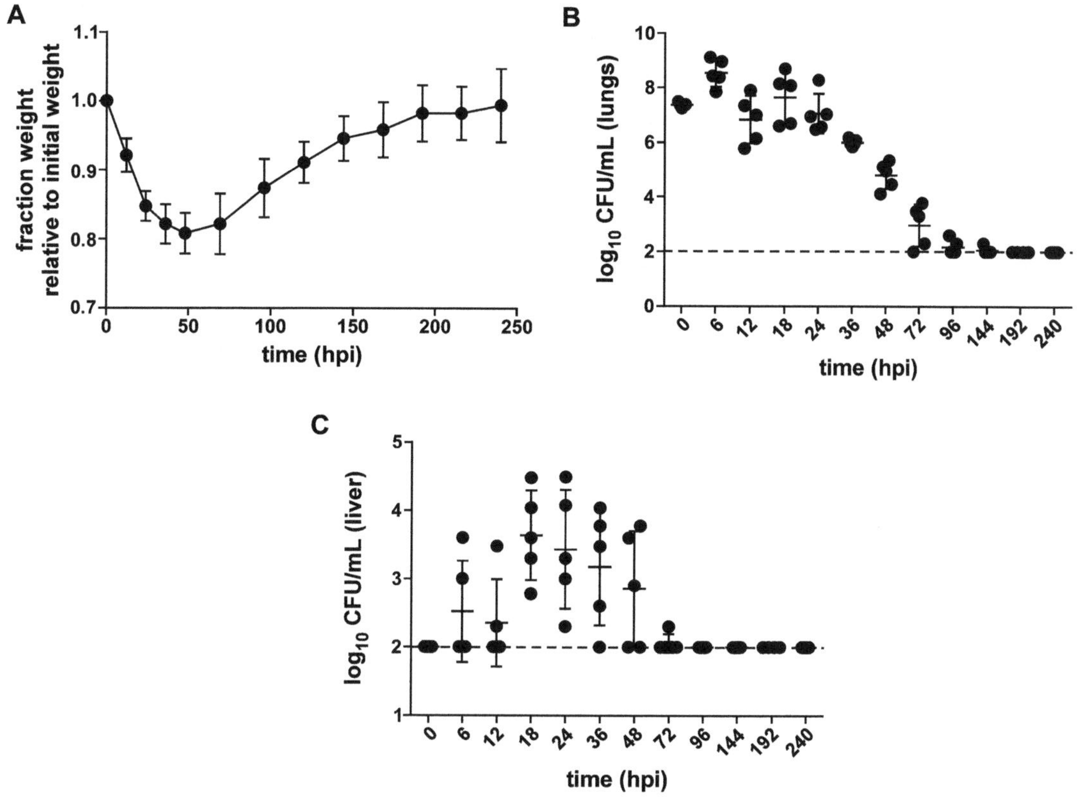

Fig. 3 Timecourse of *A. baumannii* lung infection and dissemination. Seven-week-old Swiss Webster mice were intranasally infected with 10^8 CFU *A. baumannii* and monitored over time in hours post-infection (hpi). (**a**) Weight change. Values from 6 to 216 hpi were statistically decreased ($p < 0.05$) from uninfected animals by two-way ANOVA. (**b**) Bacterial burdens in the lung. Values from 0 to 72 hpi were statistically different ($p < 0.05$) from uninfected animals by Student's *t* test. (**c**) Bacterial dissemination to the liver. Values from 18 to 36 hpi were statistically different ($p < 0.05$) from uninfected animals by Student's *t* test. Reproduced from [40] with permission from Wiley

Bullet Blender lysis tubes with 0.5 mL PBS each, placed in a 5-mL Bullet Blender lysis tube with 1 mL PBS or placed in Whirl-Pak bags.

5. *Homogenize tissues using the Bullet Blender*: Homogenize tissues in Bullet Blender lysis tubes on power 10 for 5 min at 4 °C. Homogenize lungs for two cycles. If using the 5-mL Bullet Blender lysis tubes, homogenize on power 8 for 5 min at 4 °C for two cycles.

6. *Homogenize tissues using Whirl-Pak bags (alternate method)*: Add 1 mL ice-cold PBS to each bag and seal, removing air bubbles. Tape 2–5 bags with tissues to the benchtop at the sealing end. Using a metal or plastic rolling pin, roll over tissues 70 times to homogenize or 100 times for lungs, rolling away

from the body. Visually inspect for uniform homogenization. Collect the homogenized tissues with a 2-mL pipet, and transfer to a 2-mL tube on ice.

7. Transfer 100 μL of each homogenate to the top row of a 96-well plate. Serially dilute the 10 μL tissue homogenates into 90 μL ice-cold sterile PBS (i.e., to generate dilutions 10^0–10^{-7}) using a multichannel pipettor. Using a multichannel pipettor, plate 10 μL of the serial dilutions to dried LB agar plates. For spleens and any organs with low bacterial burdens, also plate 100 μL organ homogenate to lower the limit of detection. If the experiment is to calculate the competitive index of two strains, plate to media containing the appropriate antibiotics to differentiate between the strains (i.e., wild type and mutant grow on LB, while only the mutant grows on LB containing appropriate antibiotics).
8. Incubate plates overnight at 37 °C and count total bacterial output in the lungs and other organs. For any organs with no detectable bacterial burdens, set at the limit of detection for statistical purposes.
9. *Competitive infection*: The number of the mutant strain CFU is determined by serial plating to selective medium. To determine the number of wild-type bacteria in each organ, subtract the mutant CFU counted from selective medium from the total CFU counted from LB plates. The competitive index is the ratio of mutant/wild-type CFU after normalizing for input, i.e., (mutant output CFU/mutant input CFU)/(wild-type output CFU/wild-type input CFU).
10. *Cytokine analysis*: Centrifuge the remaining homogenate at 21,000 × *g* for 5 min, and transfer the supernatant to a fresh tube to use for enzyme-linked immunosorbent assay (ELISA) or multi-analyte profiling (Luminex Corp., Austin, TX). Homogenates or supernatants can be stored at −80 °C prior to analysis. For improved precision, homogenate supernatants can be normalized for total protein content.
11. *Digestion of lungs for flow cytometry*: Place excised lungs in 1.5-mL tubes containing 1 mL ice-cold DMEM + 10% FBS, and keep on ice. Using fine-tipped scissors, chop the lungs into small pieces. Transfer lung pieces to a 15-mL conical tube, collecting any remaining lung pieces from the 1.5-mL tube with an additional 1.5 mL medium. Add 2.5 mL 2× digestion medium, and incubate at 37 °C with shaking for 15–60 min. Set a 70-μm filter on a 50-mL conical and decant the digested lung onto the filter. Push the tissue through the filter with the back of a 3-mL syringe plunger to generate single-cell

suspensions. Centrifuge at 300 × *g* for 5 min at 4 °C in a swinging-bucket rotor and discard supernatant. Resuspend the pellet in 5 mL 1× RBC lysis buffer and incubate for 5 min on ice. Stop the lysis with 25 mL DMEM or PBS. Decant the single-cell suspension through a new 70-μm filter into a new 50-mL conical tube. Centrifuge again, discarding supernatant. Resuspend cells in 1–10 mL DMEM + 10% FBS, and count cells on a hemocytometer, avoiding remaining RBCs. The single-cell suspensions can be stained and analyzed by flow cytometry using standard procedures.

3.4 Tracheal Instillation for BALF Collection or Lung Tissue Pathology

1. Euthanize mice and record body weight. Do not perform cervical dislocation prior to tracheostomy. Due to potential pulmonary hemorrhage from CO_2 inhalation [41], ketamine/xylazine anesthesia and exsanguination or overdose is preferred for these procedures. Anesthetize the mice with ketamine/xylazine for non-survival surgery. Monitor that the animal does not respond to stimuli by toe pinch.
2. Sterilize the Styrofoam board surgical surface by spraying down with 70% ethanol. Pin the mouse to the board using 21 G needles. Disinfect the mouse coat with 70% ethanol. If euthanizing the anesthetized mouse by exsanguination, open the thoracic cavity by midline and exsanguinate through cardiac puncture.
3. *Tracheostomy*: Stabilize the head and trachea of the euthanized mouse by creating a 2″ loop of suture thread, looping it around the mouse upper incisors and pinning back the loop with a 21 G needle (Fig. 4). Sterilize the surgical tools by keeping in 50-mL conical tubes filled with ethanol and then dipping in sterile water prior to use. Remove the skin from the neck with scissors, and then carefully expose the trachea by gently parting the submaxillary glands and muscle tissue with two pairs of tweezers, taking care to not rupture any blood vessels. Using fine-tipped angled tweezers, insert a 2″ suture thread under the trachea. Using fine-tipped scissors, make a small crosswise incision in the trachea, and insert a 21 G blunt needle. Tie the blunt needle in place with the suture thread.
4. *Bronchoalveolar lavage fluid*: Instill up to 0.8 mL sterile PBS into the lungs using a 1 mL syringe attached to the 21 G blunt needle or tracheostomy tool, and then withdraw to obtain BALF. BALF collection can be repeated up to three times. To quantify bacteria in the BALF, serially dilute 10 μL and plate to LB agar as described above. To measure cytokines in the BALF, freeze at −80 °C until analysis.

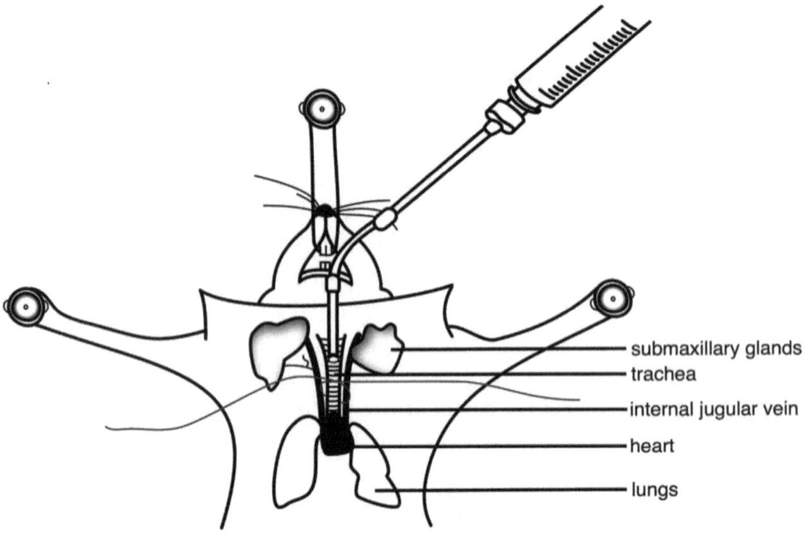

Fig. 4 Tracheal instillation. Following euthanasia, the mouse is pinned to a foam board, and the neck is stabilized by pinning a thread looped around the teeth. The neck is cut open and the tissue gently pulled aside to expose the trachea. A suture thread is inserted below the trachea. The trachea is snipped, and a blunt needle is inserted, and the suture thread is tied to secure the blunt needle. For BALF collection, PBS is instilled and withdrawn. For lung fixation, the lung is removed after instillation of the appropriate medium

5. *Formalin-fixed and paraffin-embedded lung tissue*: Inflate lung tissue prior to fixation to improve morphological analysis. Instill up to 0.8 mL 10% neutral buffered formalin in the lungs. Leaving the syringe in place, carefully open the thoracic cavity if not already open, and cut away tissue to allow removal of the trachea, lungs, and heart. Remove the syringe and needle, grip the trachea to prevent formalin loss, and remove the trachea, lungs, and heart. Place the lungs in a 50-mL conical tube containing 10 mL 10% formalin and fix for 1–4 days. Remove the heart if desired, and transfer fixed tissue to an embedding cassette dorsal side down. Embed in paraffin, cut slides 5 μm sections, stain slides with hematoxylin and eosin (H&E), and examine with a brightfield microscope according to standard procedures.
6. *Frozen lung tissue*: Instill up to 0.8 mL 25% OCT into the lungs. Cut away the heart, remove the lungs and trachea as described above, and gently wash the lungs in sterile PBS and blot dry. Place the lungs dorsal side down in a small weigh boat and freeze on dry ice. Store frozen tissue at −80 °C. Cut sections of desired thickness using a cryosectioner: 5 μm for histopathology or immunofluorescence and 10 μm for matrix-assisted laser desorption/ionization imaging mass spectrometry to image proteins [12, 40].

4 Notes

1. Each *Acinetobacter* species or strain must be tested for growth kinetics under the conditions to be used for infections. To compare strains with different growth rates in one infection experiment, subculture slower-growing strains earlier to ensure all strains are in mid-log phase when harvested by centrifugation.
2. The appropriate OD_{600} will need to be determined for each *Acinetobacter* strain or species, as differences in capsule or bacterial shape affect light scattering and therefore the OD_{600} to CFU relationship. For example, an *A. baumannii* 17978 suspension diluted 1/50 in PBS with an OD_{600} of 0.36 in a 1 cm plastic cuvette is 1×10^{10} CFU/mL. Once an appropriate OD_{600} is determined for a particular bacterial density, use the same OD_{600}, and alter the dilution factor to obtain bacterial cultures of different densities.
3. For systemic *A. baumannii* infection, inject 100 μL prepared bacterial suspension intravenously. Bacterial burdens are typically assessed at 24 hpi, as mortality or bacterial clearing is often observed after 24 h.
4. To deplete neutrophils, intraperitoneally inject mice with 3–15 mg/kg anti-mouse Ly6G clone 1A8 at 24 h prior to infection. Intraperitoneally inject the same dose of isotype control antibody to a second, control group of mice as described above. Repeat procedure 1 h prior to infection, as well as each day post-infection at 24-h intervals for the duration of the experiment. Confirm neutropenia by flow cytometric analysis using anti-mouse Gr-1 (clone RB6-8C5) and anti-mouse CD11b (clone M1/70) by standard methods.
5. When working with a new *Acinetobacter* strain or mouse strain, experimentally determine the inoculum by infecting with a range of doses, e.g., 3×10^6–1×10^9 CFU as shown in Fig. 5.

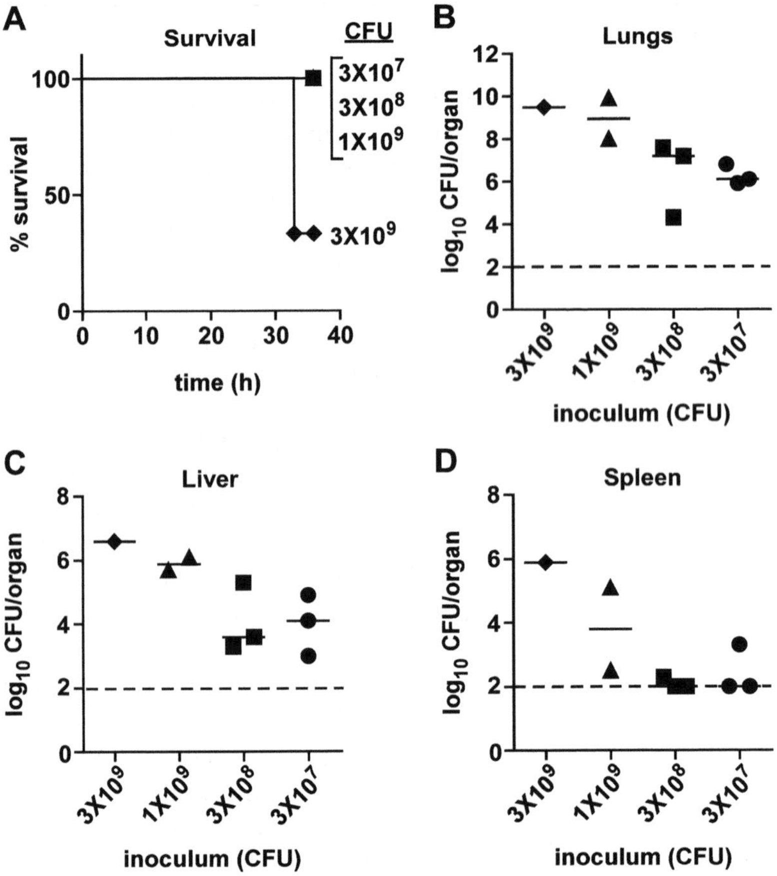

Fig. 5 Determination of appropriate inoculum. *A. nosocomialis* M2 was intranasally inoculated into 9-week-old C57BL/6J mice at the doses indicated. (**a**) Survival was monitored over 36 h. At 36 hpi, bacterial burdens were assessed in lungs (**b**), livers (**c**), and spleens (**d**). Due to mortality with 3×10^9 CFU inoculum and low dissemination to the spleen with 3×10^8 CFU inoculum, an inoculum of 1×10^9 CFU was chosen [8, 9]

Acknowledgments

We thank Michael J. Noto, Zachery R. Lonergan, and Lillian J. Juttukonda for critical reading of the manuscript. This work was supported by the National Institutes of Health (NIH) research grant AI101171 to E.P.S. and fellowships F32AI122516 to L.D.P. and T32HL094296 to L.D.P. and E.R.G.

References

1. Peleg AY, Seifert H, Paterson DL (2008) *Acinetobacter baumannii*: emergence of a successful pathogen. Clin Microbiol Rev 21 (3):538–582. https://doi.org/10.1128/CMR.00058-07
2. Talbot GH, Bradley J, Edwards JE Jr, Gilbert D, Scheld M, Bartlett JG (2006) Bad bugs need drugs: an update on the development pipeline from the Antimicrobial Availability Task Force of the Infectious Diseases

Society of America. Clin Infect Dis 42 (5):657–668. https://doi.org/10.1086/499819

3. Tacconelli E, Carrara E, Savoldi A, Harbarth S, Mendelson M, Monnet DL, Pulcini C, Kahlmeter G, Kluytmans J, Carmeli Y, Ouellette M, Outterson K, Patel J, Cavaleri M, Cox EM, Houchens CR, Grayson ML, Hansen P, Singh N, Theuretzbacher U, Magrini N, Group WHOPPLW (2018) Discovery, research, and development of new antibiotics: the WHO priority list of antibiotic-resistant bacteria and tuberculosis. Lancet Infect Dis 18:318–327. https://doi.org/10.1016/S1473-3099(17)30753-3
4. Sievert DM, Ricks P, Edwards JR, Schneider A, Patel J, Srinivasan A, Kallen A, Limbago B, Fridkin S, Facilities NHSNTNaPN (2013) Antimicrobial-resistant pathogens associated with healthcare-associated infections: summary of data reported to the National Healthcare Safety Network at the Centers for Disease Control and Prevention, 2009–2010. Infect Control Hosp Epidemiol 34(1):1–14. https://doi.org/10.1086/668770
5. Weiner LM, Webb AK, Limbago B, Dudeck MA, Patel J, Kallen AJ, Edwards JR, Sievert DM (2016) Antimicrobial-resistant pathogens associated with healthcare-associated infections: summary of data reported to the National Healthcare Safety Network at the Centers for Disease Control and Prevention, 2011–2014. Infect Control Hosp Epidemiol 37(11):1288–1301. https://doi.org/10.1017/ice.2016.174
6. Vincent JL, Rello J, Marshall J, Silva E, Anzueto A, Martin CD, Moreno R, Lipman J, Gomersall C, Sakr Y, Reinhart K, Investigators EIGo (2009) International study of the prevalence and outcomes of infection in intensive care units. JAMA 302(21):2323–2329. https://doi.org/10.1001/jama.2009.1754
7. Dexter C, Murray GL, Paulsen IT, Peleg AY (2015) Community-acquired *Acinetobacter baumannii*: clinical characteristics, epidemiology and pathogenesis. Expert Rev Anti-Infect Ther 13(5):567–573. https://doi.org/10.1586/14787210.2015.1025055
8. Harding CM, Kinsella RL, Palmer LD, Skaar EP, Feldman MF (2016) Medically relevant *Acinetobacter* species require a type II secretion system and specific membrane-associated chaperones for the export of multiple substrates and full virulence. PLoS Pathog 12(1): e1005391. https://doi.org/10.1371/journal.ppat.1005391
9. Kinsella RL, Lopez J, Palmer LD, Salinas ND, Skaar EP, Tolia NH, Feldman MF (2017) Defining the interaction of the protease CpaA with its type II secretion chaperone CpaB and its contribution to virulence in *Acinetobacter* species. J Biol Chem 292(48):19628–19638. https://doi.org/10.1074/jbc.M117.808394
10. Johnson TL, Waack U, Smith S, Mobley H, Sandkvist M (2015) *Acinetobacter baumannii* is dependent on the type II secretion system and its substrate LipA for lipid utilization and in vivo fitness. J Bacteriol 198(4):711–719. https://doi.org/10.1128/JB.00622-15
11. Harding CM, Hennon SW, Feldman MF (2018) Uncovering the mechanisms of *Acinetobacter baumannii* virulence. Nat Rev Microbiol 16(2):91–102. https://doi.org/10.1038/nrmicro.2017.148
12. Hood MI, Mortensen BL, Moore JL, Zhang Y, Kehl-Fie TE, Sugitani N, Chazin WJ, Caprioli RM, Skaar EP (2012) Identification of an *Acinetobacter baumannii* zinc acquisition system that facilitates resistance to calprotectin-mediated zinc sequestration. PLoS Pathog 8 (12):e1003068. https://doi.org/10.1371/journal.ppat.1003068
13. Mortensen BL, Rathi S, Chazin WJ, Skaar EP (2014) *Acinetobacter baumannii* response to host-mediated zinc limitation requires the transcriptional regulator Zur. J Bacteriol 196 (14):2616–2626. https://doi.org/10.1128/JB.01650-14
14. Mortensen BL, Skaar EP (2013) The contribution of nutrient metal acquisition and metabolism to *Acinetobacter baumannii* survival within the host. Front Cell Infect Microbiol 3:95. https://doi.org/10.3389/fcimb.2013.00095
15. Nairn BL, Lonergan ZR, Wang J, Braymer JJ, Zhang Y, Calcutt MW, Lisher JP, Gilston BA, Chazin WJ, de Crecy-Lagard V, Giedroc DP, Skaar EP (2016) The response of *Acinetobacter baumannii* to zinc starvation. Cell Host Microbe 19(6):826–836. https://doi.org/10.1016/j.chom.2016.05.007
16. Juttukonda LJ, Chazin WJ, Skaar EP (2016) *Acinetobacter baumannii* coordinates urea metabolism with metal import to resist host-mediated metal limitation. mBio 7(5):e01475-16. https://doi.org/10.1128/mBio.01475-16
17. Gaddy JA, Arivett BA, McConnell MJ, Lopez-Rojas R, Pachon J, Actis LA (2012) Role of acinetobactin-mediated iron acquisition functions in the interaction of *Acinetobacter baumannii* strain ATCC 19606T with human lung epithelial cells, *Galleria mellonella* caterpillars, and mice. Infect Immun 80(3):1015–1024. https://doi.org/10.1128/IAI.06279-11
18. Hood MI, Becker KW, Roux CM, Dunman PM, Skaar EP (2013) Genetic determinants of

intrinsic colistin tolerance in *Acinetobacter baumannii*. Infect Immun 81(2):542–551. https://doi.org/10.1128/IAI.00704-12

19. Jacobs AC, Hood I, Boyd KL, Olson PD, Morrison JM, Carson S, Sayood K, Iwen PC, Skaar EP, Dunman PM (2010) Inactivation of phospholipase D diminishes *Acinetobacter baumannii* pathogenesis. Infect Immun 78 (5):1952–1962. https://doi.org/10.1128/IAI.00889-09
20. Geisinger E, Isberg RR (2015) Antibiotic modulation of capsular exopolysaccharide and virulence in *Acinetobacter baumannii*. PLoS Pathog 11(2):e1004691. https://doi.org/10.1371/journal.ppat.1004691
21. Gebhardt MJ, Gallagher LA, Jacobson RK, Usacheva EA, Peterson LR, Zurawski DV, Shuman HA (2015) Joint transcriptional control of virulence and resistance to antibiotic and environmental stress in *Acinetobacter baumannii*. mBio 6(6):e01660-15. https://doi.org/10.1128/mBio.01660-15
22. Wang N, Ozer EA, Mandel MJ, Hauser AR (2014) Genome-wide identification of *Acinetobacter baumannii* genes necessary for persistence in the lung. mBio 5(3):e01163-14. https://doi.org/10.1128/mBio.01163-14
23. Subashchandrabose S, Smith S, DeOrnellas V, Crepin S, Kole M, Zahdeh C, Mobley HL (2016) *Acinetobacter baumannii* genes required for bacterial survival during bloodstream infection. mSphere 1(1):e00013-15. https://doi.org/10.1128/mSphere.00013-15
24. Peleg AY, Jara S, Monga D, Eliopoulos GM, Moellering RC Jr, Mylonakis E (2009) *Galleria mellonella* as a model system to study *Acinetobacter baumannii* pathogenesis and therapeutics. Antimicrob Agents Chemother 53(6):2605–2609. https://doi.org/10.1128/AAC.01533-08
25. Bhuiyan MS, Ellett F, Murray GL, Kostoulias X, Cerqueira GM, Schulze KE, Mahamad Maifiah MH, Li J, Creek DJ, Lieschke GJ, Peleg AY (2016) *Acinetobacter baumannii* phenylacetic acid metabolism influences infection outcome through a direct effect on neutrophil chemotaxis. Proc Natl Acad Sci U S A 113(34):9599–9604. https://doi.org/10.1073/pnas.1523116113
26. Pachon-Ibanez ME, Docobo-Perez F, Lopez-Rojas R, Dominguez-Herrera J, Jimenez-Mejias ME, Garcia-Curiel A, Pichardo C, Jimenez L, Pachon J (2010) Efficacy of rifampin and its combinations with imipenem, sulbactam, and colistin in experimental models of infection caused by imipenem-resistant *Acinetobacter baumannii*. Antimicrob Agents Chemother 54(3):1165–1172. https://doi.org/10.1128/AAC.00367-09
27. Joly-Guillou ML, Wolff M, Pocidalo JJ, Walker F, Carbon C (1997) Use of a new mouse model of *Acinetobacter baumannii* pneumonia to evaluate the postantibiotic effect of imipenem. Antimicrob Agents Chemother 41(2):345–351
28. Russo TA, Beanan JM, Olson R, MacDonald U, Luke NR, Gill SR, Campagnari AA (2008) Rat pneumonia and soft-tissue infection models for the study of *Acinetobacter baumannii* biology. Infect Immun 76 (8):3577–3586. https://doi.org/10.1128/IAI.00269-08
29. Bernabeu-Wittel M, Pichardo C, Garcia-Curiel A, Pachon-Ibanez ME, Ibanez-Martinez J, Jimenez-Mejias ME, Pachon J (2005) Pharmacokinetic/pharmacodynamic assessment of the in-vivo efficacy of imipenem alone or in combination with amikacin for the treatment of experimental multiresistant *Acinetobacter baumannii* pneumonia. Clin Microbiol Infect 11(4):319–325. https://doi.org/10.1111/j.1469-0691.2005.01095.x
30. Braunstein A, Papo N, Shai Y (2004) In vitro activity and potency of an intravenously injected antimicrobial peptide and its DL amino acid analog in mice infected with bacteria. Antimicrob Agents Chemother 48 (8):3127–3129. https://doi.org/10.1128/AAC.48.8.3127-3129.2004
31. Crandon JL, Kim A, Nicolau DP (2009) Comparison of tigecycline penetration into the epithelial lining fluid of infected and uninfected murine lungs. J Antimicrob Chemother 64 (4):837–839. https://doi.org/10.1093/jac/dkp301
32. Koomanachai P, Kim A, Nicolau DP (2009) Pharmacodynamic evaluation of tigecycline against *Acinetobacter baumannii* in a murine pneumonia model. J Antimicrob Chemother 63(5):982–987. https://doi.org/10.1093/jac/dkp056
33. Song JY, Cheong HJ, Lee J, Sung AK, Kim WJ (2009) Efficacy of monotherapy and combined antibiotic therapy for carbapenem-resistant *Acinetobacter baumannii* pneumonia in an immunosuppressed mouse model. Int J Antimicrob Agents 33(1):33–39. https://doi.org/10.1016/j.ijantimicag.2008.07.008
34. Noto MJ, Boyd KL, Burns WJ, Varga MG, Peek RM Jr, Skaar EP (2015) Toll-like receptor 9 contributes to defense against *Acinetobacter baumannii* infection. Infect Immun 83 (10):4134–4141. https://doi.org/10.1128/IAI.00410-15

35. Knapp S, Wieland CW, Florquin S, Pantophlet R, Dijkshoorn L, Tshimbalanga N, Akira S, van der Poll T (2006) Differential roles of CD14 and toll-like receptors 4 and 2 in murine *Acinetobacter* pneumonia. Am J Respir Crit Care Med 173(1):122–129. https://doi.org/10.1164/rccm.200505-730OC
36. Kale SD, Dikshit N, Kumar P, Balamuralidhar V, Khameneh HJ, Bin Abdul Malik N, Koh TH, Tan GGY, Tan TT, Mortellaro A, Sukumaran B (2017) Nod2 is required for the early innate immune clearance of *Acinetobacter baumannii* from the lungs. Sci Rep 7(1):17429. https://doi.org/10.1038/s41598-017-17653-y
37. Noto MJ, Becker KW, Boyd KL, Schmidt AM, Skaar EP (2017) RAGE-mediated suppression of interleukin-10 results in enhanced mortality in a murine model of *Acinetobacter baumannii* sepsis. Infect Immun 85(3):e00954-16. https://doi.org/10.1128/IAI.00954-16
38. Qiu H, Kuolee R, Harris G, Chen W (2009) Role of NADPH phagocyte oxidase in host defense against acute respiratory *Acinetobacter baumannii* infection in mice. Infect Immun 77 (3):1015–1021. https://doi.org/10.1128/IAI.01029-08
39. Tribromoethanol (Avertin) (2006) Cold Spring Harb Protoc 2006(1):pdb.rec701. https://doi.org/10.1101/pdb.rec701
40. Moore JL, Becker KW, Nicklay JJ, Boyd KL, Skaar EP, Caprioli RM (2014) Imaging mass spectrometry for assessing temporal proteomics: analysis of calprotectin in *Acinetobacter baumannii* pulmonary infection. Proteomics 14(7–8):820–828. https://doi.org/10.1002/pmic.201300046
41. Fisher S, Burgess WL, Hines KD, Mason GL, Owiny JR (2016) Interstrain differences in CO_2-induced pulmonary hemorrhage in mice. J Am Assoc Lab Anim Sci 55(6):811–815

Chapter 27

Computational Prediction of sRNA in *Acinetobacter baumannii*

Sankalp Arya, Vineet Dubey, Deepak Sen, Atin Sharma, and Ranjana Pathania

Abstract

Small RNAs in bacteria are noncoding RNAs that act as posttranscriptional regulators of gene expression. Over time, they have gained importance as fine-tuners of expression of genes involved in critical biological processes like metabolism, fitness, virulence, and antibiotic resistance. The availability of various high-throughput strategies enable the detection of these molecules but are technically challenging and time-intensive. Thus, to fulfil the need of a simple computational algorithm pipeline to predict these sRNAs in bacterial species, we detail a user-friendly ensemble method with specific application in *Acinetobacter* spp. The developed algorithms primarily look for intergenic regions in the genome of related *Acinetobacter* spp., thermodynamic stability, and conservation of RNA secondary structures to generate a model input for the sRNAPredict3 tool which utilizes all this information to generate a list of putative sRNA. We confirmed the accuracy of the method by comparing its output with the RNA-seq data and found the method to be faster and more accurate for *Acinetobacter baumannii* ATCC 17978. Thus, this method improves the identification of sRNA in *Acinetobacter* and other bacterial species.

Key words Computational algorithms, Noncoding RNA, Small RNA, sRNAPredict3

1 Introduction

Small RNA (sRNA) is a rather broad term encompassing various classes of noncoding RNA like microRNA (miRNA), small interfering RNA (siRNA), small nucleolar RNA (snRNA), and piwi-interacting RNA (piRNA), tRNA-derived fragments (tRFs), and small rDNA-derived RNA [1–3]. In the bacterial context, sRNAs are noncoding RNAs, 50–250 nt in length, generally transcribed from intergenic regions in the bacterial genome [4]. They are involved in regulation of gene expression by inhibition/activation of translation, altering the stability of mRNA and inhibition of protein activity as a consequence of direct binding [5]. Bacterial genomic analysis has traditionally focused on the information-rich mRNA and has overlooked these sRNAs [6]. Thus, these

Indranil Biswas and Philip N. Rather (eds.), *Acinetobacter baumannii: Methods and Protocols*, Methods in Molecular Biology, vol. 1946, https://doi.org/10.1007/978-1-4939-9118-1_27, © Springer Science+Business Media, LLC, part of Springer Nature 2019

transcripts were rarely detected by genetic screening efforts, if at all. However, in recent years, the discovery of sRNA and elucidation of their functions have gained more attention probably driven by the findings that sRNA plays diversified roles in regulation of gene expression in virulence, pathogenicity, and response to environmental stress.

Two different routes have been followed for identification of sRNA in bacteria. Using the properties of sRNA as constraints, bioinformatic programs have been written to assess the genome assemblies to predict sRNA. The other, more direct approach is to sequence the complete transcriptome and align the sequences to a reference genome [7], thereby finding sRNA in unannotated intergenic regions. Although both the approaches are followed by experimental validation by Northern analysis, the bioinformatic approach provides a simple, yet accurate, way to analyze multiple bacterial genomes in a relatively short time. Computational tools look for thermodynamic stability of RNA secondary structure by determining the free energy of the RNA loops as well as co-localization of genetic features like intergenic sequence conservation, promoters and rho-independent terminators specific to transcription of sRNA, and conserved secondary structure among closely related species. Recently, our group developed a machine-based approach in an attempt to identify regulatory sRNA in a notorious Gram-negative nosocomial pathogen, *Acinetobacter baumannii*. Using this method, we identified a novel small RNA AbsR25, the first in *A. baumannii* [8]. This small RNA was found to regulate the expression of a fosfomycin efflux pump, AbaF, one of the earliest examples of involvement of sRNA in regulation of antibiotic resistance [9].

This chapter describes the sRNA identification pipeline, originally created and later modified by Livny et al. to be used for *Pseudomonas aeruginosa* [10] and further refined by our group to make it amenable to *Acinetobacter baumannii*. Of the two general approaches in discovering sRNA—complete sequencing and manual search in intergenic regions or programming the computer to look for features complicit of RNA transcribing genes—our approach takes the more direct route, saving time compared to the manual approach. Being an ensemble method, it combines the output from different algorithms, thus increasing efficiency of predictions. In either scenario, experimental validation via Northern blotting is essential for confirmation of the predictions. Primarily, sRNAPredict3 algorithm searches the intergenic regions for sequence conservation, similar to other sRNA prediction algorithms, but it further combines the data and output from other sources to look for rho-independent terminators, promoters, and secondary structure conservation. Various public domain databases and available tools have been used in the whole process as outlined in the flowchart (Fig. 1).

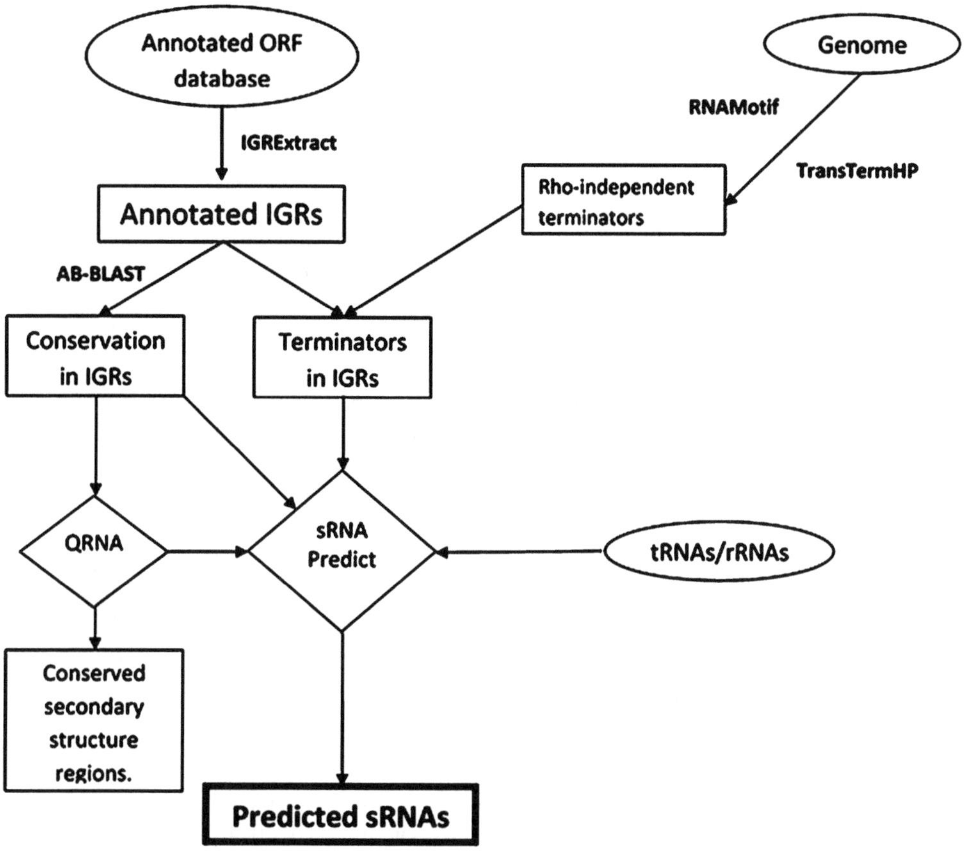

Fig. 1 Flowchart summarizing the entire process. The flowchart explains the input and output for the different tools—IGRExtract, RNAMotif, TransTermHP, and AB-BLAST—which are all essential for sRNAPredict tool to predict the final putative sRNA

2 Materials

These RNA pipeline designed in this study utilizes information from the public databases and bioinformatic tools that are freely available over the Internet. All the tasks were performed on the Linux-based, freely available Ubuntu operating system. Although the pipeline will work with other Linux systems as well, the instructions explained assume the use and a basic understanding of Ubuntu OS.

The algorithm utilizes the following bioinformatic tools.

2.1 IGRExtract

It is a C++ program used to extract the sequences of intergenic regions in FASTA format. It requires GenBank, fasta nucleic acid, and protein table files that can be downloaded from the NCBI website.

2.2 AB-BLAST

It is a general-purpose BLAST suite by Advanced Biocomputing, LLC. It is based on the WU-BLAST 2.0 which was used in the original pipeline developed by Livny et al. The AB-BLAST output format is preferred by further downstream algorithms like QRNA and sRNAPredict.

2.3 RNAMotif

It is an algorithm which allows defining and searching for RNA secondary structure. A descriptor file required for rho-independent terminators (developed by David J. Ecker) was used in this study to search for the rho-independent terminators in the intergenic regions of the *Acinetobacter baumannii* genome [11].

2.4 TransTermHP

It finds rho-independent terminators in bacterial genomes. TransTermHP utilizes a web-based service to search for the predicted terminators in the bacterial organism of interest [12].

2.5 QRNA

It is an algorithm for comparative genome sequence analysis for noncoding RNA gene search. The algorithm has three different probabilistic models for different evolutionary paths of the pattern of mutations [13]. Pairwise sequence alignment is examined, and mutations are scored dependent on if the pattern is RNA structure-constrained, coding-constrained, or position-independent.

2.6 sRNAPredict3

It is an ensemble program, written in C++ that has the flexibility to integrate different combinations of predictive features from the aforementioned programs or properly curated data obtained otherwise and use it to identify putative sRNA-encoding genes in the intergenic region of any annotated bacterial genome [14].

3 Methods

3.1 Data Gathering

3.1.1 Genome and Annotated Files

Download the genome files including the annotated GenBank (.gbk), fasta nucleic acid (.fna), and protein table (.ptt) files, in FASTA format from NCBI genome database.

3.1.2 ORF, tRNA, rRNA, and Genomes

Obtain the coordinates for the tRNA, rRNA, and open reading frame (ORF) databases from the NCBI ftp database (ftp://ftp.ncbi.nih.gov/genomes/Bacteria/). Also, download the genomes for all the bacteria from NCBI, in FASTA format for BLAST analysis. The format available at this database is directly usable with the tools utilized in this study.

3.2 Analysis

The commands and instructions provided ahead make use, as an example, of the NC_010410 genome wherever deemed necessary. In the command structure, italicized <> indicates text that has to be replaced with the instructions provided therein.

3.2.1 Phylogenetic Analysis for Closely Related Bacteria

For a meaningful comparative analysis, only the closely related bacterial genomes are required to be taken into account. Thus, the choice for selection of other genomes (up to three for each comparative analysis) should be made on the basis of 16S rRNA gene sequences and subsequent construction of a phylogenetic tree. To increase the confidence in the prediction strategy, two parallel studies for sRNA prediction by using the same methodology can be performed, for instance, one with *Acinetobacter* spp. genomes and another with closely related bacterial genomes from other species. We reasoned that overlapping results from the two studies would help narrow down the range of putative sRNAs for experimental verification.

3.2.2 Construction of Intergenic Region Database Using IGRExtract

The following are the steps and commands for compiling and running IGRExtract:

1. In the console, navigate to the folder where IGRExtract was unpacked using the cd command.
2. Compile the program using the following command to create the executable file *a.out*. Figure 2 illustrates the use and output of this command:

```
g++ -o a.out IGRExtract2.cpp.
```

3. In the same folder as the executable file *a.out*, put the downloaded GenBank (.gbk), fasta nucleic acid (.fna), and protein table (.ptt) files. For example, if we are interested in NC_010410 genome, then we put the NC_010410.gbk, NC_010410.fna, and NC_010410.ptt files downloaded from NCBI in IGRExtract folder.
4. Run the program via the executable file, and follow the instructions that appear on the screen. Figure 3 provides a view of the screen and an example of the options to enter at each prompt:

```
./a.out
```

   This will create an output <genomeName>_IGR.txt file with the intergenic regions of the genome. In the example from the screenshot above, the output file will be NC_010410_IGR.txt.

3.2.3 Construction of Databases by Using AB-BLAST

Before performing BLAST analysis, we use the *xdformat* command of AB-BLAST to create database from the intergenic regions predicted in the above step. Download and install AB-BLAST.

1. Set the PATH variable to point to AB-BLAST installation folder:

```
PATH=<Path to the ab-blast folder>
```

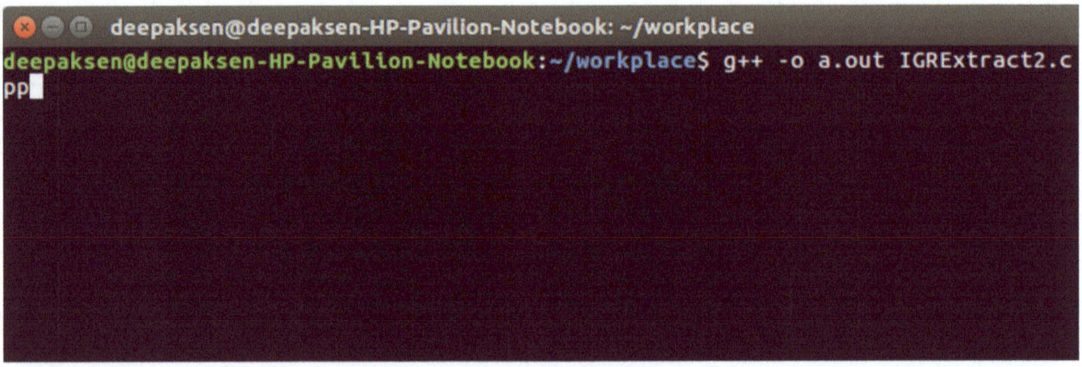

Fig. 2 IGRExtract compilation command. The command shown in the console window compiles the program and produces the *a.out* executable file in the folder

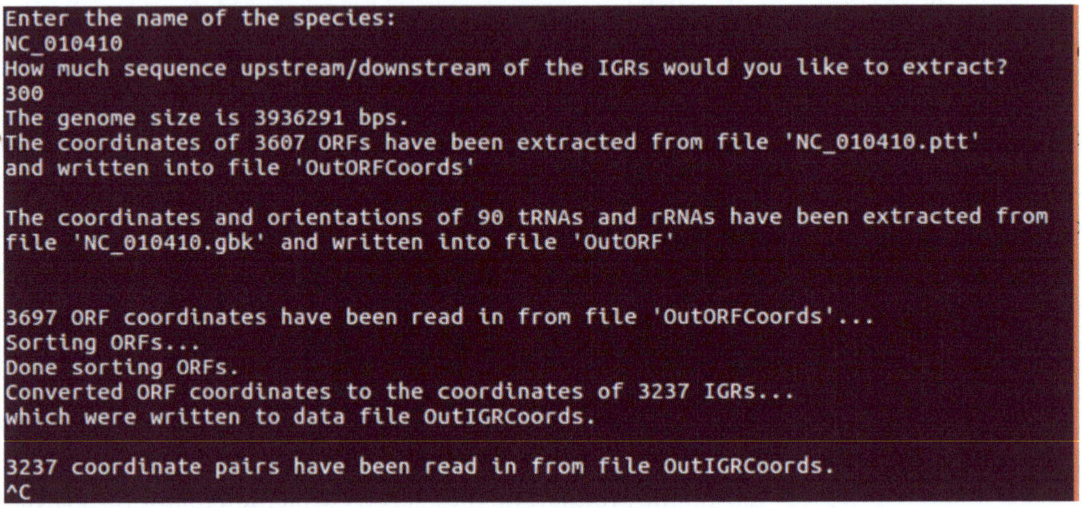

Fig. 3 Sample output of IGRExtract execution. The output of IGRExtract program, obtained after executing the *a.out* file. The program asks the user to input the name of the species and the length of sequence up-/downstream of IGRs to be extracted. As an example, we have entered NC_010410 as name and 300 as length

2. Copy/paste the output file from IGRExtract (NC_010410_IGR.txt from the above example) and also the .fna file of same genome to AB-BLAST folder.
3. In the console navigate to the AB-BLAST folder, and run the following command:

```
xdformat -n -o <database name> -I <IGRExtract output
filename>
```

This will create a database of the IGR sequences with the name provided in the command. In Fig. 4, it is data010410.

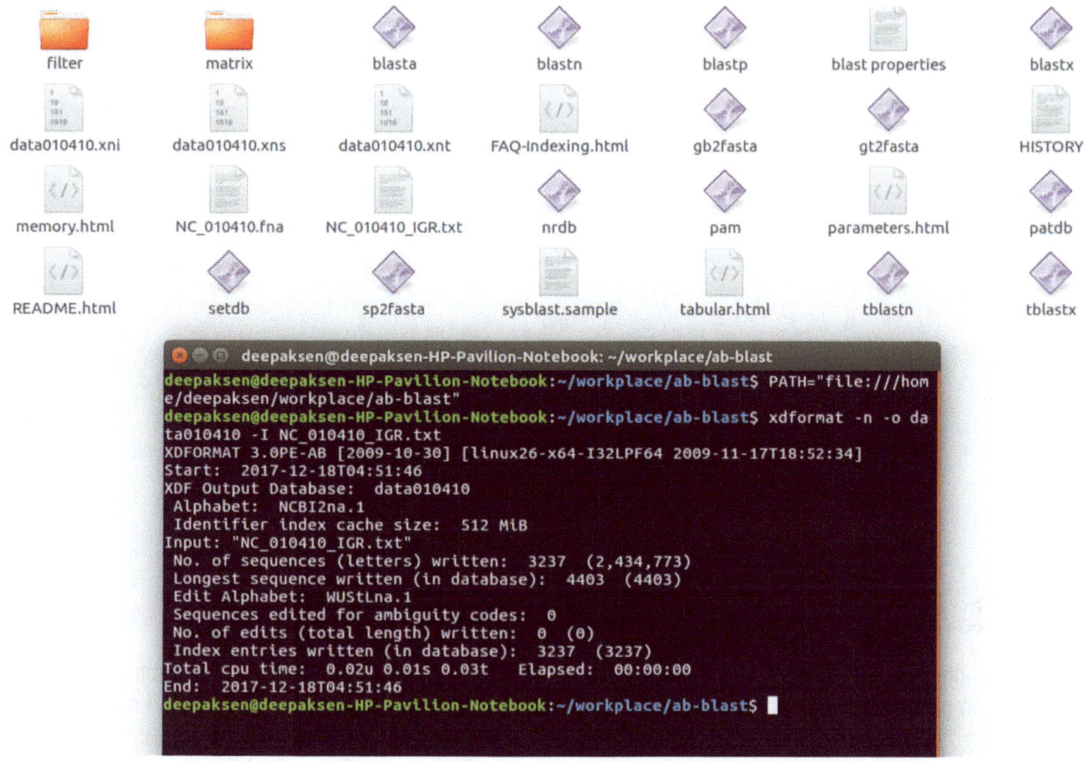

Fig. 4 Example command and output for databases. Example of the commands given in Subheading 3.2.3 can be seen in the console and the files created as databases—data010410.xnl, data010410.xns, and data010410.xnt—in the folder. The name data010410 is defined in the command itself

3.2.4 Running BLAST

Once we have the database file ready, we can now perform BLAST.

1. Run the following commands to set up the filter variable required by BLASTN utility. The default location of filter folder is in the AB-BLAST folder:

```
declare -x ABBLASTFILTER=<path to 'filter' folder>
```

2. Create the combined FASTA file for the closely related genomes against which BLAST has to be done. This can be done by simply copying and pasting the full contents of the FASTA file of each genome downloaded from NCBI to a single file.
3. Run the following commands for performing BLASTN. An example of the command is illustrated in Fig. 5:

```
Blastn<name of database><name of fasta file> -filter=dust
-hspsepQmax 1e-5 -V 2000 -B 500 -nogaps -wstrict -o
blastfile
```

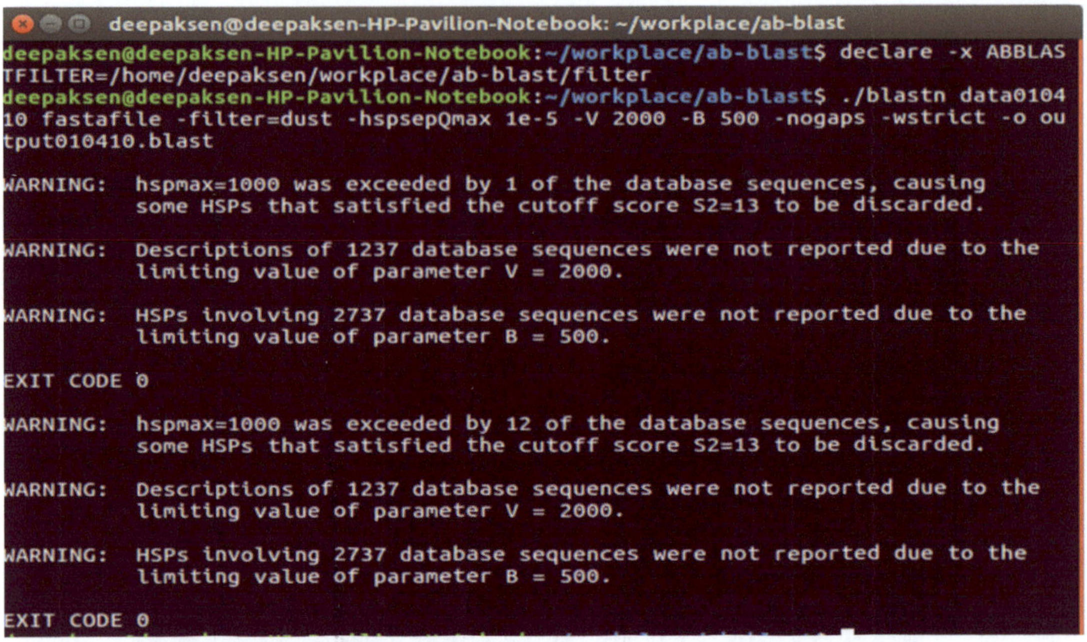

Fig. 5 Example of commands for BLAST. This is an example of the commands described in Subheading 3.2.4. The first command declares a system variable ABBLASTFILTER which lets the AB-BLAST tool know where to look for the filter tables necessary for performing BLAST. The second command runs the BLASTN tool to perform nucleotide BLAST with the parameters necessary for RNA match. The output filename is also defined in this command as output010410.blast

4. Once BLAST is done, we will have the output file with extension .blast. This file needs a little modification before we can use it for further steps.
5. Open the .blast file in a text editor, and replace "_< NCBI reference>_" with " " (blank space). In our example, this would require replacing all occurrences of "_NC_010410_" with "<>."

3.2.5 Predicting Consensus Secondary Structure by Comparative Sequence Analysis Using QRNA

1. Navigate to the scripts directory inside the QRNA directory.
2. Copy/paste the updated BLAST file from the previous step to this directory.
3. In the terminal console, navigate to same directory, and type in the following command:

```
./blastn2qrnadepth.pl -g <Name of organism>blastfile
```

4. Move the newly created .q file to the src directory of QRNA folder. Figure 6 gives an example of the command and highlights the output file required for further steps.

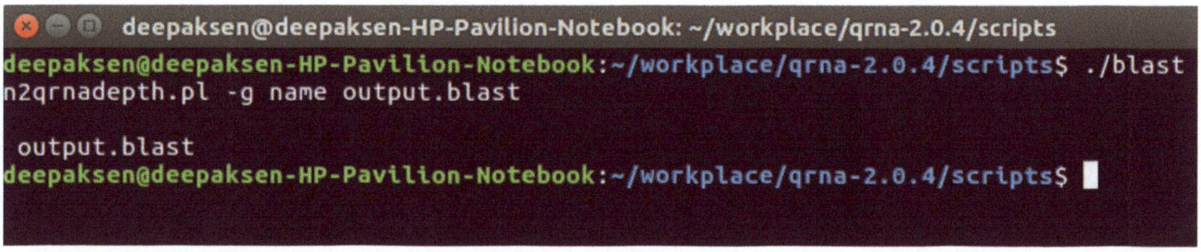

Fig. 6 QRNA scripts directory and command. Using output.blast as the input file, obtained from the previous step, we run the blastn2qrnadepth perl script and obtain the output.blast.E0.01.D1.q file, which is required in **steps 5** and **6** of Subheading 3.2.5

5. Declare the QRNADB variable to point toward the lib directory of QRNA:

```
declare -x QRNADB=<Path to qrna/lib directory>
```

6. Navigate to the src directory, and run the following command to get the final output file:

```
eqrna -w 150 -x 50 -o <qrna output><.q filename>
```

3.2.6 Rho-Independent Terminators Using RNAMotif

The descriptor file used here for searching rho-independent terminators was provided by D.J. Ecker, the creator of the program RNAMotif, to Jonathan Livny for his study of sRNAs of *P. aeruginosa* and *V. cholerae*. This file was available for download with the sRNAPredict3 program.

1. In the terminal, navigate to src directory under RNAMotif directory.
2. Type the following command to declare the EFNDATA variable, setting it to point to the efndata directory under RNAMotif directory:

```
declare -x EFNDATA=<Path to efndata directory>
```

3. Run RNAMotif using the following command. Output filename can be changed by replacing output.txt with desired filename. Figure 7 shows an example of the command and highlights the output file in the directory:

```
rnamotif -descr RNAMotif_descr.txt <.fna filename>>output.txt
```

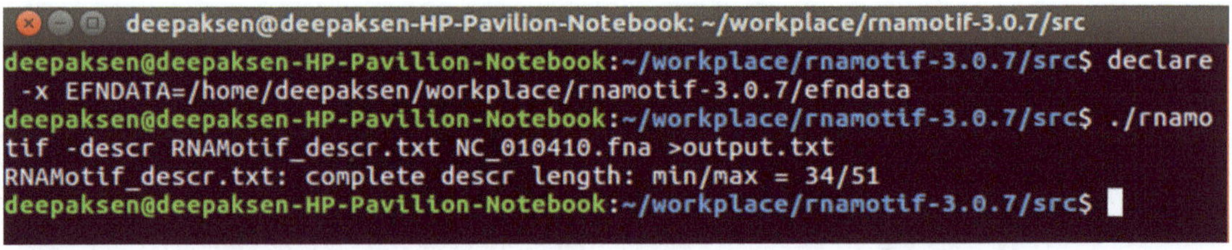

Fig. 7 RNAMotif commands and directory. This is a screenshot of the commands in console and the files in src directory inside RNAMotif parent directory. The first command declares the EFNDATA environment variable, which is used by the RNAMotif tool to find the efndata directory inside RNAMotif parent directory. The second command executes the RNAMotif program with descriptor file RNAMotif_descr.txt on nucleotide sequences in NC_010410.fna file and writes the output of the program in output.txt file

3.2.7 Prediction of Rho-Independent Terminators by Using TransTermHP

Although TransTermHP can be used via its web-based service very easily, here we also provide the instructions to download and run the stand-alone software package on your own computer. The compilation instructions can be found in the downloaded files.

1. Download and unpack the TransTermHP software from its website.
2. In the TransTermHP directory, move the .fna, .gbk, and .ptt files.
3. In the terminal navigate to the TransTermHP folder, and run the following command:

```
./transterm -p expterm.dat <NCBI reference>.fna<NCBI re-
ference>.ptt> tt.txt
```

4. This will create the output file output<NCBI reference number>.txt. Figure 8 shows an example of the above command and the output file in the created TransTermHP directory.

3.2.8 Prediction of Putative sRNA by Using sRNAPredict3

1. Navigate to the sRNAPredict folder, and type the following command to compile the program:

```
g++ -o srnapredict.out sRNAPredict3.cpp
```

2. Move the files obtained in all the previous steps by AB-BLAST, QRNA, RNAMotif, and TransTermHP along with the genome, gbk, and ptt files to sRNAPredict folder.
3. Open the file input.txt present in this folder in any text editing software, and edit the filenames accordingly. Example of this file is shown in Fig. 9.

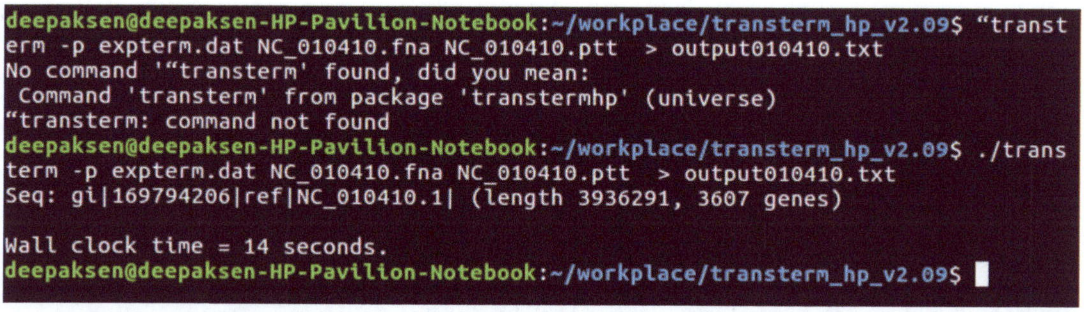

Fig. 8 TransTermHP directory and commands. The command in the example on execution uses the FASTA format nucleic acid sequences and protein table files to predict rho-independent terminators in the output010410.txt file. The output filename can be changed by modifying the name in the command

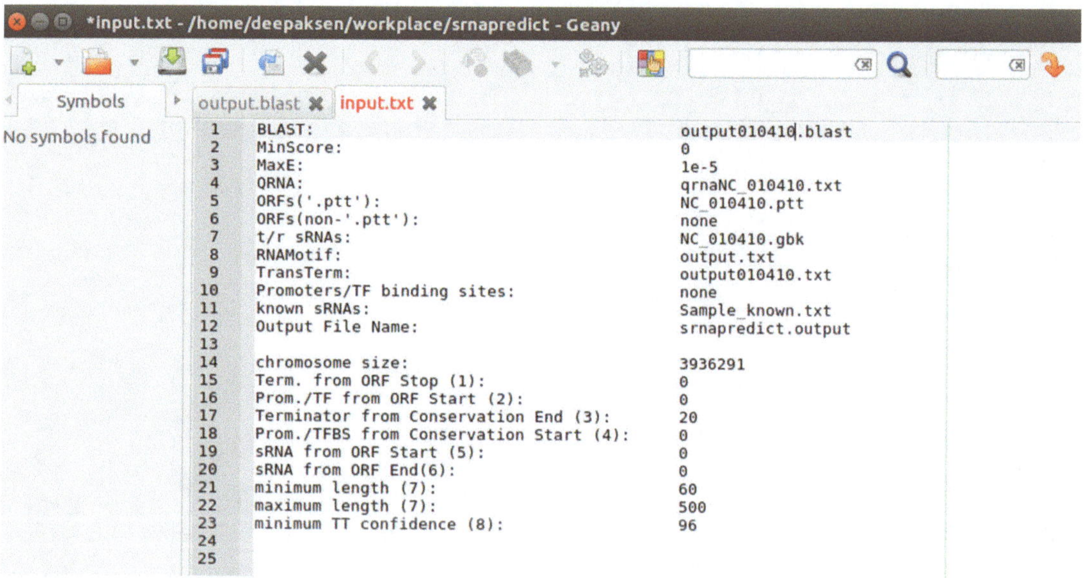

Fig. 9 The input.txt file. This figure shows a screenshot of the input.txt file and provides the sample data to be entered in the file. sRNAPredict tool searches its directory for the files mentioned in this file, as well as sets the parameters for its search as defined in this file. The parameters in this figure are specific for *Acinetobacter* spp. sRNA

4. In the terminal type the following command for running sRNAPredict3 software, and follow the instructions, as illustrated in the screenshot in Fig. 10:

```
./srnapredict.out
```

```
----------------------------------------------------------------------------------------
                  Welcome to the 'sRNAPredict2' program
----------------------------------------------------------------------------------------

Enter the number of searches you would like to conduct:
1
Enter the name of input file #1:
input.txt

Reading in data from file 'input.txt'.
********************************************************************************

11368 have been extracted from 'output' and written to file 'OutBlast'.

11368 coordinate pairs have been read in from data file 'BLAST_Out'
The 11368 blast coordinate pairs have been parsed to the
coordinates of 2837 conserved segments which have been written to file 'OutBlastParsed'.

Finished parsing BLAST coordinates...
********************************************************************************

Extracting QRNA coordinate pairs from file 'qrnaNC_010410.txt'...
and writting them into file 'QRNA_out'.

0 coordinate pairs of 'winner = RNA' segments have been written to file 'QRNA_out'
Finished extracting QRNA coordinates...
********************************************************************************

The genome size is 3936291 bps.
The coordinate pairs, orientations, numbers, and names of 3607 ORFs have been
extracted from file 'NC_010410.ptt' and written into file 'OutORF'

3607 ORF coordinates, orientations, numbers, and names have been read in from data file 'NC_01
t'...
The coordinates and orientations of 90 tRNAs and rRNAs have been extracted from file 'NC_01041
and written into file 'OutORF'

3697 ORF coordinates, orientations, numbers, and names have been read in from data file 'OutOR
Sorting ORFs...
Done sorting ORFs.
The sorted ORF coordinates are being converted to IGR coordinates...
Coordinates of 3237 intergenic regions have been written to data file OutIG.

The total amount of intergenic sequence is 489637.
Finished converting ORF coordinates to IG coordinates...
********************************************************************************

6453 terminators have been have been read in from file 'rnamotif.txt'.
1072 terminators have been have been read in from file 'tt.txt'
and have been written into file 'OutTerm_temp'.

7525 terminator coordinates have been read in from data file 'OutTerm_temp'
pruned to the coordinates of 2673 terminators and written to file 'OutTerms'.

Finished processing terminators...
********************************************************************************

The coordinate pairs of 2837 conserved seqments have been read in from file 'OutBlastParsed'.
```

Fig. 10 Output from executing sRNAPredict. This is a screenshot of the welcome message displayed on executing sRNAPredict. It asks the user to input the number of searches to be performed and then asks for the input files corresponding to the number of searches. For our purposes, this number is 1 and has only 1 input file: input.txt

4 Notes

1. The output of sRNAPredict3 was compared with the data obtained from the RNA-seq of *A. baumannii* ATCC 17978.

18 out of 31 predicted candidate sRNAs were confirmed by RNA-seq.

2. Some programs might take a long time to execute depending on the computational power or in some cases the length of sequences.
3. The parameters for BLAST were set as follows:
   (a) *nogaps*—does not create gapped alignments.
   (b) *wstrict*—will prevent the software from exhaustively searching diagonals that are found to contain HSPs in an effort to find other HSPs that would not be seeded by neighborhood word hits.
   (c) *filter*—for nucleotide-level (BLASTN) searches, supported filter programs include dust and seg. By default, no filtering is performed.
   (d) *V*—sets the maximum number of one-line descriptions of significant database sequences to report in the first section of program output to *v*. The default limit is 500.
   (e) *B*—sets the maximum number of database sequences for which *any* alignments will be reported to *b*. The default limit is 250. The maximum number of alignments that may be saved and reported per database sequence is governed by other parameters.
   (f) *hspsepQmax*—maximum allowed separation along the *subject* (database) sequence between two HSPs (gapped or ungapped) that will be clustered into a consistent set. We have set it to 1e−5.
4. For programs like QRNA, RNAMotif, and TransTerm_hp, default parameters were used.
5. There are other approaches to predicting sRNA, some of them specific to a particular type of sRNA, for example, protract [15] to search specifically for piRNA. The sRNAPredict pipeline was later turned into SIPHT by Livny [16]. It is a web-based portal which uses similar algorithms as detailed in this chapter.

References

1. Gomes AQ, Nolasco S, Soares H (2013) Non-coding RNAs: multi-tasking molecules in the cell. Int J Mol Sci 14(8):16010–16039. https://doi.org/10.3390/ijms140816010
2. Lee YS, Shibata Y, Malhotra A, Dutta A (2009) A novel class of small RNAs: tRNA-derived RNA fragments (tRFs). Genes Dev 23 (22):2639–2649. https://doi.org/10.1101/gad.1837609
3. Wei H, Zhou B, Zhang F, Tu Y, Hu Y, Zhang B, Zhai Q (2013) Profiling and identification of small rDNA-derived RNAs and their potential biological functions. PLoS One 8 (2):e56842
4. Barman RK, Mukhopadhyay A, Das S (2017) An improved method for identification of small non-coding RNAs in bacteria using support vector machine. Sci Rep 7:46070. https://doi.org/10.1038/srep46070
5. Storz G, Vogel J, Wassarman KM (2011) Regulation by small RNAs in bacteria: expanding frontiers. Mol Cell 43(6):880–891. https://doi.org/10.1016/j.molcel.2011.08.022

6. Mercer TR, Dinger ME, Mattick JS (2009) Long noncoding RNAs: insights into functions. Nat Rev Genet 10:155–159
7. Gómez-Lozano M, Marvig RL, Molin S, Long KS (2014) Identification of bacterial small RNAs by RNA sequencing. In: Filloux A, Ramos JL (eds) Pseudomonas methods and protocols. Methods in molecular biology (methods and protocols), vol 1149. Humana Press, New York, NY
8. Sharma R, Arya S, Patil SD, Sharma A, Jain PK, Navani NK, Pathania R (2014) Identification of novel regulatory small RNAs in Acinetobacter baumannii. PLoS One 9(4):e93833
9. Sharma A, Sharma R, Bhattacharyya T, Bhando T, Pathania R (2016) Fosfomycin resistance in *Acinetobacter baumannii* is mediated by efflux through a major facilitator superfamily (MFS) transporter—AbaF. J Antimicrob Chemother 72(1):68–74
10. Livny J, Brencic A, Lory S, Waldor MK (2006) Identification of 17 Pseudomonas aeruginosa sRNAs and prediction of sRNA-encoding genes in 10 diverse pathogens using the bioinformatic tool sRNAPredict2. Nucleic Acids Res 34(12):3484–3493. https://doi.org/10.1093/nar/gkl453
11. Macke TJ, Ecker DJ, Gutell RR, Gautheret D, Case DA, Sampath R (2001) RNAMotif, an RNA secondary structure definition and search algorithm. Nucleic Acids Res 29 (22):4724–4735
12. Kingsford CL, Ayanbule K, Salzberg SL (2007) Rapid, accurate, computational discovery of Rho-independent transcription terminators illuminates their relationship to DNA uptake. Genome Biol 8(2):R22. https://doi.org/10.1186/gb-2007-8-2-r22
13. Rivas E, Eddy SR (2001) Noncoding RNA gene detection using comparative sequence analysis. BMC Bioinformatics 2:8. https://doi.org/10.1186/1471-2105-2-8
14. Livny J, Fogel MA, Davis BM, Waldor MK (2005) sRNAPredict: an integrative computational approach to identify sRNAs in bacterial genomes. Nucleic Acids Res 33 (13):4096–4105. https://doi.org/10.1093/nar/gki715
15. Rosenkranz D, Zischler H (2012) proTRAC—a software for probabilistic piRNA cluster detection, visualization and analysis. BMC Bioinformatics 13:5. https://doi.org/10.1186/1471-2105-13-5
16. Livny J (2012) Bioinformatic discovery of bacterial regulatory RNAs using SIPHT. In: Keiler K (ed) Bacterial regulatory RNA. Methods in molecular biology (methods and protocols), vol 905. Humana Press, Totowa, NJ

Chapter 28

Global Metabolic Analyses of *Acinetobacter baumannii*

Mohd Hafidz Mahamad Maifiah, Tony Velkov, Darren J. Creek, and Jian Li

Abstract

Acinetobacter baumannii is rapidly emerging as a multidrug-resistant pathogen responsible for nosocomial infections including pneumonia, bacteremia, wound infections, urinary tract infections, and meningitis. Metabolomics provides a powerful tool to gain a system-wide snapshot of cellular biochemical networks under defined conditions and has been increasingly applied to bacterial physiology and drug discovery. Here we describe an optimized sample preparation method for untargeted metabolomics studies in *A. baumannii*. Our method provides a significant recovery of intracellular metabolites to demonstrate substantial differences in global metabolic profiles among *A. baumannii* strains.

Key words *Acinetobacter baumannii*, Multidrug-resistant, Metabolomics, Polymyxins

1 Introduction

Acinetobacter baumannii is one of the six significant ESKAPE 'superbugs' identified by the Infectious Diseases Society of America (IDSA) which required urgent attention due to their resistance to most of current antibiotics [1]. The WHO has classified *A. baumannii* under the "Critical" level giving the priority to discover novel antibiotics [2]. It also exhibits significant tolerance to a wide range of environmental conditions and can cause pneumonia, bacteremia, meningitis, urinary tract infection, and wound infection [3]. Worryingly, infections caused by *A. baumannii* have contributed to high morbidity and mortality rates worldwide [4]. Effective treatment for *A. baumannii* infections is a great challenge, as the pathogen can rapidly develop different resistance mechanisms to almost all clinically available antibiotics [5]. Polymyxins (i.e., polymyxin B and colistin, also known as polymyxin E) have been widely used as the only effective therapeutic option for multidrug-resistant (MDR) *A. baumannii* [6, 7]. Even though polymyxin resistance is not common, *A. baumannii* can develop resistance by modifications of the lipid A component of lipopolysaccharides (LPS) in the bacterial outer membrane, which is the

Indranil Biswas and Philip N. Rather (eds.), *Acinetobacter baumannii: Methods and Protocols*, Methods in Molecular Biology, vol. 1946, https://doi.org/10.1007/978-1-4939-9118-1_28,

initial polymyxin binding site [8–10]. We reported that *A. baumannii* ATCC 19606 spontaneously acquired colistin resistance via the loss of LPS due to single random mutations in the lipid A biosynthesis genes, *lpxA*, *lpxC*, and *lpxD* [11].

Metabolomics is able to profile cellular metabolites and metabolic pathways at the systems level [12, 13]. Metabolites are small molecules typically with a molecular weight of less than 1000 Da generated as a result of complex cellular interactions of the genome, transcriptome, proteome, and environment [14]. Untargeted metabolomics, also known as global metabolite profiling, involves the comprehensive measurement and identification of cellular metabolites and is commonly used to generate new hypotheses [15]. Alternatively, targeted metabolomic approaches aim to analyze a predetermined set of metabolites of interest in specific pathways [16]. We have developed and optimized a sample pretreatment method which provides extensive metabolite recovery for analyzing global metabolic profiles of *A. baumannii*. This method has been extensively used in our antimicrobial systems pharmacology studies, including a comparative global metabolomic analysis of two different pairs of polymyxin-susceptible and polymyxin-resistant *A. baumannii* strains [17]. Our results demonstrated significant metabolic differences between the paired polymyxin-susceptible ATCC 19606 and polymyxin-resistant 19606R. Notably, compared to the polymyxin-susceptible ATCC 19606, the resistant 19606R strain showed significant metabolic perturbations in amino acids, carbohydrates, nucleotides, glycerophospholipids, and peptidoglycan biosynthesis [17].

2 Materials

2.1 Bacterial Strains

A. baumannii ATCC 19606 was purchased from the American Type Culture Collection (ATCC). The LPS-deficient polymyxin-resistant *A. baumannii* strain 19606R (MIC >128 mg/L) is an *lpxA* mutant of ATCC 19606 [11].

2.2 Culture Media

1. Nutrient agar.
2. Mueller-Hinton agar supplemented with 10 mg/L polymyxin B.
3. Cation-adjusted Mueller-Hinton broth (CaMHB; 20–25 mg/L Ca^{2+} and 10–12.5 mg/L Mg^{2+}).

2.3 Cellular Metabolite Extraction

1. Quenching solution: Add dry ice into ethanol solution in a basin container.
2. Washing solution: 0.9% NaCl (4 °C). Weigh 9 g of NaCl and add distilled water to a volume of 1 L. Autoclave and then keep the washing solution at 4 °C until use.

3. Extraction solvent: Chloroform, methanol, and distilled water are prepared in the ratio of 1:3:1 (*v/v*), respectively. Add 30 mL chloroform, 90 mL methanol, and 30 mL distilled water to a glass bottle to make a final volume of 150 mL. Keep the extraction solvent on dry ice during the experiment to maintain the temperature below −40 °C.
4. Generic internal standards: Prepare stock solutions of the buffer salts CHAPS, CAPS, PIPES, and TRIS at 1 mM of each in 50% methanol/water (*v/v*) (*see* **Note 1**). Add the internal standards into the extraction solvent at a final concentration of 1 μM.
5. Liquid nitrogen. Prepare an enough volume of liquid nitrogen to rapidly freeze sample tubes for the freeze-thaw process during metabolite extraction.
6. Ice.

2.4 LC-MS Analysis

1. LC-MS instrument: Dionex high-performance liquid chromatography (HPLC) system (RSLCU3000, Thermo Fisher) coupled to a Q-Exactive Orbitrap mass spectrometer (Thermo Fisher).
2. Column: ZIC-pHILIC (5 μm, polymeric, 150 × 4.6 mm, SeQuant, Merck).
3. LC solvent A: 20 mM ammonium carbonate in Milli-Q water. Filter the solvent before use.
4. LC solvent B: LC-MS grade acetonitrile.
5. LC system: a multistep gradient system from 80% B to 50% B over 15 min, then to 5% B at 18 min, followed by wash with 5% B for 3 min, return to 80% B over 3 min. Re-equilibrate for 8 min with 80% B at a flow rate of 0.3 mL/min [18]. The run time is 32 min and the injection sample volume is 10 μL.
6. Operation of the Q-Exactive Orbitrap mass spectrometer: 35,000 resolution in both positive and negative electrospray ionization (ESI) mode (rapid switching) and a detection range of 85–1275 *m/z*.
7. Metabolite standards: Pooled mixtures (up to eight individual vials) of over 250 authentic metabolite standards that cover a wide range of metabolic pathways (Mass Spectrometry Metabolite Library of Standards, IROA Technologies).

3 Methods

3.1 Bacterial Culture Preparation

1. Inoculate the frozen stock (−80 °C) of *A. baumannii* ATCC 19606 onto a nutrient agar plate, and incubate it for 16–18 h at 37 °C.

2. For the polymyxin-resistant strain, *lpxA* mutant 19606R, inoculate the bacteria onto a Mueller-Hinton agar plate supplemented with polymyxin B (10 mg/L) (*see* **Note 2**).
3. For each *A. baumannii* strain, pick up a single colony, and inoculate in 10 mL CaMHB then incubate it overnight (16–18 h) at 37 °C with constant shaking (180 rpm). For the polymyxin-resistant strain 19606R, the CaMHB contains no polymyxin B. Prepare at least four biological replicate culture samples from different colonies of *A. baumannii* ATCC 19606 and *lpxA* mutant 19606R strains.
4. Prepare four reservoirs of 50 mL CaMHB in shake flasks for each biological replicate.
5. Inoculate each reservoir with 500 μL (1:100 dilution) overnight culture of the bacterial strains ATCC 19606 and polymyxin-resistant 19606R, and grow at 37 °C with shaking at 180 rpm to an OD_{600} ~0.5 (mid-exponential growth phase) (*see* **Note 3**).
6. Prepare a 50 mL CaMHB reservoir without bacterial inoculation as a blank control.

3.2 Sample Preparation for the Metabolomic Study

1. Collect 10 mL of mid-exponential culture (OD_{600} ~0.5) of *A. baumannii* ATCC 19606 and polymyxin-resistant 19606R in 15 mL tubes (*see* **Note 4**). The volume of bacterial culture should be adjusted based on the amount of cell pellet obtained.
2. Rapidly and briefly quench the samples in a dry ice/ethanol bath until temperature reaches 4 °C, and preserve them on ice for all the following steps. As a precaution, do not let bacterial culture frozen during the quenching step (*see* **Note 5**).
3. Centrifuge the tubes at 3220 × *g* at 4 °C for 5 min, and keep the supernatant for extracellular metabolite (footprint) analysis (refer **steps 11–13** for footprint sample preparation) (*see* **Note 6**).
4. Wash the cell pellets twice with 0.5 mL of 0.9% NaCl (4 °C) to remove any residue of extracellular metabolites and medium components (*see* **Note 7**).
5. Centrifuge the tube at 3220 × *g* at 4 °C for 3 min after each wash, and discard the supernatant. Collect as much supernatant as possible using a fine-tip pipette after the final wash to remove the residual washing solution (*see* **Note 8**).
6. Resuspend the cell pellets in 0.5 mL extraction solvent consisting of chloroform/methanol/water (CMW; 1:3:1, *v/v*; −40 °C) (*see* **Note 9**); the solvent mixture contained the internal standards (CHAPS, CAPS, PIPES, and TRIS; 1 μM of each). Vortex the tubes in 5 s to mix. The volume of extraction solvent (CMW) can be decreased depending on the size of the

pellet (i.e., 0.3 mL is appropriate in most experiments) (*see* **Note 10**).

7. Freeze the cell pellet extraction mixtures by submersion of tubes in liquid nitrogen, and thaw on ice. Repeat the step three times in order to permeabilize the cells and release intracellular metabolites.
8. Centrifuge the mixture at 3220 × *g* at 4 °C for 10 min.
9. Collect 300 μL of the supernatant containing the extracted metabolites in 1.5-mL tubes.
10. For LC-MS analysis, centrifuge the tube at 14,000 × *g* for 10 min at 4 °C, and transfer 200 μL of particle-free supernatant into the glass injection vial. Keep the vials at −80 °C until further use.
11. For footprint samples, filter an aliquot of approximately 1.5 mL of the culture supernatant (from **step 3**) through a 0.22-μm filter into a 1.5-mL tube.
12. For LC-MS analysis, prepare 250 μL extraction solvent (CMW, 1:3:1, *v/v*) in 1.5 mL tubes. Add 10 μL of the footprint samples into the tubes and mix by vortexing.
13. Centrifuge the tubes at 14,000 × *g* for 10 min at 4 °C, and collect 200 μL supernatant into the injection vial. Keep the vials at −80 °C until further use.
14. Mix equal volumes from each of the samples (10 μL each), for both intra- and extracellular metabolites, in an injection vial for the pooled quality control (QC) sample for LC-MS analysis (*see* **Note 11**).

3.3 LC-MS Analysis

1. Untargeted metabolomic analysis utilizes a hydrophilic interaction liquid chromatography (HILIC)-high-resolution mass spectrometry (HRMS) method.
2. Separate the intracellular and extracellular metabolite samples into two batches, and randomize the sequence of all the samples to be analyzed in each batch (*see* **Note 12**).
3. Analyze the mixtures of pure metabolite standards at the start of the batch to aid in the identification of metabolites.
4. Include the pooled QC sample at regular intervals (every five to ten samples) during the analysis sequence.
5. Include one sample of blank extraction solvent as a blank control.
6. Run the LC-MS analysis on a high-performance liquid chromatography (HPLC) system (RSLCU3000) using a ZIC-pHILIC column (5 μm, polymeric, 150 × 4.6 mm) coupled to a Q-Exactive Orbitrap mass spectrometer.

7. Monitor the chromatographic peaks, signal reproducibility, and analyte stability by assessment of the pooled QC sample, internal standards, and total ion chromatograms for each sample.
8. Perform data processing, bioinformatics, and statistical analyses using a free software IDEOM (http://mzmatch.sourceforge.net/ideom.php) [19].

4 Notes

1. These common buffer salts are selected as the internal standards as they are physicochemically diverse small molecules that are not naturally occurring in any microorganism and can be spiked at known concentrations to determine the analytical performance of the method used.
2. Mueller-Hinton agar supplemented with polymyxin B (10 mg/L) is to induce the selection pressure in order to maintain the resistance phenotype of the polymyxin-resistant strain, *lpxA* mutant 19606R.
3. Bacterial culture at OD_{600nm} ~0.5 at mid-exponential growth phase (10^8 CFU/mL) is to obtain enough bacterial cells for metabolite extraction.
4. Four technical replicates from each sample are recommended for cellular metabolite analysis. However, one technical replicate (with at least four biological replicates) is sufficient if the recovery of metabolites is reproducible.
5. When necessary, depending on the nature of the metabolomic study conducted (e.g., comparison between different bacterial strains, treatment of drug), after quenching the bacterial cells, bacterial cultures are normalized by OD_{600nm} ~0.5 to ensure samples within a particular study have comparable biomass.
6. Five min centrifugation is sufficient, as longer time may cause intracellular content leakage.
7. The washing step is essential as CaMHB is nutrient rich. Our results showed that the culture media components are sufficiently removed during the washing step with 0.9% NaCl [17]. There is no substantial effect on the recovery of intracellular metabolites (those not present in the broth), as only little metabolite leakage is detected in the supernatant.
8. Any remaining washing solution interferes the ratio of organic and aqueous extraction solvents used during the extraction process. If the 1:3:1 ratio is imbalanced, phase partitioning of solvents can be observed, and extractions are not reproducible.

9. CMW was demonstrated to be the most efficient and reproducible solvent for the extraction of a wide coverage of different metabolite classes, compared to the other solvents tested including chloroform/methanol (CM; 1:2, *v/v*), 60% ethanol, and absolute methanol [17].
10. We recommend that the cell pellet extraction should be processed and finished in a single session to avoid any bias and inconsistency between sample pretreatment procedures.
11. This pooled QC sample is used to estimate a composite sample profile representing all the analytes that will be encountered during the LC-MS analysis [20].
12. Sample randomization can reduce the impact of within-batch variation of LC-MS signal responses between samples.

Acknowledgments

Jian Li, Darren J. Creek, and Tony Velkov are supported by the National Institute of Allergy and Infectious Diseases of the National Institutes of Health (R01 AI111965 and AI132681 [JL and TV]). Mohd Hafidz Mahamad Maifiah is an academic staff from the International Institute for Halal Research and Training (INHART), International Islamic University Malaysia (IIUM). Darren J. Creek and Tony Velkov are Australian National Health and Medical Research Council (NHMRC) Dareer Development Fellows. Jian Li is an Australian NHMRC Senior Research Fellow.

References

1. Boucher HW, Talbot GH, Bradley JS, Edwards JE, Gilbert D, Rice LB, Scheld M, Bartlett J (2009) Bad bugs, no drugs: no ESKAPE! An update from the Infectious Diseases Society of America. Clin Infect Dis 48(1):1–12
2. World Health Organization (WHO) (2017) Global priority list of antibiotic-resistant bacteria to guide research, discovery, and development of new antibiotics. Available at: http://www.who.int/medicines/publications/global-priority-list-antibiotic-resistant-bacteria/en/. Accessed 10 Dec 2018
3. Fishbain J, Peleg AY (2010) Treatment of Acinetobacter infections. Clin Infect Dis 51 (1):79–84
4. Dijkshoorn L, Nemec A, Seifert H (2007) An increasing threat in hospitals: multidrug-resistant *Acinetobacter baumannii*. Nat Rev Microbiol 5(12):939–951
5. Peleg AY, Seifert H, Paterson DL (2008) *Acinetobacter baumannii*: emergence of a successful pathogen. Clin Microbiol Rev 21 (3):538–582
6. Li J, Nation RL, Turnidge JD, Milne RW, Coulthard K, Rayner CR, Paterson DL (2006) Colistin: the re-emerging antibiotic for multidrug-resistant Gram-negative bacterial infections. Lancet Infect Dis 6(9):589–601
7. Nation RL, Li J, Cars O, Couet W, Dudley MN, Kaye KS, Mouton JW, Paterson DL, Tam VH, Theuretzbacher U, Tsuji BT, Turnidge JD (2015) Framework for optimization of the clinical use of colistin and polymyxin B: the Prato polymyxin consensus. Lancet Infect Dis 15(2):225–234
8. Beceiro A, Llobet E, Aranda J, Bengoechea JA, Doumith M, Hornsey M, Dhanji H, Chart H, Bou G, Livermore DM, Woodford N (2011) Phosphoethanolamine modification of lipid A in colistin-resistant variants of *Acinetobacter baumannii* mediated by the pmrAB

two-component regulatory system. Antimicrob Agents Chemother 55(7):3370–3379

9. Arroyo LA, Herrera CM, Fernandez L, Hankins JV, Trent MS, Hancock RE (2011) The *pmrCAB* operon mediates polymyxin resistance in *Acinetobacter baumannii* ATCC 17978 and clinical isolates through phosphoethanolamine modification of lipid A. Antimicrob Agents Chemother 55 (8):3743–3751
10. Pelletier MR, Casella LG, Jones JW, Adams MD, Zurawski DV, Hazlett KR, Doi Y, Ernst RK (2013) Unique structural modifications are present in the lipopolysaccharide from colistin-resistant strains of *Acinetobacter baumannii*. Antimicrob Agents Chemother 57 (10):4831–4840
11. Moffatt JH, Harper M, Harrison P, Hale JDF, Vinogradov E, Seemann T, Henry R, Crane B, St Michael F, Cox AD, Adler B, Nation RL, Li J, Boyce JD (2010) Colistin resistance in *Acinetobacter baumannii* is mediated by complete loss of lipopolysaccharide production. Antimicrob Agents Chemother 54 (12):4971–4977
12. Chen C, Gonzalez FJ, Idle JR (2007) LC-MS-based metabolomics in drug metabolism. Drug Metab Rev 39(2–3):581–597
13. Mastrangelo A, Armitage EG, Garcia A, Barbas C (2014) Metabolomics as a tool for drug discovery and personalised medicine. A review. Curr Top Med Chem 14(23):2627–2636
14. Johnson CH, Ivanisevic J, Siuzdak G (2016) Metabolomics: beyond biomarkers and towards mechanisms. Nat Rev Mol Cell Biol 17(7):451–459
15. Theodoridis G, Gika HG, Wilson ID (2008) LC-MS-based methodology for global metabolite profiling in metabonomics/metabolomics. TrAC-Trends Anal Chem 27(3):251–260
16. Patti GJ, Yanes O, Innovation SG (2012) Metabolomics: the apogee of the omics trilogy. Nat Rev Mol Cell Biol 13(4):263–269
17. Mahamad Maifiah MH, Cheah SE, Johnson MD, Han ML, Boyce JD, Thamlikitkul V, Forest A, Kaye KS, Hertzog P, Purcell AW, Song J, Velkov T, Creek DJ, Li J (2016) Global metabolic analyses identify key differences in metabolite levels between polymyxin-susceptible and polymyxin-resistant *Acinetobacter baumannii*. Sci Rep 6:22287
18. Zhang T, Creek DJ, Barrett MP, Blackburn G, Watson DG (2012) Evaluation of coupling reversed phase, aqueous normal phase, and hydrophilic interaction liquid chromatography with Orbitrap mass spectrometry for metabolomic studies of human urine. Anal Chem 84 (4):1994–2001
19. Creek DJ, Jankevics A, Burgess KE, Breitling R, Barrett MP (2012) IDEOM: an Excel interface for analysis of LC-MS-based metabolomics data. Bioinformatics 28 (7):1048–1049
20. Gika HG, Theodoridis GA, Wingate JE, Wilson ID (2007) Within-day reproducibility of an HPLC-MS-based method for metabonomic analysis: application to human urine. J Proteome Res 6(8):3291–3303

Chapter 29

Reverse Vaccinology Approach to Potential Vaccine Candidates Against *Acinetobacter baumannii*

Fatima Shahid, Shifa Tariq Ashraf, and Amjad Ali

Abstract

Acinetobacter baumannii is a rapidly evolving pathogen that largely inhabits intensive care units (ICU). This opportunistic, gram-negative organism has shown noteworthy taxonomic variations during the past three decades. *A. baumannii* functions as a catalase-positive, oxidase-negative obligate, aerobic, nonmotile, highly infectious, and multidrug-resistant bacterium. Therefore, the infection caused by this bacterium tends to have a fairly higher incidence rate in immune-compromised individuals ranging from 26.5% to 91%, as it colonizes in skin tissues and secretions of the respiratory tract. Recently, it has been globally labeled as a "red alert" pathogen, setting alarms throughout the medical community, arising mainly due to its widespread antibiotic resistance continuum. There is a dire need for alternative therapeutic intervention to combat *A. baumannii*-associated infections and the growing resistance. This chapter focuses upon the reverse vaccinology-based steps and strategies to identify novel potential vaccine candidates against this emerging pathogen.

Key words *A. baumannii*, Reverse vaccinology, PVCs, Antibiotic resistance, In silico

1 Introduction

1.1 Global Epidemiology of A. baumannii Infection

Studies have reported numerous incidences of *A. baumannii* infections in various geographical locations. *A. baumannii* infections may vary according to the geographical vicinity and the socioeconomic status of the patients. In a Moroccan study carried out in the year 2002–2005, this particular bacterium accounted for 13.63% of isolates obtained from ICUs' blood samples; and in another study, this pathogen was found to be 6.74% of the total gram-negative bacilli [1, 2]. An international study depicted the continent vise distribution of A. *baumannii* infections in the healthcare setting, where infection rate was 19.2% in Asian patients, 17.1% in Eastern Europeans, 13.8% in Central and South Americans, 5.6% in Western Europeans, and 14.8% in Africans while 4.4% in Oceania and 3.7% in North American patients [3]. It has been found to infect

Indranil Biswas and Philip N. Rather (eds.), *Acinetobacter baumannii: Methods and Protocols*, Methods in Molecular Biology, vol. 1946, https://doi.org/10.1007/978-1-4939-9118-1_29,

15% of South African individuals that are HIV-positive and 13% of the Canadian burn care unit patients [4, 5].

1.2 Prevalence of Antibiotic Resistance in A. baumannii

A. baumannii is widely reckoned for harboring intrinsic antibiotic resistance genes and is prone to quickly obtain and integrate mobile genetic elements such as plasmids, jumping genes, and specific integron cassettes along with various insertion sequences that manifest antibiotic resistance [6]. According to the available literature, multidrug-resistant (MDR) *A. baumannii* is posing a massive worldwide threat [7]. Undeniably, this organism is resistant to quite a few available antibiotics. The rate of resistance, according to several studies, varies from approximately 32% to 92% to ceftazidime, from 8.8% to 90% to imipenem, from 28.8% to 91.6% to fluoroquinolones; from 12% to 90% to tazobactam/piperacillin, and from 30% to 90% to aminoglycosides [5, 8–10]. Furthermore, colistin is considered as the most effective treatment alternative, but some *Acinetobacter* strains have developed resistance to colistin [5, 9–11]. Colistin resistance was calculated to be around 5.3% in the United States [11], 1.2% in India [10], 2.7% in South Africa [5], 0.5% in Saudi Arabia [12], and 0.9% in Tunisia [9]. The rate of resistance in Morocco was 0.3–69% for ceftazidime, in the range of 24–43% for imipenem, while approximately around 17.0–77.5% for aminoglycosides, along with 65–68% for ciprofloxacin; but none of the clinical isolates have been found to be colistin resistant [1, 2, 13].

2 Employing Reverse Vaccinology Approaches

With the advent of genome sequencing technology, the need to culture bacteria has subsided and taken over by reverse vaccinology techniques. Various predictive and analytical software programs such as Vaxign, NERVE, Jenner-predict, VaxiJen, and Vacceed are available that employ the reverse vaccinology approach [14]. These software programs have been created to identify potential vaccine candidates. These pipelines are readily available online, but hardly any of these comprise completely accessible packages to comprehensively identify candidate substrates. In addition, these pipelines take limited numbers and sizes as input files and are prone to time deferrals. In this short note, we focus on in silico methods utilizing reverse vaccinology approaches to identify and specifically filter out potential vaccine candidates from the sequenced bacterial genomes.

The acquisition of completely sequenced proteome of an organism, against which the vaccine has to be formulated, is the prerequisite to employ the vaccine development techniques via reverse vaccinology tools. The general steps that are followed during reverse vaccinology development are shown in Fig. 1, and the tools employed to execute these steps are listed in Table 1.

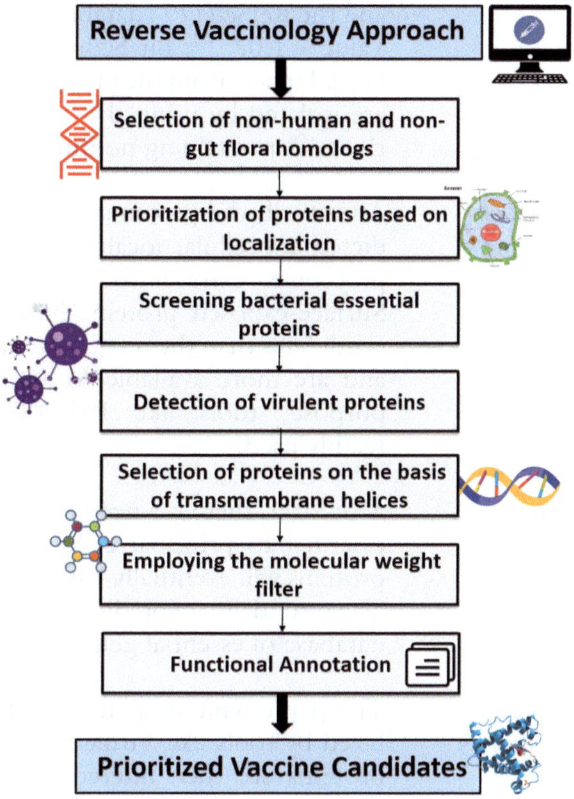

Fig. 1 A methodology flowchart to be followed in the reverse vaccinology approach for prioritization of vaccine candidates against *A. baumannii*

Table 1
Tools employed for executing reverse vaccinology approaches

Parameters	Tools/databases
Nonhuman homologs	BlastP against RefSeq and Swiss-Prot [17]
Non-gut flora homologs	BlastP against RefSeq and Swiss-Prot
Proteins in localization selected	CELLO2GO, CELLO, PSORT [20]
Bacterial essential proteins	DEG [21]
Virulent proteins	VFDB, MvirDB, PRINTs, TFvac [22]
Helices <2	HMMTOP version 2.0 [23]
Molecular weight <110 kDa	Compute p*I*/MW, peptide molecular weight calculated by GenScript
Annotation	UniProt, String [24, 25]

2.1 Selection of Nonhuman and Non-gut Flora Homologs

The primary task of this approach is to identify nonhost homologs (nonhuman homologs). This step is executed to reduce the chances of autoimmunity [15, 16]. The proteome of the organism (*A. baumannii*) is subjected to BLASTp against the Swiss-Prot and the RefSeq [17] databases where the BLAST filters are set as, *E*-Value $<1.0e^{(-5)}$, Bit Score >100, and percentage identity >35% [18]. Hence, from the prioritized proteins, non-gut flora homologs are isolated by using the same cutoff values and procedure mentioned for obtaining nonhuman homologs.

2.2 Prioritization of Proteins Based on Localization

Afterward, the proteins are distributed on the basis of their predicted subcellular localization comprising of the secretome/exoproteome, extracellular region, and outer membrane proteins. Surface-exposed proteins have a tendency to be better vaccine candidates than the rest, as they are chiefly involved in pathogenesis and are more available to antibody development [19]. For this purpose, tools like PSPORTb and CELLO2GO [20] are readily used.

2.3 Screening Bacterial Essential Proteins

Moreover, there is a need to explore and filter the proteins that are essential to principal cellular functions. Attacking these specific proteins will eventually have a fatal effect on the microbial viability. The essentiality of predicted proteins is validated through the use of database of essential genes (DEG) [21].

2.4 Detection of Virulence Factors

The pathogenic propensity of the candidate proteins is next analyzed by tools like Virulence Factors Database (VFDB) [22]. The proteins are then filtered on the basis of their transmembrane topology.

2.5 Selection of Proteins on the Basis of Transmembrane Helices

Software such as HMMTOP version 2.0 [23] is then used to select all proteins with less than two helices in the transmembrane region in order to avoid complications during purification of proteins during vaccine development. Afterward, the proteins are filtered on the basis of their molecular weight and are functionally annotated via UniProt [24].

2.6 Employing the Molecular Weight Filter

The proteins with a weight of <110 kDa are best suited for vaccine development, since small proteins are easily purified and can be handled effectively. This leads to the final identification of the potential vaccine candidates.

2.7 Functional Annotation

Functional annotation to depict biological, cellular, and molecular functional significance of the filtered proteins is subsequently performed by STRING analysis.

3 Our Contribution Toward Reverse Vaccinology

There are nearly 100 completely sequenced genomes available for *A. baumannii* on NCBI database (last accessed on March 2018). The average genome size is estimated at approximately 4.0 Mb,

Table 2
Predicted potential vaccine candidates identified in *A. baumannii* strain AB030

Parameters	No. of proteins
Total proteins	3857
Nonhuman homologs	3479
Non-gut flora homologs	3388
Proteins in localization selected	1180
Bacterial essential proteins	91
Virulent factors	16
Helices <2	14
Molecular weight <110 kDa	13
Filtered proteins	**13**

while the GC content is about 39%. Each strain on average possesses a total of 3894 genes and 3691 proteins. We have established a pipeline, called VACSOL [14], by amalgamating the abovementioned key steps of reverse vaccinology that is comparably less laborious and takes minimal time to perform the task. A total of 3875 proteins from *A. baumannii* strain AB030 were subjected to reverse vaccinology analysis using VACSOL. This generated nearly 3479 nonhuman homologous proteins. These hits were further screened to obtain 3388 non-gut flora homologs. Subsequent analysis revealed that about 1180 proteins were localized in the extracellular matrix, periplasm, and outer membrane and are the most probable locations to harbor vaccine candidates. Afterward, 91 essential proteins were identified that were required for the survival of the pathogen and hence are potential targets for further development. Additional screening revealed 16 proteins to be highly virulent, out of which 14 possessed less than two transmembrane helices. A total of 13 easy to purify, small proteins were finally screened out on the basis of their molecular weight (less than 110 kDa). These 13 proteins can serve as potential vaccine candidates. The predicted vaccine candidates identified are presented in Table 2, categorized according to the filters mentioned in the previous section, and their functional annotations are enlisted in Table 3.

3.1 Future Prospects

There are a few success stories where reverse vaccinology strategy has employed and predicted several potential vaccine epitopes against meningococcal strains, group A *Streptococcus*, pneumococcal strains, virulent *Escherichia coli* isolates, as well as antibiotic-resistant *Staphylococcus aureus*. The subsequent task would be to

Table 3
Functional annotation of the potential vaccines candidates identified in *A. baumannii* AB030

Sr. no.	Protein	Annotation
1	F911_00130	Peptidyl-prolyl cis–trans isomerase
2	MrcB	Murein polymerase; cell wall formation. Synthesis of cross-linked peptidoglycan from the lipid intermediates
3	HMPREF0010_03292	Peptidyl-prolyl cis–trans isomerase
4	F911_00513	Penicillin-binding protein, 1A family
5	F911_03727	Metallo-β-lactamase domain protein
6	Pal	Peptidoglycan-associated lipoprotein
7	F911_02226	Gram-negative pili assembly chaperone domain protein
8	F911_00852	UPF0339 protein
9	panD	Aspartate alpha-decarboxylase
10	AcrA	Acriflavine resistance protein A
11	OprM	Outer membrane efflux protein
12	PepN	Aminopeptidase N
13	HMPREF0010_00856	Fimbrial usher protein

identify the best possible vaccine candidates that are suitable for vaccine development platforms. The main goal of the analysis performed in this chapter was to detect highly immunogenic potential vaccine candidates for *A. baumannii* using AB030 as model strain with the reverse vaccinology approaches. Henceforth, in vivo and in vitro studies could be carried out to examine immune responses against the predicted proteins identified in this in silico study. Moreover, molecular modeling analysis and energy calculations could be executed to establish the binding efficiency of these proteins. The tools and databases employed in the analysis are publically available and easy to use by a microbiologist.

References

1. Elouennass M, Sahnoun I, Zrara A, Bajjou T, Elhamzaoui S (2008) Épidémiologie et profil de sensibilité des isolats d'hémoculture dans un service de réanimation (2002–2005). Médecine Mal Infect [Internet] 38(1):18–24 [cited 2018 Mar 26]. http://www.ncbi.nlm.nih.gov/pubmed/18065180
2. Lahsoune M, Boutayeb H, Zerouali K, Belabbes H, El Mdaghri N (2007) Prévalence et état de sensibilité aux antibiotiques d'Acinetobacter baumannii dans un CHU marocain. Médecine Mal Infect [Internet] 37 (12):828–831 [cited 2018 Mar 26]. http://www.ncbi.nlm.nih.gov/pubmed/17669611
3. Vincent J-L, Rello J, Marshall J, Silva E, Anzueto A, Martin CD et al (2009) International study of the prevalence and outcomes of infection in intensive care units. JAMA [Internet] 302(21):2323 [cited 2018 Mar 26]. http://www.ncbi.nlm.nih.gov/pubmed/19952319

4. Simor AE, Lee M, Vearncombe M, Jones-Paul L, Barry C, Gomez M et al (2002) An outbreak due to multiresistant Acinetobacter baumannii in a burn unit: risk factors for acquisition and management. Infect Control Hosp Epidemiol [Internet] 23(5):261–267 [cited 2018 Mar 26]. http://www.ncbi.nlm.nih.gov/pubmed/12026151
5. Ntusi NBA, Badri M, Khalfey H, Whitelaw A, Oliver S, Piercy J et al (2012) ICU-associated Acinetobacter baumannii colonisation/infection in a high HIV-prevalence resource-poor setting. PLoS One [Internet] 7(12):e52452 [cited 2018 Mar 26]. http://www.ncbi.nlm.nih.gov/pubmed/23300673
6. Fournier P-E, Vallenet D, Barbe V, Audic S, Ogata H, Poirel L et al (2006) Comparative genomics of multidrug resistance in Acinetobacter baumannii. PLoS Genet [Internet] 2(1):e7 [cited 2018 Mar 26]. http://www.ncbi.nlm.nih.gov/pubmed/16415984
7. Uwingabiye J, Frikh M, Lemnouer A, Bssaibis F, Belefquih B, Maleb A et al (2016) Acinetobacter infections prevalence and frequency of the antibiotics resistance: comparative study of intensive care units versus other hospital units. Pan Afr Med J [Internet] 23:191 [cited 2018 Mar 26]. http://www.ncbi.nlm.nih.gov/pubmed/27347280
8. Xu J, Sun Z, Li Y, Zhou Q (2013) Surveillance and correlation of antibiotic consumption and resistance of Acinetobacter baumannii complex in a Tertiary Care Hospital in Northeast China, 2003–2011. Int J Environ Res Public Health [Internet] 10(4):1462–1473 [cited 2018 Mar 26]. http://www.ncbi.nlm.nih.gov/pubmed/23571451
9. Ben Haj Khalifa A, Khedher M (2010) Profil de sensibilité aux antibiotiques des souches d'Acinetobacter baumannii isolées dans la région de Mahdia. Médecine Mal Infect [Internet] 40(2):126–128 [cited 2018 Mar 26]. http://www.ncbi.nlm.nih.gov/pubmed/19796896
10. Jaggi N, Sissodia P, Sharma L (2012) Acinetobacter baumannii isolates in a tertiary care hospital: antimicrobial resistance and clinical significance. J Microbiol Infect Dis [Internet] 2(2):57–63 [cited 2018 Mar 26]. http://www.jmidonline.org/upload/sayi/8/JMID-00562.pdf
11. Queenan AM, Pillar CM, Deane J, Sahm DF, Lynch AS, Flamm RK et al (2012) Multidrug resistance among Acinetobacter spp. in the USA and activity profile of key agents: results from CAPITAL Surveillance 2010. Diagn Microbiol Infect Dis [Internet] 73(3):267–270 [cited 2018 Mar 26]. http://www.ncbi.nlm.nih.gov/pubmed/22579652
12. Al-Mously N, Hakawi A, Author C (2013) Acinetobacter baumannii bloodstream infections in a tertiary hospital: antimicrobial resistance surveillance. Int J Infect Control [Internet] 9(9):2 [cited 2018 Mar 26]. www.ijic.info
13. Elouennass M, Bajou T, Lemnouer A, Foissaud V, Hervé V, Baaj A (2003) Acinetobacter baumannii: étude de la sensibilité des souches isolées à l'hôpital militaire d'instruction Mohammed V, Rabat, Maroc. Médecine Mal Infect [Internet] 33(7):361–364 [cited 2018 Mar 26]. http://linkinghub.elsevier.com/retrieve/pii/S0399077X03002051
14. Rizwan M, Naz A, Ahmad J, Naz K, Obaid A, Parveen T et al (2017) VacSol: a high throughput in silico pipeline to predict potential therapeutic targets in prokaryotic pathogens using subtractive reverse vaccinology. BMC Bioinformatics [Internet] 18(1):106 [cited 2018 Mar 26]. http://www.ncbi.nlm.nih.gov/pubmed/28193166
15. Kotb M (1995) Infection and autoimmunity: a story of the host, the pathogen, and the copathogen. Clin Immunol Immunopathol [Internet] 74(1):10–22 [cited 2018 Mar 26]. http://www.ncbi.nlm.nih.gov/pubmed/7994918
16. Naz A, Awan FM, Obaid A, Muhammad SA, Paracha RZ, Ahmad J et al (2015) Identification of putative vaccine candidates against Helicobacter pylori exploiting exoproteome and secretome: a reverse vaccinology based approach. Infect Genet Evol [Internet] 32:280–291 [cited 2018 Mar 26]. http://www.ncbi.nlm.nih.gov/pubmed/25818402
17. Pruitt KD, Tatusova T, Maglott DR (2007) NCBI reference sequences (RefSeq): a curated non-redundant sequence database of genomes, transcripts and proteins. Nucleic Acids Res [Internet] 35(Database):D61–D65 [cited 2018 Mar 26]. http://www.ncbi.nlm.nih.gov/pubmed/17130148
18. Pertsemlidis A, Fondon JW (2001) Having a BLAST with bioinformatics (and avoiding BLASTphemy). Genome Biol [Internet] 2(10):REVIEWS2002. http://www.ncbi.nlm.nih.gov/pubmed/11597340
19. Kaufmann SHE, Lambert PH (2005) The grand challenge for the future: vaccines for poverty-related diseases from bench to field. Birkhäuser, Basel, 290 p
20. Yu C-S, Cheng C-W, Su W-C, Chang K-C, Huang S-W, Hwang J-K et al (2014) CELLO2GO: a web server for protein subCELlular LOcalization prediction with functional gene ontology annotation. PLoS One [Internet] 9

(6):e99368 [cited 2018 Mar 26]. http://dx.plos.org/10.1371/journal.pone.0099368

21. Luo H, Lin Y, Gao F, Zhang C-T, Zhang R (2014) DEG 10, an update of the database of essential genes that includes both protein-coding genes and noncoding genomic elements: Table 1. Nucleic Acids Res [Internet] 42(D1):D574–D580 [cited 2018 Mar 26]. http://www.ncbi.nlm.nih.gov/pubmed/24243843
22. Chen L, Yang J, Yu J, Yao Z, Sun L, Shen Y et al (2004) VFDB: a reference database for bacterial virulence factors. Nucleic Acids Res [Internet] 33(Database issue):D325–D328 [cited 2018 Mar 26]. http://www.ncbi.nlm.nih.gov/pubmed/15608208
23. Tusnády GE, Simon I (2001) The HMMTOP transmembrane topology prediction server. Bioinformatics [Internet] 17(9):849–850 [cited 2018 Mar 26]. http://www.ncbi.nlm.nih.gov/pubmed/11590105
24. UniProt Consortium (2007) The Universal protein resource (UniProt). Nucleic Acids Res [Internet] 36(Database):D190–D195 [cited 2018 Mar 26]. http://www.ncbi.nlm.nih.gov/pubmed/18045787
25. von Mering C, Huynen M, Jaeggi D, Schmidt S, Bork P, Snel B (2003) STRING: a database of predicted functional associations between proteins. Nucleic Acids Res [Internet] 31(1):258–261 [cited 2018 Mar 27]. http://www.ncbi.nlm.nih.gov/pubmed/12519996

Index

A

B

C

D

E

F

G

H

L

Indranil Biswas and Philip N. Rather (eds.), *Acinetobacter baumannii: Methods and Protocols*, Methods in Molecular Biology, vol. 1946, https://doi.org/10.1007/978-1-4939-9118-1, © Springer Science+Business Media, LLC, part of Springer Nature 2019

MIX
Papier aus verantwortungsvollen Quellen
Paper from responsible sources
FSC® C105338
FSC
www.fsc.org

If you have any concerns about our products,
you can contact us on
ProductSafety@springernature.com

In case Publisher is established outside the EU,
the EU authorized representative is:
Springer Nature Customer Service Center GmbH
Europaplatz 3, 69115 Heidelberg, Germany

Printed by Libri Plureos GmbH
in Hamburg, Germany